One-dimensional Stable Distributions

TRANSLATIONS OF MATHEMATICAL MONOGRAPHS

VOLUME 65

One-dimensional Stable Distributions

V. M. ZOLOTAREV

American Mathematical Society · Providence · Rhode Island

ВЛАДИМИР МИХАЙЛОВИЧ ЗОЛОТАРЕВ

ОДНОМЕРНЫЕ УСТОЙЧИВЫЕ РАСПРЕДЕЛЕНИЯ

«НАУКА», МОСКВА, 1983

Translated from the Russian by H. H. McFaden
Translation edited by Ben Silver

1980 *Mathematics Subject Classification* (1985 *Revision*). Primary 60E07.

ABSTRACT. The class of stable distributions, which includes normal distributions and Cauchy distributions, is one of the most important classes in probability theory. In recent years there has been an intensive expansion of the circle of practical problems in which stable distributions appear in a natural way (such mathematical models can be found in engineering, physics, astronomy, and economics). The present book is the first—not only in this country but also abroad—to be specifically devoted to a systematic exposition of the essential facts now known about properties of stable distributions and methods of statistical treatment of them. Also included here are some of the practically useful models connected with stable distributions. This book is intended for experts in the area of probability theory and its applications, for engineers, and for students in university graduate courses.
Illustrations: 7
Bibliography: 254 titles

Library of Congress Cataloging-in-Publication Data

Zolotarev, V. M.
One-dimensional stable distributions.
(Translations of mathematical monographs, ISSN 0065-9282; v. 65)
Translation of: Odnomernye ustoĭchivye raspredelenii͡a.
Bibliography: p. 263
Includes index.
I. Distribution (Probability theory) I. Title. II. Series.
QA273.6.Z6413 1986 519.2 86-10943
ISBN 0-8218-4519-5

The paper used in this book is acid-free and falls within the guidelines
established to ensure permanence and durability.

Contents

Foreword

I begin my first book with words of gratitude and deep respect for my father, Mikhail Ivanovich Zolotarev, whose whole life has been connected with the Soviet Army from the moment of its formation. This monograph is dedicated to him.

More than 50 years have passed since the appearance of the concept of a stable distribution in Paul Lévy's 1925 book *Calcul des probabilités*. Our knowledge about the properties of these remarkable probability laws has by now become so much richer that it could fill several monographs. However, no monograph dealing specifically with stable laws has yet appeared. There are numerous and diverse results relating to stable laws scattered in journal articles or, at best, appearing as auxiliary sections or chapters in books on other branches of probability theory. For example, information about limit theorems for sums of independent random variables when the limit distributions are stable laws can be found in the well-known monographs of Gnedenko and Kolmogorov [26], Feller [22], Ibragimov and Linnik [35], and Petrov [65]. The main part of the results about characterizing stable laws is in the book of Kagin, Linnik, and Rao [38].

A number of facts reflecting the analytic properties of stable laws are contained in the Feller and Ibragimov-Linnik monographs, the book of Lukacs [54], and the well-known survey article of Holt and Crow [31]. Some properties of homogeneous stable processes with independent increments are included in Skorokhod's book [77].

Nevertheless, this information about stable laws is of a nonsystematic, fragmentary nature and does not permit one to form a sufficiently complete picture of the contemporary level of knowledge in any particular direction. This is apparently explained and to a certain extent justified by the fact that, with rare exceptions, stable laws did not find applications for a long time. However, the situation changed in the 1960s after the appearance of a series of papers by Mandelbrot and his successors, who sketched the use of stable laws in certain economic models. And there is now a basis for believing that the role of these laws in the areas of economics, sociology, and biology where the Zipf-Pareto distribution appears will grow in the future. For that to happen, of course, a systematization and more thorough exposition of the known facts are needed.

By way of a preliminary classification, these facts can, in our view, be divided into the following four groups.

1. Limit theorems for sums of independent (or dependent in a special way) random variables, along with various refinements of them such as estimates of the rate of convergence to limiting stable distributions in diverse metrics, asymptotic expansions, large deviations, etc., as well as properties of stable processes with independent increments.

2. Characterization of stable distributions.

3. Analytic properties of stable distributions.

4. Statistical problems associated with stable distributions (estimators of the parameters determining these distributions, problems involving hypothesis testing, and so on).

By now a fairly large number of results have accumulated in the first three groups, and they permit us to speak of the need for a systematic presentation of them.

But the realization of this program would require a volume three or four times the size of the present book, and that exceeds my scope by far. This circumstance prompted me to undertake the more modest task of selecting only one of the groups of results for systematic exposition.

The choice of the third group as primary had to do with two considerations. First, the material relevant here was already sufficiently established, and the influx of new facts is not great at present, which certainly cannot be said of, for example, the first group. Second, I hope that the systematic exposition of the analytic properties of one-dimensional stable laws will stimulate analogous investigations for multidimensional stable laws, of whose properties very little is known to us.

In addition to an Introduction and the second and third chapters, which contain the main bulk of information about the analytic properties of stable laws, the book also includes examples of the occurrence of stable distributions in applied problems (Chapter 1) and a chapter connected with the problem of statistical estimation of the parameters determining stable laws. However, the inclusion of this chapter was dictated not so much by the desire to reflect the material of the fourth group to some extent, as by the desire to demonstrate the possibility of exploiting the analytic properties of stable distributions in solving statistical problems.

The structure of the book is such that only the Introduction and the second and third chapters are interconnected. The first chapter is not necessary for understanding the remaining material, and the fourth chapter makes only minimal use of the information in Chapters 2 and 3.

Although in preparing the material for the main chapters I tried to encompass known results as completely as possible, it is not excluded that some facts accidentally escaped my view.

The same can be said of the first three sections of the references at the end of the book. The fourth section, which deals with limit theorems connected with stable laws, was put together only for a rough orientation of the reader, and thus I did not even try to make it complete.

Information of a historical nature or concerning priority is reduced to a minimum in the main text and is relegated to the section entitled Comments.

The material in the Introduction and the second and third chapters is fundamental to the book. A part of this material contains new results, while another part of the collected material was essentially revised in the preparation process, both in the formulations of the theorems and in their proofs.

A characteristic feature of the exposition is that to describe the properties of stable laws we use not one but several formally different ways of expressing the characteristic functions corresponding to these laws. The fact of the matter is that there simply is no single form of expression leading to the simplest formulations of the properties presented for stable laws. The use of different ways of expressing the characteristic functions enables us to concentrate all the formulational complexities not of fundamental importance in the formulas for passing from one form to another.

In this work I used the advice and all-around help of my friends and colleagues, of whom I should mention first and foremost V. V. Senatov and I. S. Shiganov. I. A. Ibragimov became familiar with the book and made a number of reasonable comments leading to its improvement. I received valuable advice from Yu. B. Sindler and L. A. Khalfin on certain parts of Chapter 1, and from I. V. Ostrovskiĭ on isolated questions in Chapter 2. The material in Chapter 4 benefited greatly from a discussion of its content with È. V. Khmaladze and D. M. Chibisov. N. Kalinauskaitė helped complete the bibliography.

Professor J. H. McCulloch of Ohio State University was invaluable in supplying me with copies of several articles from American publications not easily accessible to me.

G. D. Smychnikov and L. L. Petrov assisted me in the preparation of the manuscript.

Here I express to all these people my sincere gratitude.

Moscow, 1981 *V. M. Zolotarev*

Introduction

I.1. Despite its very close kinship with the normal law, the family of stable distributions has never enjoyed the great interest of mathematicians in probability theory. The laws in this family, which came to light due to the talent of Paul Lévy and received their name from him, attracted only moderate attention from the leading experts, though there were also enthusiasts, of whom our own remarkable mathematician Aleksandr Yakovlevich Khintchine should be mentioned first of all. The reserve met by stable laws was apparently due to the fact that they did not find any applications for a long time. However, there was one exception here. In 1919, several years before the appearance of Lévy's monograph introducing the concept of a stable law, the Danish astronomer Holtsmark published a paper [114] where he found a probabilistic principle obeyed by random fluctuations of the gravitational field of stars in space under certain natural assumptions about their distribution and masses. It turned out that the three-dimensional Fourier transform of the density $p(x)$ of this probability distribution law has a very simple form:

$$\int_{R^3} \exp\{i(t,x)\}p(x)\,dx = \exp(-\lambda|t|^{3/2}), \qquad t \in R^3,$$

where λ is a positive constant determined by the physical characteristics of the object under consideration.

As became clear much later (the article did not at first attract the attention of mathematicians), the distribution found by Holtsmark and now bearing his name belongs to the collection of spherically symmetric stable laws with parameter $\alpha = 3/2$.

In our times there has been a sharp increase in the interest in stable laws, due to their appearance in certain socio-economic models.

Returning to the prehistory of the appearance of stable laws, we should mention that Lévy's creation did not arise from nothing. Already a century before, Poisson and later Cauchy turned their attention to the distribution with density

$$p_\lambda(x) = \lambda^{-1}p_1(x\lambda^{-1}) = \frac{\lambda}{\pi(x^2+\lambda^2)}, \qquad \lambda > 0,\ x \in R^1,$$

which has the property that

$$p_{\lambda_1} * p_{\lambda_2} = p_\lambda, \tag{$*$}$$

where λ is uniquely determined by λ_1 and λ_2. Cauchy, who in the context of his statistical investigations was interested in finding out when the error in an individual observation can be comparable with the average error in a series of many independent observations, discovered that such is the situation if the errors in the observations have probability distribution with density p_λ. The density p_λ is included (for the value $\alpha = 1$) in the set $\{f_\lambda^{(\alpha)}\}$ of functions whose Fourier transforms have the simple form

$$\exp(-\lambda|t|^\alpha), \qquad \alpha > 0,\ \lambda > 0. \tag{$**$}$$

Cauchy knew [8] that each of the functions $f_\lambda^{(\alpha)}$ has the property $(*)$, but except for $f_\lambda^{(1)} = p_\lambda$ he could not distinguish among them the nonnegative functions (each such function is clearly the density of some probability distribution). Only in the beginning of our century did the famous Hungarian mathematician Pólya succeed in proving that $f_\lambda^{(\alpha)} \geq 0$ in the case $0 < \alpha < 1$ (I.1).† He proved that among the functions $f_\lambda^{(\alpha)}$ there is only one that is the density of a distribution with finite variance. And the value corresponding to it is $\alpha = 2$, i.e., the density of an unbiased normal law is such a function. The fact that the functions $f_\lambda^{(\alpha)}$ are the densities of probability distributions only for $0 < \alpha \leq 2$ was established in full by Lévy in his first monograph [50]. For a long time, stable distributions with characteristic functions of the form $(**)$ were called Cauchy laws. Unfortunately, the tradition later died out, and distributions of this type (except for p_λ) became known simply as symmetric distributions.

The concept of stable distributions took full shape in 1937 with the appearance of Lévy's monograph [51] and Khintchine's monograph [44]. It then became clear that, with certain exceptions, stable laws do not have explicit expressions for their densities or distribution functions. The question therefore arose of indirectly investigating the analytic properties of stable distributions. In past years there have been many investigations dealing with this problem. As our knowledge has accumulated about the analytic properties of stable distributions (which form a four-parameter family of functions), it has become

†Raised numbers in this form refer to the corresponding notes in the Comments at the end of the book.

clear that the complexity involved in investigating various groups of properties depends in large part on how fortuitously the parameter system is chosen. Expressing analytic relations between stable distributions, and proving them, can thus be simpler or more complicated depending on what parametrization we employ. This circumstance makes it not only possible but even desirable to use several parametrizations on an equal basis, choosing in each specific situation the one ensuring the least complexity in statements and proofs. With this approach we free ourselves of additional complications that in a reasonably chosen parameter system are put aside and concentrated in the formulas for passing from one system of parameters to another.

I.2. It can be said without exaggerating that contemporary probability theory has acquired its independence, its authority as a mathematical discipline important for applications, and a large part of its arsenal of methods in the course of the solution of the problem of approximating distributions of sums of independent random variables. The results here, which are grouped around the central limit theorem, have come to form in our time a great theory with many branches. Associated with the creation of this theory, which was roughly completed only in the 1930's and which required in all more than a century and a half of effort by several generations of mathematicians, are such illustrious names as Laplace, Gauss, Poisson, Tchebycheff, Lyapunov, Markov, Lévy, Khintchine, Kolmogorov, and Gnedenko. The theory of limit theorems for sums of independent random variables, which has already become classical in our time, continues to nourish many new areas of probability theory, including the theory of Markov chains, stationary processes, martingales, distributions on groups, and so on.

In the theory of summation of independent random variables Khintchine [44] obtained a fundamental result which underlines the first stage of the investigations. It consists in the following.

Consider a sequence $\{X_{nj}, j = 1, 2, \dots, k_n\}$, $n = 1, 2, \dots$, of series of independent (in the individual series) random variables, and form the sums

$$Z_n = X_{n1} + \cdots + X_{nk_n} - A_n, \tag{I.1}$$

centered by some numbers A_n. It is assumed that the terms in the sums are uniformly infinitesimal, i.e., for each $\varepsilon > 0$,

$$\lim_{n\to\infty} \max\{\mathsf{P}(|X_{nj}| > \varepsilon) \colon j = 1, \dots, k_n\} = 0. \tag{I.2}$$

Denote by F_n and F_{nj} the distribution functions of the respective random variables Z_n and X_{nj}, and by $\mathfrak{G}$ the set of all distribution functions G that can be weak limits of such functions F_n as $n \to \infty$.

THEOREM A. *A distribution function G belongs to $\mathfrak{G}$ if and only if the characteristic function $\mathfrak{g}$ corresponding to it can be written in the form*

$$\mathfrak{g}(t) = \exp\left\{ita - bt^2 + \int_{x\neq 0} (e^{itx} - 1 - it\sin x)\, dH(x)\right\}, \tag{I.3}$$

where $a, b \geq 0$ are real numbers, and the function $H(x)$, which is defined on the whole x-axis except for the point $x = 0$, is nondecreasing on the semi-axis $x < 0$ and $x > 0$, tends to zero as $|x| \to \infty$, and satisfies the condition

$$\int_{0<|x|<1} x^2\, dH(x) < \infty. \tag{I.4}$$

The representation (I.3) is called the canonical form for the characteristic function $\mathfrak{g}$, and the function H is called the spectral function of the infinitely divisible distribution G ([I.2]).

The following result of Gnedenko [25] was the second important achievement of the theory. Let

$$a_n(y) = \sum_{j=1}^{k_n} \mathsf{E}(X_{nj} I(|X_{nj}| < y)),$$

where $I(A)$ is the indicator function of an event A and y is a fixed positive number such that y and $-y$ are points of continuity of H, and let

$$\sigma_n^\varepsilon = \sum_{j=1}^{k_n} \operatorname{Var}(X_{nj} I(|X_{nj}| < \varepsilon)), \qquad \varepsilon > 0.$$

THEOREM B. *Let G be a distribution function in the class $\mathfrak{G}$ whose characteristic function in the canonical form* (I.3) *is determined by the triple a, b, H. In the sums* (I.1) *choose the centering constants*

$$A_n = a_n(y) - a - \int_{|u|<y} u\, dH(u) + \int_{|u|\geq y} u^{-1}\, dH(u).$$

Then the distribution functions F_n converge weakly to G as $n \to \infty$ if and only if:

1. *at each point x of continuity of H*

$$\lim_{n\to\infty} \sum_{j=1}^{k_n} \left(F_{nj}(x) - \frac{1}{2} - \frac{1}{2}\operatorname{sgn} x\right) = H(x); \tag{I.5}$$

2.

$$\lim_{\varepsilon\to 0} \limsup_{n\to\infty} \sigma_n^\varepsilon = \lim_{\varepsilon\to 0} \liminf_{n\to\infty} \sigma_n^\varepsilon = 2b.$$

Proofs of Theorems A and B (possibly formulated differently) can be found in any solid course in probability theory; for example, in [22] or in the well-known monograph [26], which deals especially with the problematics of limit approximations of the distributions of sums of independent random variables.

The simplest variant of the scheme (I.1) is a sequence of linearly normalized sums of independent and identically distributed random variables:

$$Z_n = (X_1 + \cdots + X_n)B_n^{-1} - A_n, \qquad n = 1, 2, \ldots, \tag{I.6}$$

where $B_n > 0$ and A_n are real constants.

For condition (I.2) to hold in the scheme (I.6) it suffices that

$$B_n \to \infty \quad \text{as } n \to \infty. \tag{I.7}$$

It turns out that this property holds whenever the distribution functions $F_n(x) = \mathsf{P}(Z_n < x)$ converge weakly to a nondegenerate distribution G.

Indeed, suppose that (I.7) fails to hold. Then, clearly, there is a subsequence $n_k \to \infty$ such that $B_{n_k} = O(1)$. The convergence $F_n \to G$ implies that for each t in any bounded interval

$$|\mathfrak{f}(t/B_{n_k})|^{n_k} = |\mathfrak{g}(t)|(1 + o(1)) \quad \text{as } k \to \infty,$$

where $\mathfrak{f}$ and $\mathfrak{g}$ are the characteristic functions of the distributions $F(x) = \mathsf{P}(X_1 < x)$ and $G(x)$. Consequently, for each sufficiently small value of t

$$|\mathfrak{f}(t)| = |\mathfrak{g}(tB_{n_k})|^{1/n_k}(1 + o(1)).$$

But this relation is possible only if $|\mathfrak{f}(t)| \equiv 1$. This means, in turn, that $|\mathfrak{g}(t)| \equiv 1$, i.e., the distribution G is degenerate, contrary to assumption.

An analysis of the limit behavior of the distributions F_n in the scheme (I.6), carried out (under strong restrictions on the distribution of the terms) near the end of the 18th century by Laplace and Gauss, led to the emergence of the first and, as it later turned out, most important distribution in probability theory and its applications—the normal distribution (sometimes also called the Gaussian law):

$$\Phi_{a,\sigma}(x) = (2\pi\sigma^2)^{-1/2} \int_{-\infty}^{x} \exp\left(-\frac{1}{2\sigma^2}(t - a)^2\right) dt.$$

Once the central limit theorem was proved in a very broad treatment a hundred years later through the efforts of Lyapunov and Markov (who used completely different methods), the experts could switch their attention to the analysis of the variety of limit distributions connected with the scheme (I.6) and its generalizations. Lévy made the first fundamentally new step in this direction. We begin our acquaintance with this result and its subsequent development by a formal definition of stable laws, to the study of whose properties we direct all our attention below.

DEFINITION. A distribution function G is said to be *stable* if it can occur in the scheme (I.6) as a weak limit of the distribution functions F_n as $n \to \infty$.

The set of all such functions G is called *the family of stable laws* and denoted by $\mathfrak{S}$.

Since the scheme (I.6) is a special case of the scheme (I.1), the family of stable laws is a subset of the set $\mathfrak{G}$ of infinitely divisible distribution laws.

There are many different criteria for a distribution function to belong to the family $\mathfrak{S}$, and, if desired, any one of them can be taken as the original definition of stable laws. Some of these criteria will be considered below. We begin with the following one, which, by established tradition, is used particularly often as a definition. This tradition stems from work of Lévy [51] and Khintchine [44].

CRITERION 1. *A distribution function G belongs to the family $\mathfrak{S}$ if and only if it has the following property.*

For any positive numbers b_1 and b_2 there exist a positive number b and a real number a such that

$$G(x/b_1) * G(x/b_2) = G((x-a)/b). \tag{I.8}$$

See §I.4 for a proof.

In Lévy's first monograph [50] the definition of stable laws used is not (I.8) but the more restrictive relation

$$G(x/b_1) * G(x/b_2) = G(x/b), \tag{I.9}$$

which corresponds only to a certain subfamily $\mathfrak{W}$ of $\mathfrak{S}$. Some years later, in the second monograph [51], Lévy did start out from the definition (I.8), but called the laws in the complement $\mathfrak{S}\backslash\mathfrak{W}$ *quasistable laws.* Unfortunately, this term did not take root in the literature, possibly because an overwhelming number of experts did not feel that it was natural to single out the subclass $\mathfrak{W}$ of $\mathfrak{S}$ in the scheme of the problems they were considering. Here Feller was a characteristic exception. Following the terminology in his book [22], we shall call the distributions in $\mathfrak{W}$ *strictly stable* distribution laws.

If in the scheme (I.6) we trace the situation leading to limit distributions in the class $\mathfrak{W}$, then it turns out to correspond to the case when the linear normalization does not need to be centered, i.e., when we can choose the constants $A_n = 0$.

I.3. Before passing to the proof of Criterion 1 and to a discussion of other criteria, we find it convenient first to obtain by purely methodical considerations a description, analogous to that given above for the class $\mathfrak{G}$ of infinitely divisible distributions, for the family $\mathfrak{S}$ of stable laws and the subset $\mathfrak{W}$ of it.

THEOREM C.1. *For each stable distribution G the spectral function H corresponding to it has the form*

$$H(x) = \begin{cases} -C_1 x^{-\alpha} & \text{if } x > 0, \\ C_2(-x)^{-\alpha} & \text{if } x < 0, \end{cases} \tag{I.10}$$

where C_1, C_2, and α are nonnegative numbers, and $0 < \alpha < 2$.

PROOF. We can obviously confine ourselves to the case of a nondegenerate distribution G. Denote by $\mathfrak{X}_H$ the set of points of continuity of H. Since $G \in \mathfrak{S}$, there exists a sequence of independent random variables $X_1, X_2, \dots$ with a common distribution function F and a sequence of pairs A_n, B_n ($B_n \to \infty$) of normalizing constants such that $F_n(x) = \mathsf{P}(Z_n < x)$ converges weakly to $G(x)$ as $n \to \infty$. According to the necessary convergence condition (I.5), this implies that as $n \to \infty$

$$n\left(F(B_n x) - \tfrac{1}{2} - \tfrac{1}{2}\operatorname{sgn} x\right) \to H(x), \qquad x \in \mathfrak{X}_H. \tag{I.11}$$

Consider the case when $x > 0$. If $H(x) \not\equiv 0$ (otherwise choose $C_1 = 0$) on the semi-axis $x > 0$, then there is a point $x_0 \in \mathfrak{X}_H$ for which $q = -H(x_0) > 0$. For each $t > 0$ we now define the positive integer

$$n = n(t) = \min(k\colon B_k x_0 \le t < B_{k+1} x_0).$$

The function $U(x) = 1 - F(x)$ is nonincreasing on the positive semi-axis, and hence

$$\frac{U(B_{n+1}x_0 x)}{U(B_n x_0)} \le \frac{U(tx)}{U(t)} \le \frac{U(B_n x_0 x)}{U(B_{n+1}x_0)} \tag{I.12}$$

for any $x > 0$. Since $n(t) \to \infty$ as $t \to \infty$, it follows from (I.11) that the left-hand and right-hand sides of the inequalities in (I.12) tend as $t \to \infty$ to the common limit

$$L(x) = -H(x_0 x)/q$$

if $x_0 x \in \mathfrak{X}_H$. Consequently, on the one hand

$$U(txy)/U(t) \to L(xy)$$

as $t \to \infty$ for any positive x and y such that $x_0 x$, $x_0 y \in \mathfrak{X}_H$. On the other hand, we can compute the same limit differently:

$$\frac{U(txy)}{U(t)} = \frac{U(txy)}{U(ty)} \cdot \frac{U(ty)}{U(t)} \to L(x)L(y).$$

In summary, we get the equation

$$L(xy) = L(x)L(y), \tag{I.13}$$

valid for all positive x and y except possibly for the points in some countable set.

The function $L(x)$ is defined at all points of the positive semi-axis, where it is also nonnegative and nonincreasing with increasing x, and it has the values $L(1) = 1$ and $L(\infty) = 0$. Under these conditions a solution of equation (I.13) necessarily has the form $L(x) = x^{-\alpha}$, where α is a positive constant (see the monograph [30] for details on solution of the equation of the form (I.13) or of equations reducing to this equation).

The function $H(x)$ must therefore have the following form for $x > 0$:

$$H(x) = -qL(x/x_0) = x_0^\alpha H(x_0)x^{-\alpha} = -C_1 x^{-\alpha}.$$

The additional condition (I.4), which is imposed on any spectral function of an infinitely divisible (hence also of a stable) distribution, constrains the values of the parameter α to the interval $0 < \alpha < 2$. The constant $C_1 \geq 0$ is not restricted by any additional conditions. It is not hard to see this if together with the sequence of normalized sums Z_n we consider another sequence of normalized sums of the form CZ_n, $C > 0$. Both sequences have limiting stable distributions, with respective spectral functions equal to $H(x) = -C_1 x^{-\alpha}$ and $H_C(x) = -C_1 C^\alpha x^{-\alpha}$.

The case $x < 0$ is analyzed similarly and leads to the conclusion that $H(x) = C_2(-x)^{-\delta}$, where $C_2 \geq 0$ and $0 < \delta < 2$. Let us show that $\alpha = \delta$.

In the course of analyzing the case $x > 0$ it was established that

$$\frac{U(tx)}{U(t)} = \frac{1 - F(tx)}{1 - F(t)} \sim x^{-\alpha} \quad \text{as } t \to \infty.$$

This means (see [22]) that $1 - F(x) = x^{-\alpha}h_1(x)$, where $h_1(x)$ is a slowly varying function. Consequently, by (I.11),

$$n(1 - F(B_n x)) = nB_n^{-\alpha}h_1(B_n x)x^{-\alpha} \to C_1 x^{-\alpha}$$

as $n \to \infty$ for any $x > 0$. From this, using properties of slowly varying functions, we find that

$$nB_n^{-\alpha}h_1(B_n) \sim C_1. \tag{I.14}$$

The representation $F(x) = (-x)^{-\delta}h_2(-x)$ for $x < 0$ is obtained in a completely analogous way, and so

$$nB_n^{-\delta}h_2(B_n) \sim C_2. \tag{I.15}$$

If we assume that the spectral function $H(x)$ vanishes on neither semi-axis $x > 0$ nor $x < 0$, i.e., that $C_1 > 0$ and $C_2 > 0$, then the relations (I.14) and (I.15) are possible only if $\alpha = \delta$.

REMARK. We note that the last part of the arguments in the proof allow us to determine the behavior of the normalizing coefficients B_n as $n \to \infty$ in the case when F_n in the scheme (I.6) converges to a nondegenerate distribution G with nonzero spectral function H (i.e., $C_1 + C_2 > 0$). Indeed, let $C_1 > 0$. Then, according to (I.14), $n \sim C_1 B_n^\alpha h_1^{-1}(B_n)$, where $0 < \alpha < 2$. The basic order of variation of the right-hand side is obviously connected with B_n^α. Therefore, the basic order of variation of B_n is a power, and the quantity $h_1(B_n)$, as a function of n, varies more slowly than any power: it is a slowly varying function $h_*(n)$. Consequently, as $n \to \infty$

$$B_n \sim n^{1/\alpha}(C_1^{-1} h_*(n))^{1/\alpha} = n^{1/\alpha} h(n), \tag{I.16}$$

where $h(n)$ is a slowly varying function. In the theory of regularly varying functions the relation (I.16) would be a simple consequence of the fact that the function inverse to a regularly varying (but not slowly varying) function is again a regularly varying function.

A description of the family $\mathfrak{S}$, to which we now turn, is obviously also a criterion for membership in $\mathfrak{S}$.

THEOREM C.2. *A nondegenerate distribution G belongs to the family $\mathfrak{S}$ if and only if the logarithm of its characteristic function $\mathfrak{g}$ can be represented in the form*

$$\log \mathfrak{g}(t) = \lambda(it\gamma - |t|^\alpha + it\omega_A(t, \alpha, \beta)), \tag{A}$$

where the real parameters vary within the limits $0 < \alpha \le 2$, $-1 \le \beta \le 1$, $-\infty < \gamma < \infty$, $\lambda > 0$, and

$$\omega_A(t, \alpha, \beta) = \begin{cases} |t|^{\alpha-1}\beta \tan(\pi\alpha/2) & \text{if } \alpha \neq 1, \\ -\beta(2/\pi)\log|t| & \text{if } \alpha = 1. \end{cases}$$

REMARK. The index A is used to distinguish the given form of expression for characteristic functions of stable laws from the other forms of expression we deal with below.

PROOF. Since $\mathfrak{S}$ is a subset of $\mathfrak{G}$, to make more precise the form of $\mathfrak{g}$ we substitute the expression (I.10) for its spectral function H into the general form (I.3) and compute the resulting integrals. For $0 < \alpha < 2$ we have that

$$\begin{aligned} \log \mathfrak{g}(t) &= ita - bt^2 - C_1 \int_0^\infty (e^{itx} - 1 - it\sin x)\, dx^{-\alpha} \\ &\quad - C_2 \int_0^\infty (e^{-itx} - 1 + it\sin x)\, dx^{-\alpha} \\ &= ita - bt^2 + C_1 Q_\alpha + C_2 \overline{Q}_\alpha. \end{aligned} \tag{I.17}$$

Consider the function

$$\varphi_\alpha(s, p) = p \int_0^\infty (e^{-sx} - e^{-px}) x^{-\alpha}\, dx, \qquad 0 < \alpha < 2,$$

in the domain $\operatorname{Re} s > 0$, $\operatorname{Re} p > 0$. It is not hard to see that $\varphi_\alpha(s,p)$ is continuous with respect to α in the interval $(0,2)$ for fixed s and p and is continuous on the boundary of the domain (i.e., on the straight lines $\operatorname{Re} s = 0$ and $\operatorname{Re} p = 0$) for a fixed value of α. A comparison of the integral Q_α with the function φ_α shows that if we set $s = i$ and single out the real part of $\varphi_\alpha(i,p)$, assuming here that p is positive, and then pass to complex p and replace p by $-it$, then we get the integral Q_α, i.e.,

$$Q_\alpha = \operatorname{Re} \varphi_\alpha(i,p)|_{\,p=-it}.$$

It is not complicated to compute an explicit expression for $\varphi_\alpha(s,p)$. Assuming first that $\alpha \neq 1$, we integrate by parts and find that

$$\begin{aligned} \varphi_\alpha(s,p) &= \frac{p}{1-\alpha}\int_0^\infty (se^{-sx} - pe^{-px})x^{1-\alpha}\,dx \\ &= p\frac{\Gamma(2-\alpha)}{1-\alpha}(s^{\alpha-1} - p^{\alpha-1}). \end{aligned}$$

We then pass to the limit as $\alpha \to 1$ and get the equality

$$\varphi_1(s,p) = p\log(p/s).$$

This procedure leads us to the following expression for Q_α:

$$Q_\alpha = \begin{cases} -|t|^\alpha\Gamma(1-\alpha)\cos(\pi\alpha/2) & \\ \qquad - it(1-|t|^{\alpha-1})\Gamma(1-\alpha)\sin(\pi\alpha/2) & \text{if } \alpha \neq 1, \\ -|t|\pi/2 - it\log|t| & \text{if } \alpha = 1. \end{cases}$$

Let

$$d = (C_1 + C_2)\Gamma(2-\alpha)\frac{\sin(\pi(1-\alpha)/2)}{1-\alpha}$$

(in the case $\alpha = 1$ the value of d is understood in the sense of the limit as $\alpha \to 1$); then

$$|\mathfrak{g}(t)| = \exp(-bt^2 - d|t|^\alpha).$$

We show that the numbers b and d cannot simultaneously be nonzero. Since $G \in \mathfrak{S}$, there exist a characteristic function $\mathfrak{f}(t)$ and sequences of normalizing constants A_n and B_n ($\to \infty$) such that

$$\mathfrak{f}^n(t/B_n)e^{-itA_n} \to \mathfrak{g}(t) \quad \text{as } n \to \infty.$$

Consequently, $|\mathfrak{f}(tB_n^{-1})|^n \to |\mathfrak{g}(t)|$. Assume that $d > 0$. In this case $B_n = n^{1/\alpha}h(n)$ according to (I.16). Therefore, $B_n/B_{nk} \to k^{-1/\alpha}$ as $n \to \infty$ for any integer $k \geq 1$. This fact then gives us that

$$|\mathfrak{f}(tB_{nk}^{-1})|^{nk} = \left|\mathfrak{f}\left(t\frac{B_n}{B_{nk}} \cdot B_n^{-1}\right)\right|^{nk} \to |\mathfrak{g}(tk^{-1/\alpha})|^k.$$

On the other hand, this same limit must obviously be equal to $|\mathfrak{g}(t)|$. Thus,

$$\exp(-bt^2k^{1-2/\alpha} - d|t|^\alpha) = \exp(-bt^2 - d|t|^\alpha),$$

for any integer $k \geq 1$. But this is possible only if $b = 0$.

Define [I.4]

$$\lambda = \begin{cases} d & \text{if } C_1 + C_2 > 0, \\ b & \text{if } C_1 + C_2 = 0, \end{cases}$$

$$\beta = \begin{cases} (C_1 - C_2)/(C_1 + C_2) & \text{if } C_1 + C_2 > 0, \\ 0 & \text{if } C_1 + C_2 = 0, \end{cases}$$

$$\gamma = (a + \tilde{a})/\lambda,$$

where

$$\tilde{a} = \begin{cases} (C_2 - C_1)\Gamma(1-\alpha)\sin(\pi\alpha/2) & \text{if } \alpha \neq 1, \\ 0 & \text{if } \alpha = 1. \end{cases}$$

The right-hand side of (I.17) can now be written in the form (A) we need.

The canonical representation (A) of the characteristic functions of stable laws has one disagreeable feature. The functions $\mathfrak{g}$, and hence also the distributions G associated with them, are not continuous functions of the parameters determining them, since, as is clear from the definition of the function $\omega_A(t, \alpha, \beta)$, they have discontinuities at all points of the form $\alpha = 1$, $\beta \neq 0$. Taking the limits $\alpha^* \to 1$ $(\alpha^* \neq 1)$, $\beta^* \to \beta \neq 0$, $\lambda^* \to \lambda$, and $\gamma^* \to \gamma$ not only does not yield the stable law with the parameters $\alpha = 1$, β, γ, and λ, but does not even yield a proper distribution in the limit. The whole measure goes to infinity. Moreover, we ourselves created this situation by concealing the compensating shift $\tilde{a}$ in the parameter γ. In the original expression for the functions $\mathfrak{g}$ the integrals Q_α, as was especially emphasized, are continuous functions of the variable α. Thus, we can remove the discontinuity by selecting the shift $\tilde{a}\lambda^{-1} = -\beta\tan(\pi\alpha/2)$. The modification of formula (A) in this case is as follows:

$$\log \mathfrak{g}(t) = \lambda(it\gamma - |t|^\alpha + it\omega_M(t, \alpha, \beta)), \tag{M}$$

where

$$\omega_M(t, \alpha, \beta) = \begin{cases} (|t|^{\alpha-1} - 1)\beta\tan(\pi\alpha/2) & \text{if } \alpha \neq 1 \\ -\beta(2/\pi)\log|t| & \text{if } \alpha = 1. \end{cases}$$

It is easy to see that the function ω_M is jointly continuous in its variables. The domain of variation of the parameters in the form (M) is the same as in the form (A). The parameters of the forms (A) and (M) are connected by the following relations (the subscripts indicate that the parameters are associated with the corresponding form):

$$\alpha_A = \alpha_M, \quad \beta_A = \beta_M, \quad \gamma_A = \gamma_M - \beta\tan(\pi\alpha/2), \quad \lambda_A = \lambda_M.$$

There are many different forms of expression for the characteristic function of stable laws. One of them (whose use can be justified by considerations of an analytic nature) is the following.

THEOREM C.3. *The characteristic function* $\mathfrak{g}$ *of nondegenerate distributions* G *in* $\mathfrak{S}$ *can be written in the form*

$$\log \mathfrak{g}(t) = \lambda(it\gamma - |t|^\alpha \omega_B(t, \alpha, \beta)), \tag{B}$$

where

$$\omega_B(t,\alpha,\beta) = \begin{cases} \exp(-i(\pi/2)\beta K(\alpha)\operatorname{sgn} t) & \text{if } \alpha \neq 1, \\ \pi/2 + i\beta \log|t| \operatorname{sgn} t & \text{if } \alpha = 1, \end{cases}$$

$K(\alpha) = \alpha - 1 + \operatorname{sgn}(1-\alpha)$, *and the parameters have the same domain of variation as in the form* (A).

The connection between the parameters (which is not hard to establish by equating the right-hand sides of (A) and (B)) gives us the following relations (the main parameter α is the same in both forms): if $\alpha = 1$, then

$$\beta_A = \beta_B, \quad \gamma_A = 2\gamma_B/\pi, \quad \lambda_A = \pi\lambda_B/2, \tag{I.18}$$

and if $\alpha \neq 1$, then

$$\begin{aligned} \beta_A &= \cot(\pi\alpha/2)\cdot\tan(\pi\beta_B K(\alpha)/2), \\ \gamma_A &= \gamma_B(\cos(\pi\beta_B K(\alpha)/2)^{-1}, \\ \lambda_A &= \lambda_B\cos(\pi\beta_B K(\alpha)/2). \end{aligned} \tag{I.19}$$

In the form (B), as in (A), stable laws are not continuous at points of the form $\alpha = 1$. However, contrary to (A), the limit distribution as $\alpha^* \to 1$, $\beta^* \to \beta$, $\gamma^* \to \gamma$, and $\lambda^* \to \lambda$ exists if α^* always remains greater than or equal to 1. Moreover, the limit is a stable law with characteristic function $\mathfrak{g}$ of the form

$$\log \mathfrak{g}(t) = \lambda(it(\gamma \pm \sin(\pi\beta/2)) - |t|\cos(\pi\beta/2)),$$

where "+" is used if $\alpha^* > 1$, and "−" is used if $\alpha^* < 1$.

I.4. In what follows, the characteristic functions $\mathfrak{g}$ will be accompanied by parameter values, i.e., $\mathfrak{g}(t) = \mathfrak{g}(t,\alpha,\beta,\gamma,\lambda)$. In a special case we use the briefer notation $\mathfrak{g}(t,\alpha,\beta) = \mathfrak{g}(t,\alpha,\beta,0,1)$. If it must be underscored that a particular form $((A),(B)$, etc.) is being used to express the functions, then $\mathfrak{g}$ (and the other quantities and functions connected with it) will be given the corresponding subscript. For example, using this convention, we write the following easily verified property:

$$|\mathfrak{g}_A(t,\alpha,\beta,\gamma,\lambda)| = \exp(-\lambda|t|^\alpha). \tag{I.20}$$

It follows from this property of $\mathfrak{g}_A$ that the corresponding distribution G has a density $\mathfrak{g}_A$ which exists and is uniformly bounded on the whole axis, as is any derivative of it. Indeed, by using the inversion formula

$$g_A(x) = \frac{1}{2\pi}\int e^{-itx}\mathfrak{g}_A(t)\,dt$$

and property (I.20) it is easy to see that

$$|g_A^{(n)}(x)| \le \frac{1}{2\pi}\int |t|^n|\mathfrak{g}_A(t)|\,dt = \frac{\Gamma((n+1)/\alpha)}{\pi\alpha}\lambda^{-(n+1)/\alpha}.$$

The distribution function and density of the stable law with characteristic function $\mathfrak{g}(t,\alpha,\beta,\gamma,\lambda)$ are denoted by $G(x,\alpha,\beta,\gamma,\lambda)$ and $g(x,\alpha,\beta,\gamma,\lambda)$, respectively, and, in shortened variants,

$$G(x,\alpha,\beta) = G(x,\alpha,\beta,0,1),$$
$$g(x,\alpha,\beta) = g(x,\alpha,\beta,0,1).$$

Everywhere in what follows we use the respective notation $Y(\alpha,\beta,\gamma,\lambda)$ or $Y(\alpha,\beta)$ for random variables with stable distributions $G(x,\alpha,\beta,\gamma,\lambda)$ or $G(x,\alpha,\beta)$, and in general the symbol Y is reserved solely for denoting random variables with distributions in $\mathfrak{S}$.

We should stipulate one more rule which we follow throughout the book.

In all equalities connecting functions of random variables with the meaning that they have the same distribution (the symbol $\overset{d}{=}$ will be used for such equalities) the random variables on one side of an equality (even when written the same) are understood as being independent.

For example, with the use of this rule, Criterion 1 can be formulated as follows.

A distribution of random variables $X_1 \overset{d}{=} X_2 \overset{d}{\neq}$ const *belongs to* $\mathfrak{S}$ *if and only if for any positive numbers* b_1 *and* b_2 *there exist a positive number* b *and a real number* a *such that*

$$b_1X_1 + b_2X_2 \overset{d}{=} bX_1 + a. \tag{I.21}$$

Here no special mention is made of the fact that the X_1 and X_2 on the left-hand side of (I.21) are independent.

This way of expressing relations between distributions turns out fairly often to be preferable to an expression using the distributions themselves, the latter being more cumbersome and less intuitive. There will be more than one opportunity below to see the truth of this statement.

Let us turn to the proof of Criterion 1 in the form (I.21). Induction arguments, beginning with the relation $X_1 + X_2 \overset{d}{=} b_2X_1 + a_2$, enable us to verify that

$$X_1 + \cdots + X_n \overset{d}{=} b_nX_1 + a_n, \qquad n \ge 2.$$

Consequently, for any $n \geq 2$

$$\frac{X_1 + \cdots + X_n}{b_n} - \frac{a_n}{b_n} \stackrel{d}{=} X_1.$$

Obviously, the limit distribution of the left-hand side as $n \to \infty$ exists and coincides with the distribution of X_1, which must belong to $\mathfrak{S}$. This proves that the criterion is sufficient. We show that it is necessary.

Let $X_1 \stackrel{d}{=} X_2 \stackrel{d}{=} Y(\alpha, \beta, \gamma, \lambda)$, and choose positive numbers b_1 and b_2. In terms of the characteristic functions $\mathfrak{g}$ of the random variables Y, (I.21) is equivalent to the equality

$$\mathfrak{g}(b_1 t)\mathfrak{g}(b_2 t) = \mathfrak{g}(bt)e^{ita}. \tag{I.22}$$

By using the form (A) it is easy to verify this equality for the choices $b = (b_1^\alpha + b_2^\alpha)^{1/\alpha}$ and

$$a = \begin{cases} \lambda\gamma(b_1 + b_2 - b) & \text{if } \alpha \neq 1, \\ \lambda\beta(2/\pi)(b_1 \log(b_1/b) + b_2 \log(b_2/b)) & \text{if } \alpha = 1. \end{cases} \tag{I.23}$$

CRITERION 2. *The distribution of the random variables* $X_1 \stackrel{d}{=} X_2 \stackrel{d}{=} \cdots \stackrel{d}{=} X_n \stackrel{d}{=} \cdots \stackrel{d}{\neq}$ const *belongs to* $\mathfrak{S}$ *if and only if for any* $n \geq 2$ *there exist a positive number* b_n *and a real number* a_n *such that*

$$X_1 + X_2 + \cdots + X_n \stackrel{d}{=} b_n X_1 + a_n. \tag{I.24}$$

In outward appearance this criterion seems to be a weakening of the condition (I.21). It can be proved in the same way as Criterion 1. In turn, Criterion 2 can be weakened in outward appearance ([I.5]).

CRITERION 3. *The distribution of the random variables* $X_1 \stackrel{d}{=} X_2 \stackrel{d}{=} X_3 \stackrel{d}{\neq}$ const *belongs to* $\mathfrak{S}$ *if and only if the relation* (I.24) *holds for* $n = 2$ *and* $n = 3$.

The proof of necessity is analogous to that for Criterion 1 and does not present difficulties.

We prove sufficiency, i.e., we prove that if

$$X_1 + X_2 \stackrel{d}{=} b_2 X_1 + a_2, \qquad X_1 + X_2 + X_3 \stackrel{d}{=} b_3 X_1 + a_3,$$

where b_1 and b_2 are some real numbers, then the X_j have a stable distribution. Denote by $\mathfrak{f}(t)$ the characteristic function of the random variable X_1. Our goal is to prove that $\mathfrak{f}(t)$ satisfies for each integer $n \geq 2$ the equation

$$\mathfrak{f}^n(t) = \mathfrak{f}(b_n t)\exp(ia_n t), \qquad t \in R^1,$$

which is equivalent to the condition (I.24).

The validity of this equation for $n = 2$ obviously also implies its validity for any of the values $n = 2^m$ with constants b_n and a_n of the form $b_n = b_2^m$,

$a_n = a_2(1 + b_2 + \cdots + b_2^{m-1})$. From this it now follows (see, for example, [54]) that the distribution of X_1 is infinitely divisible, and hence $\mathfrak{f}(t) \neq 0$ for any t.

Define the function $\chi(t) = \log \mathfrak{s}(t)$, $t \in R^1$. The condition of the criterion can be written in terms of χ as follows:

$$2\chi(t) = \chi(b_2 t) + ia_2 t, \qquad 3\chi(t) = \chi(b_3 t) + ia_3 t.$$

Consequently, for any $j, k = 0, \pm 1, \pm 2, \ldots$

$$2^j 3^k \chi(t) = \chi(b_2^j b_3^k t) + ia_{jk},$$

where the a_{jk} are real numbers.

The set $\{2^j 3^k \colon j, k = 0, \pm 1, \pm 2, \ldots\}$ of numbers is dense in the set of all positive numbers. This follows from the well-known fact that the set of numbers of the form $j + \omega k$, $j, k = 0, \pm 1, \pm 2, \ldots$, where ω is irrational, is dense in the real line. Thus, for any n there exists a sequence of numbers $r_m = 2^{j_m} 3^{k_m}$ such that $r_m \to n$ as $m \to \infty$. We let $b_n(m) = b_2^{j_m} b_3^{k_m}$ and show that this sequence of numbers is bounded. From the functional equation for $\chi(t)$,

$$r_m \operatorname{Re} \chi(t) = \operatorname{Re} \chi(b_n(m) t)$$

for any integer m and real t. If the sequence $b_n(m)$ were unbounded, then it would have a subsequence $b_n(m')$ such that $|b_n(m')| \to \infty$ as $m' \to \infty$. Let us perform the change of variable $t' = t b_n(m')$ in the last equation. Then, considering that $r_{m'} \to n$ as $m' \to \infty$, we have that

$$\operatorname{Re} \chi(t') = r_{m'} \operatorname{Re} \chi(t'/b_n(m')) \to 0.$$

Consequently, $|\mathfrak{f}(t)| \equiv 1$. But this contradicts the condition that the distribution of X_1 be nondegenerate.

The sequence $b_n(m)$ is bounded, so it has a subsequence $b_n(m')$ converging to some number b_n as $m' \to \infty$. Hence,

$$\begin{aligned} a_{j_{m'} k_{m'}} &= it^{-1}[\chi(b_n(m')t) - r_{m'} \chi(t)] \\ &\to it^{-1}[\chi(b_n t) - n\chi(t)] = a_n \end{aligned}$$

as $m' \to \infty$. Consequently, for any n and any real t

$$n\chi(t) = \chi(b_n t) + ita_n.$$

But this equation is equivalent to (I.24), i.e., we have reduced the sufficiency of Criterion 3 to that of Criterion 2.

I.5. Let us proceed to a description of the class $\mathfrak{W}$ of strictly stable laws. The definition (I.9) can be rephrased as follows in terms of random variables.

The distribution of the random variables $X_1 \stackrel{d}{=} X_2 \stackrel{d}{\neq} \text{const}$ belongs to $\mathfrak{W}$ if and only if for any positive numbers b_1 and b_2 there is a positive number b such that

$$b_1 X_1 + b_2 X_2 \stackrel{d}{=} b X_1. \tag{I.25}$$

We begin with the necessary conditions. Let $X_1 \stackrel{d}{=} X_2 \stackrel{d}{=} Y(\alpha, \beta, \gamma, \lambda)$. In this case (I.25) is equivalent to the equality

$$\mathfrak{g}(b_1 t)\mathfrak{g}(b_2 t) = \mathfrak{g}(bt),$$

which is a particular case of (I.22). It is clear from the expression for a in (I.23) that this quantity is zero only if

$$\gamma = 0 \quad (\text{if } \alpha \neq 1) \quad \text{or} \quad \beta = 0 \quad (\text{if } \alpha = 1). \tag{I.26}$$

Although this conclusion relates to the form (A), it is the same (as is clear from the transition formulas (I.18) and (I.19)) in the case when we use the form (B).

The following is a criterion for a nondegenerate distribution to belong to the class $\mathfrak{W}$.

CRITERION 4. *Let $c_j = \exp(-d_j)$, $j = 1, \ldots, k$, $k \geq 2$, be numbers such that*

a) $c_1^2 + \cdots + c_k^2 \leq 1$, *and*

b) *among the numbers $d_1, \ldots, d_k$ there are two with an irrational ratio.*

The distribution of the random variables $X_1 \stackrel{d}{=} X_2 \stackrel{d}{=} \ldots \stackrel{d}{=} X_k \stackrel{d}{\neq} \text{const}$ belongs to $\mathfrak{W}$ if and only if

$$c_1 X_1 + c_2 X_2 + \cdots + c_k X_k \stackrel{d}{=} X_1. \tag{I.27}$$

The necessity in the criterion can be verified without difficulty in the same way as for the preceding criterion (by means of characteristic functions, with membership in $\mathfrak{W}$ taken into account in the form of the condition (I.26)). Furthermore, in the course of the proof the following facts become obvious.

a) The parameter α of the random variables $X_1 \stackrel{d}{=} X_2 \stackrel{d}{=} \cdots \stackrel{d}{=} X_k \stackrel{d}{=} Y(\alpha, \beta, \gamma, \lambda)$, satisfying the condition (I.27) is determined as the unique positive solution of the equation $c_1^s + \cdots + c_k^s = 1$.

b) The parameters β, γ, and λ (from the set of those not bound by the condition (I.26)) can vary within the limits of their domains of variation.

The sufficiency of Criterion 4 is proved by fairly complicated analytic arguments, and we shall not present the proof here. Those wishing to go through it will find it in the monograph [38], where Theorem 5.4.2 contains the sufficiency assertion in Criterion 4 as a particular case ([I.6]).

The class $\mathfrak{W}$ is formally described by four parameters, although it is actually a three-parameter set of distributions (because of (I.26)). This is clearly inconvenient from an analytic point of view. We can pass to a system of three parameters by changing the original system $(\alpha, \beta, \gamma, \lambda)$, and it is most convenient to connect the transition with the form (B).

THEOREM C.4. *The characteristic functions* $\mathfrak{g}$ *of the distributions in* $\mathfrak{W}$ *have the representation*

$$\log \mathfrak{g}(t) = -\lambda |t|^{\alpha} \exp(-i(\pi/2)\theta\alpha \operatorname{sgn} t), \tag{C}$$

where the parameters α, θ, *and* λ *vary within the limits*

$$0 < \alpha \leq 2, \quad |\theta| \leq \theta_{\alpha} = \min(1, 2/\alpha - 1), \quad \lambda > 0.$$

The case $\alpha = 1$ occupies a special position in the parametrization system, because, in particular, the values of the parameters $\alpha = 1, \theta$, and λ with $|\theta| = 1$ correspond to a degenerate distribution at the point $\theta\lambda$. If we exclude this case, then the parameters α, β, γ, and λ of the same characteristic function $\mathfrak{g}$ in the form (B) are connected with the parameters α_C, θ, and λ_C by the equalities

$$\begin{gathered} \alpha_C = \alpha_B, \\ \theta = \beta_B K(\alpha)/\alpha, \quad \lambda_C = \lambda_B \quad \text{if } \alpha \neq 1, \\ \theta = \tfrac{2}{\pi} \arctan(2\gamma B/\pi), \quad \lambda_C = \lambda_B(\pi^2/4 + \gamma_B^2)^{1/2} \quad \text{if } \alpha = 1. \end{gathered} \tag{I.28}$$

Together with the parametrization system α, θ, λ_C we can use another parametrization system α, ρ, λ_C close to it for the laws in $\mathfrak{W}$, where $\rho = (1 + \theta)/2$.

It is reasonable not to single this out as an independent form, and to call it also the form (C), since it is after all only a modification of (C). We need not worry about confusion if we use the notation α, θ, λ for the parameters in the case of the form (C), and the notation α, ρ, λ for the parameters in the case of the modification. In the modified variant the characteristic function $\mathfrak{g}$ has the form

$$\mathfrak{g}(t) = \exp\{-\lambda(it)^{\alpha} \exp(-i\pi\alpha\rho \operatorname{sgn} t)\}.$$

We complete the description of the various forms of expression for the characteristic functions of stable laws by one more modification of the form (C) (it is more significant than the preceding one) which is needed in solving the problem of statistical estimation of the parameters of these laws. The logarithm of the characteristic function $\mathfrak{g}$ of each distribution in $\mathfrak{W}$ can be written in the form

$$\log \mathfrak{g}(t) = -\exp\{\nu^{-1/2}(\log|t| + \tau - i(\pi/2)\theta \operatorname{sgn} t) + \mathbb{C}(\nu^{-1/2} - 1)\}, \tag{E}$$

where $\mathbb{C} = 0.577\ldots$ is the Euler constant, and the parameters vary within the limits $\nu \geq 1/4$, $|\theta| \leq \min(1, 2\sqrt{\nu} - 1)$, $|\tau| < \infty$ and are connected with the parameters α, θ, and λ by the equalities

$$\nu = \alpha^{-2}, \quad \theta_E = \theta_C, \quad \tau = (1/\alpha)\log\lambda_C + \mathbb{C}(1/\alpha - 1). \tag{I.29}$$

Of course, one may puzzle over the question: why such an abundance of different forms for expressing the characteristic functions of stable laws? In studying the analytic properties of the distributions of stable laws we encounter groups of properties with their own diverse features. The expression of analytic relations connected with stable distributions can be simpler or more complicated depending on how felicitous the choice of the parameters determining the distributions for our problem turns out to be. By associating with a particular group of properties the parametrization form most natural for it we thereby minimize the complexity involved in expressing these properties. In this approach the extraneous complexity is, as it were, isolated from the problem and relegated to the formulas for passing from one form of expressing the characteristic functions $\mathfrak{g}$ to another ([I.3]).

We make one more important remark concerning the forms given above for expressing the functions $\mathfrak{g}$. In the form (A) the structure of the parameter β_A, i.e., ([I.4])

$$\beta_A = (C_1 - C_2)/(C_1 + C_2)$$

enables us to connect the extreme values $\beta_A = 1$ and $\beta_A = -1$ with two distinctive cases in the scheme (I.6). Indeed, suppose that the sums Z_n in this scheme are formed from positive random variables $X_1, X_2, \ldots$ such that the limit distribution G is not normal. Then the left-hand side of the limit expression (I.11) of the spectral function H corresponding to G is identically zero for $x < 0$. Therefore, H is also equal to zero on the semi-axis $x < 0$, i.e., $C_2 = 0$. Consequently, G has the parameter $\beta_A = 1$.

In exactly the same way we find that the limit distribution G for the normalized sums Z_n has the parameter $\beta_A = -1$ in the case of negative terms X_j. This connection between the properties of the terms X_j and the extreme values of the parameter β_A, though not reversible (i.e., for example, the X_j are not necessarily positive when $\beta_A = 1$ for the limit law), nonetheless serves as a simple reference point in describing the properties of stable distributions. This suggests that it is desirable to have the same rule in the other forms of expression for the characteristic functions.

It holds in the forms (A), (M), and (B), i.e., $\beta_A = 1$ corresponds to $\beta_M = 1$ and $\beta_B = 1$. In the forms (C) and (E), however, $\beta_A = 1$ corresponds to the value $\theta = \theta_\alpha$ if $\alpha < 1$ and to the value $\theta = -\theta_\alpha$ if $\alpha > 1$. Of course, it is easy to redefine the expressions (C) and (E) so that $\theta = \theta_\alpha$ also in the case

$\alpha > 1$. But here we must choose between two incompatible properties of the form (B):

1) $\beta_A = 1 \leftrightarrow \theta = \theta_\alpha$.

2) The distributions in $\mathfrak{W}$ are jointly continuous with respect to the defining parameters in the whole domain of their admissible values.

And in this situation the preference should undoubtedly be given to the second property ([I.7]). In the light of this it is worth turning our attention to the form (M), which has both the indicated properties.

I.6. The definition given above for the family $\mathfrak{S}$ of one-dimensional stable laws can naturally be extended to the case of finite-dimensional and even infinite-dimensional spaces. We consider a sequence of independent and identically distributed random variables $X_1, X_2, \ldots$ with values in the k-dimensional Euclidean space R^k and form the sequence of sums

$$\overline{X}_n = \sigma_n(X_1 + \cdots + X_n) - a_n, \qquad n = 1, 2, \ldots,$$

normalized by some sequences of positive numbers σ_n and nonrandom elements a_n of R^k. The set $\mathfrak{S}^k$ of all weak limits of the distributions of such sequences $\overline{X}_n$ as $n \to \infty$ is called the family of stable distributions on R^k or the family of Lévy-Feldheim distributions. This is not the only way of generalizing the distributions in $\mathfrak{S}$. If the sums $X_1 + \cdots + X_n$ are normalized by nonsingular matrices σ_n and not by positive numbers, then the concept of stable laws becomes essentially broader. Very little is known at present about the properties of multidimensional stable laws (in particular, about their analytic properties). Neither the amount nor the diversity of the facts known here can compare in any way with what is known about the distributions in $\mathfrak{S}$. For this reason a summation of facts on the properties of multidimensional stable laws would certainly be a premature undertaking at this time. We confine ourselves here to presenting and commenting on a canonical representation of the characteristic function $\mathfrak{g}(t)$, $t \in R^k$, of finite-dimensional Lévy-Feldheim laws.

It turns out that the distributions in $\mathfrak{S}^k$ now form a nonparametric set. The corresponding characteristic functions have the form

$$\mathfrak{g}(t) = \exp\{i(t, a) - \psi_\alpha(t)\}, \qquad 0 < \alpha \leq 2, \tag{I.30}$$

where $a \in R^k$ and the functions $\psi_\alpha(t)$, which are determined by the parameter α and by a certain finite measure $M(d\xi)$ on the sphere $S = \{\xi\colon |\xi| = 1\}$, are as follows:

If $\alpha = 2$, then $\psi_\alpha(t) = (\sigma t, t)$, where σ is the so-called covariance matrix. If $0 < \alpha < 2$, then

$$\psi_\alpha(t) = \int_S |(t, \xi)|^\alpha \omega_\alpha(t, \xi) M(d\xi), \tag{I.31}$$

where

$$\omega_\alpha(t,\xi)=\begin{cases}1-i\tan(\pi\alpha/2)\operatorname{sgn}(t,\xi) & \text{if } \alpha\neq 1,\\ 1+i(2/\pi)\log|(t,\xi)|\operatorname{sgn}(t,\xi) & \text{if } \alpha=1.\end{cases}$$

The representation (I.30)–(I.31) is an analogue of formula (A) for expressing characteristic functions of one-dimensional stable laws.

This analogue is not the only one. If we use a spherical system of coordinates in R^k and write a vector t in the form $t=r\eta$, where $r=|t|\neq 0$ and $\eta=t/|t|$, then it is not difficult to give the representation (I.30)–(I.31) the following form:

$$\log \mathfrak{g}(r\eta)=\begin{cases}\lambda[ir\gamma-r^\alpha(1-i\beta\tan(\pi\alpha/2))] & \text{if } \alpha\neq 1,\\ \lambda[ir\gamma-r(1+i(2/\pi)\beta\log r)] & \text{if } \alpha=1,\end{cases} \tag{I.32}$$

where $0<\alpha\leq 2$ and β, γ, and λ are real functions defined on the unit sphere S and determined by the equalities

$$\lambda=\lambda(\eta)=\int_S |(\eta,\xi)|^\alpha M(d\xi), \qquad \eta\in S,$$

$$\lambda\beta=\lambda\beta(\eta)=\int_S |(\eta,\xi)|^\alpha \operatorname{sgn}(\eta,\xi)M(d\xi),$$

$$\lambda\gamma=\lambda\gamma(\eta)=\begin{cases}(\eta,a) & \text{if } \alpha\neq 1,\\ (\eta,a)-\dfrac{2}{\pi}\displaystyle\int_S(\eta,\xi)\log|(\eta,\xi)|M(d\xi) & \text{if } \alpha=1.\end{cases}$$

We note some properties of the functions β, γ, and λ.

1. They are continuous on S, and for a given value of α they uniquely determine the shift a and the measure $M(d\xi)$ in the representation (I.30)–(I.31). In particular, for a given value $\alpha\neq 1$ the functions β and λ uniquely determine the measure M. This implies that the integral transform

$$\Delta(\eta)=\int_S(\eta,\xi)^\alpha I((\eta,\xi)\geq 0)M(d\xi), \qquad \eta\in S,$$

of M on S (which is a function on S) uniquely determines this measure.

2. The domain of variation for the values of the function γ is the whole real axis.

3. The following relations hold for any $\eta\in S$:

$$\beta(-\eta)=-\beta(\eta), \qquad \lambda(-\eta)=\lambda(\eta),$$

$$|\beta(\eta)|\leq 1, \qquad 0\leq\lambda(\eta)\leq M_0,$$

where M_0 is the value of the complete measure $M(d\xi)$ on S. Here all the inequalities are strict unless $M(d\xi)$ is concentrated entirely on some subspace of R^k. This leads, in particular, to the conclusion that

$$\lambda_0=\inf\{\lambda(\eta)\colon \eta\in S\}>0, \qquad |\mathfrak{g}(t)|\leq\exp(-\lambda_0|t|^\alpha),$$

and hence the corresponding stable distribution has density $p(x, \alpha, a, M)$ bounded by the quantity

$$\frac{\Gamma(1+k/\alpha)}{\Gamma(1+k/2)}(2\sqrt{\pi}\lambda_0^{1/\alpha})^{-k}.$$

Each of the forms (M) and (B) also has two analogues obtained by transforming the corresponding analogues of (A). Namely, if $\alpha \neq 1$, then

$$\begin{aligned} \log \mathfrak{g}(t) &= i(t,a) - \int_S \{|(t,\xi)|^\alpha - i(t,\xi)\tan(\pi/2)\alpha(|(t,\xi)|^{\alpha-1} - 1)\} M(d\xi), \\ \log \mathfrak{g}(r\eta) &= \lambda[ir\gamma - r^\alpha + i\beta r(r^{\alpha-1} - 1)\tan(\pi\alpha/2)]. \end{aligned} \tag{I.33}$$

In the case $\alpha = 1$ the functions $\log \mathfrak{g}(t)$ and $\log \mathfrak{g}(r\eta)$ are defined just as in (I.30)–(I.32). As a result, these representations turn out to be continuous functions of α in the whole domain $0 < \alpha \leq 2$ in which α varies.

The first analogue of the form (B) is the representation (I.30)–(I.31) with $\omega_\alpha(t,\xi)$ replaced by the function

$$\tilde{\omega}_\alpha(t,\xi) = \begin{cases} \exp(-i\frac{\pi}{2}K(\alpha)\operatorname{sgn}(t,\xi)) & \text{if } \alpha \neq 1, \\ \frac{\pi}{2} + i\log|(t,\xi)|\operatorname{sgn}(t,\xi) & \text{if } \alpha = 1, \end{cases}$$

and $M(d\xi)$ replaced by the measure

$$\tilde{M}(d\xi) = \begin{cases} |\cos(\pi\alpha/2)|^{-1}M(d\xi) & \text{if } \alpha \neq 1, \\ (2/\pi)M(d\xi) & \text{if } \alpha = 1. \end{cases}$$

The second analogue of (B) is obtained from (I.32) by the same transformations used in the one-dimensional case for getting (B) from (A). This representation has the form

$$\log \mathfrak{g}(r\eta) = \begin{cases} \lambda[ir\gamma - r^\alpha \exp(-i(\pi/2)K(\alpha)\beta)] & \text{if } \alpha \neq 1, \\ \lambda[ir\gamma - r((\pi/2) + i\beta\log r)] & \text{if } \alpha = 1. \end{cases} \tag{I.34}$$

The elements α, β, γ and λ in (I.34) determining the stable distributions are connected with the set of analogous determining elements in (I.32) by relations (I.18) and (I.19).

In those cases when the measure $\tilde{M}(d\xi)$ happens to be concentrated on a half S^* of the sphere S, the first analogue of (B) allows us to write the Laplace transform of the corresponding stable distribution $p(x, \alpha, a, \tilde{M})$ by a componentwise analytic extension of $\mathfrak{g}$ (the substitution $t = is$). Namely,

$$\begin{aligned} &\int_{R^k} \exp\{-(s,x)\} p(x,\alpha,a,\tilde{M})\,dx \\ &\quad = \begin{cases} \exp\{-(s,a) - \varepsilon(\alpha)\int_{S^*}(s,\xi)^\alpha \tilde{M}(d\xi)\} & \text{if } \alpha \neq 1, \\ \exp\{-(s,a) + \int_{S^*}(s,\xi)\log(s,\xi)\tilde{M}(d\xi)\} & \text{if } \alpha = 1, \end{cases} \end{aligned}$$

where $\varepsilon(\alpha) = \operatorname{sgn}(1-\alpha)$, and the vector s takes values in the half-space L^* containing S^*.

The fact that the Laplace transform of the density p exists for the values $s \in L^k$ indicates, in particular, that the function $p(x, \alpha, a, \tilde{M})$ decreases exponentially in the case when $x \notin L^k$ and $|x| \to \infty$.

This observation can be refined. Namely, let $T \subseteq S^*$ be the support of the measure $\tilde{M}$. The set T generates in R^k a corresponding cone C contained in the half-space corresponding to the hemisphere S^*. Denote by $\overline{C}$ the convex hull of C.

In this situation it is relatively simple to see the following properties of the density function p:

a) If $0 < \alpha < 1$, then for all $x \notin \overline{C}$

$$p(x + a, \alpha, a, \tilde{M}) = p(x, \alpha, 0, \tilde{M}) = 0,$$

and $p(x, \alpha, 0, \tilde{M}) > 0$ for all interior points x of $\overline{C}$.

b) If $1 \leq \alpha \leq 2$, then $p(x, \alpha, a, \tilde{M}) > 0$ for all points $x \in R^k$; however, the function $p(x + a, \alpha, a, \tilde{M}) = p(x, \alpha, 0, \tilde{M})$ decreases exponentially in the case when $x \notin \overline{C}$ and $|x| \to \infty$.

The representations given for the characteristic functions of stable distributions can serve as a starting point for various analytic investigations of properties of one-dimensional stable laws. At present the forms (I.30)–(I.31) are those used most in the literature, apparently due to the manifestation of a peculiar "inertia" created by the first investigations.

The analogues of the forms (B) and (M) have not yet been used in general, although they unquestionably have some merits that may in time make them no less popular. For example, these representations make it possible to connect the density $p(x, \alpha, a, M)$ with the densities $g(u, \alpha, \beta, \gamma, \lambda)$ of the one-dimensional stable laws. This connection has the simplest form in the case when the dimension of the space is odd: $\kappa = 2m + 1$. Namely, for all $x \in R^k$

$$p(x, \alpha, a, M) = \frac{(-1)^m}{2(2\pi)^{2m}} \int_S g^{(2m)}((\tau, x), \alpha, \beta, \gamma, \lambda)\, d\tau,$$

where β, γ, and λ are the functions in the representation of the characteristic function of the distribution $p(x, \alpha, a, M)$, and where the density $g(u, \alpha, \beta, \gamma, \lambda)$ corresponds to the form (A) or (B), depending on which of the representations (I.32) or (I.34) we use.

The functions β, γ, and λ (independently of what representation they are associated with) are, respectively, the generalized asymmetry, shift, and scale characteristics of the distribution, as in their interpretation in the one-dimensional case. However, it should be borne in mind here that in carrying various concepts associated with one-dimensional distributions over to the multidimensional case we inevitably encounter a variety of possible generalizations. For instance, the analogue of the symmetric laws in $\mathfrak{S}$, which have characteristic functions of the form

$$\mathfrak{g}(t) = \exp(-\lambda|t|^{\alpha}),$$

can be taken to be the spherically symmetric distributions on R^k having the same form of characteristic functions (with $\lambda = \text{const}$), but the analogues of the symmetric laws can also be taken (and this is more natural) to be the stable distributions with the functions $\beta(\eta) = \gamma(\eta) = 0$ for all $\eta \in S$, which is equivalent to the equality

$$\log \mathfrak{g}(t) = -\lambda(\eta) r^{\alpha} = -\int_S |(t, \xi)|^{\alpha} M(d\xi), \tag{I.35}$$

where $M(d\xi)$ is a centrally symmetric measure.

It is still comparatively rare to encounter multidimensional stable laws in applications (we saw some examples of their occurrence in §1.1). However, there is reason to expect that this situation will change in the near future (due first and foremost to applications in economics).

CHAPTER 1

Examples of the Occurrence of Stable Laws in Applications(1.5)*

§1.0. Introduction

The main and most prominent user of stable laws thus far has been probability theory itself. Stable distributions can be encountered in the most diverse areas in the contemporary mathematical literature. First and foremost, this involves limit theorems for sums of independent and weakly dependent random variables, and a variety of refinements of such theorems. Further, we can name the theory of random processes, in particular, the theory of branching processes, the theory of random determinants, and a number of other areas where problems connected with stable laws are encountered to some extent. The two volumes of Feller [22] can provide a first and fairly thorough acquaintance with such problems, and have justly received worldwide recognition.

The pertinent material here is so extensive that we have not attempted a comprehensive compilation. We have nevertheless regarded it as useful to reflect a part of this material, connected mainly with limit theorems, in the fourth section of the Bibliography.

As illustrations we present three examples of the occurrence of stable laws in mathematical problems.

EXAMPLE 1. The following problem is a classic in probability theory and is given in many publications (for example, in the first volume of Feller cited above).

We consider a sequence of tosses of a symmetric coin and compute the difference Δ_n between the number of heads and tails in n tosses.

This procedure can be represented schematically as a random walk along the integer points of the real axis under the assumptions that we are initially at the point $x = 0$ and that in any position a shift by 1 takes place to the

* Raised numbers in this form refer to the corresponding notes in the Comments at the end of the book.

right or to the left with probability $\frac{1}{2}$, independently of the position and independently of how this position was reached.

The probability that a point $r > 0$ is first reached after n steps is denoted by $p_{r,n}$. Of course, $p_{r,n} = 0$ if the numbers r and n have different parity. If they have the same parity, then

$$p_{r,n} = \frac{r}{n}\binom{n}{(n+r)/2}2^{-n}.$$

The probability $q_n(N)$ that the point r is first reached in no more than N steps is obtained by summing the probabilities $p_{r,n}$:

$$q_r(N) = \sum(p_{r,n} \colon 1 \le n \le N).$$

Let $N = 2[r^2t]$, where $[\cdot]$ denotes the integer part, and the number $t > 0$ is fixed. Then a simple asymptotic analysis shows that as $r \to \infty$

$$q_r(N) \to \sqrt{2/\pi}\int_{1/\sqrt{2t}}^{\infty} \exp(-x^2/2)\,dx = G(t, \tfrac{1}{2}, 1),$$

i.e., the limit distribution of the first hit of the point r within $2[tr^2]$ steps coincides with the Lévy distribution (see §2.2).

EXAMPLE 2. We consider a random branching process with discrete time and with a single type of particle. Each particle is transformed in one of three possible ways at the end of a unit of time: it disappears with probability $p > 0$, is transformed into the same particle with probability $1 - 2p$, or is transformed into two particles with probability p. The generating function $F(s)$ of the transformation probabilities for a single particle (this function determines the random process) has the form

$$F(s) = p + (1 - 2p)s + ps^2.$$

Denote by $\nu_0(t)$ the number of particles in the $(t-1)$st generation that do not have descendants (i.e., that disappear at the appearance of the tth generation), and by $\nu = \sum_{k=1}^{\infty} \nu_0(k)$ the total number of such particles that appear in the process during the entire time $0 < t < \infty$ of its evolution ($\nu_0(0) = 0$, because precisely one particle was considered at the beginning of the process). The generating function $\varphi(u) = \sum_0^{\infty} q_n u^n$ of the distribution $q_n = \mathsf{P}\{\nu = n\}$ is connected with the function $F(s)$ by the equation $\varphi(u) = F(\varphi(u)) + p(u-1)$, which easily yields the following explicit expression for $\varphi(u)$:

$$\varphi(u) = 1 - \sqrt{1-u} = \sum_{n=1}^{\infty} \frac{1}{2\sqrt{\pi}} \frac{\Gamma(n-1/2)u^n}{n!}.$$

This implies that

$$q_n = \frac{1}{2\sqrt{\pi}}\frac{\Gamma(n-1/2)}{\Gamma(n+1)} \sim \frac{1}{2\sqrt{\pi}}n^{-3/2} \quad \text{as } n \to \infty.$$

This asymptotic expression indicates that the distribution of the random variable ν belongs to the domain of normal attraction of the stable law with parameters (in the form (B)) $\alpha = 1/2, \beta = 1, \gamma = 0$, and $\lambda = 1$.

Thus, if at the present time there are not one but n particles, then the total number of particles without descendants is expressed by a sum $\nu_1 + \cdots + \nu_n$ of independent random variables with the same distributions as ν. After normalization, the sum $(\nu_1 + \cdots + \nu_n)n^{-2}$ has a distribution that is approximated by the stable distribution $G(x, \frac{1}{2}, 1)$ as n increases.

It is worth remarking that examples of stable laws with various values of the main parameter α can be found among this kind of problems involving branching processes. For instance, [74] mentions situations in which stable laws with $\alpha = 2^{-r}, r = 1, 2, \ldots$, appear as limit laws.

EXAMPLE 3. This example is taken from the theory of random matrices (concerning this see [24]). We consider systems of linear algebraic equations of the form

$$\Xi_n \mathbf{x}_n = \mathbf{W}_n, \qquad n = 1, 2, \ldots, \tag{1.0.1}$$

where $\Xi_n = (X_{ij}^{(n)})$ is an $n \times n$ matrix with random elements $X_{ij}^{(n)}$, and $\mathbf{W}_n = (W_i^{(n)})$ is a random vector. As we know, there is a unique solution of (1.0.1) if $\det \Xi_n \neq 0$, and

$$\mathbf{x}_n = (x_j^{(n)}) = \Xi_n^{-1} \mathbf{W}_n.$$

In the case when $\det \Xi_n = 0$, a solution of (1.0.1) can fail to exist, and in this situation we agree to assign $\mathbf{x}_n$ the value $\mathbf{x}_n = \mathbf{0}$. For large values of n, when solving linear equations becomes a very laborious computational problem, the following limit approximation can supply a certain amount of information.

Assume that for each value of n the random variables $X_{ij}^{(n)}$ and $W_j^{(n)}$, $i, j = 1, \ldots, n$ are mutually independent, $\mathsf{E}X_{ij}^{(n)} = \mathsf{E}W_j^{(n)} = 0, \operatorname{Var} X_{ij}^{(n)} = \operatorname{Var} W_j^{(n)} = 1$, and the quantities $\sup_{n,i,j} \mathsf{E}(|X_{ij}^{(n)}|^5 + |W_j^{(n)}|^5)$ are finite. Then the following limit relations hold for any $1 \leq i, j \leq n, i \neq j$:

$$\begin{aligned} \lim \mathsf{P}(x_i^{(n)} < \xi) &= \lim_{n \to \infty} \mathsf{P}(x_i^{(n)}/x_j^{(n)} < \xi) \\ &= \tfrac{1}{2} + \tfrac{1}{\pi} \arctan \xi = G_A(\xi, 1, 0). \end{aligned} \tag{1.0.2}$$

It turns out that an analogous limit distribution arises if we consider the joint distributions of any finite number of components of the solution x_n. Namely, under the same assumptions

$$\begin{aligned} &\lim_{n \to \infty} \mathsf{P}(x_{i_1}^{(n)} < \xi_1, \ldots, x_{i_k}^{(n)} < \xi_k) \\ &\quad = \pi^{-(k+1)/2} \Gamma\left(\frac{k+1}{2}\right) \int_{u_j < \xi_j} (1 + |\mathbf{u}|^2)^{-(k+1)/2} \, d\mathbf{u} \end{aligned}$$

for any fixed k, i.e., the limit law here turns out to be a k-dimensional Cauchy distribution.

It should be mentioned that the Cauchy distribution (both one-dimensional and multidimensional) occurs in various problems more frequently than other stable laws, yielding only to the normal law in this respect. The special position of the Cauchy law, like that of the normal law ($^{1.1}$), can be observed also in the analytic setting.

To these laws we could also add the Laplace distribution, assigning the value $\alpha = 0$ to it conditionally (there are well-known grounds for this; see, for example, the relation (2.9.1)). The symmetric laws with integer values of α thereby distinguish themselves from other stable laws by their significance.

The Lévy law with $\alpha = 1/2, \beta = 1$, and $\gamma = 0$ can also occur fairly often in problems. Both the Cauchy law and the Lévy law are closely connected with the normal law. This is clear from the fact that the ratio N_1/N_2 of independent random variables with the standard normal law has a Cauchy distribution, while the random variable N_1^{-2} has a Lévy distribution* (in the final analysis, it is the first circumstance that explains the appearance of the limit relation (1.0.2)).

There are more than a few examples in which the mechanism for the appearance of the Cauchy law reminds us little of our usual mechanism for summation of random variables. We illustrate this by another example taken from [22].

EXAMPLE 4. Consider a point source of radiation on the plane of a foil located parallel to a screen at a distance equal to 1. The radioactive emission causes point flashes of light on the screen. Our problem is to compute the distribution of these flashes on the screen. First of all, choose a Cartesian system of coordinates (x, y, z) in space such that the (x, y)-plane coincides with the plane of the screen, while the radiation source has coordinates $(0, 0, 1)$.

The points at which we see flashes are randomly located on the plane. Denote the coordinates of one of these flashes by $(U, V, 0)$. Of course, the geometric symmetry in the very statement of the problem implies that the random vector (U, V) has a circular distribution, for the determination of which it suffices to find the distribution of one of its coordinates, say U. With this aim we project the whole process on the (x, z)-plane. In the planar model the radiation source has coordinates $(0, 1)$, and the random positions of the flashes on the x-axis have the coordinate U.

Denote by $F(x)$ the distribution function of the random variable U; $F(x)$ is the probability that a ray hits the axis to the left of the point with coordinate

*These facts follow from relations in Chapter 3.

$(x, 0)$. All the rays satisfying this condition are included in an angle equal to $\pi/2 + \arctan x$, measured counterclockwise from the line $z = 1$. All the rays that can hit the screen are obviously included in the angle π. Moreover, it is not hard to see that the angle φ corresponding to the random point $(U, 0)$ has a uniform distribution between the limits 0 and π. Consequently,

$$F(x) = \tfrac{1}{2} + \tfrac{1}{\pi}\arctan x = G_A(x, 1, 0),$$

i.e., the component U (as well as V) has a Cauchy distribution.

Since the vector (U, V) has a circular distribution, it coincides with a two-dimensional Cauchy distribution.

The distributions in the Cauchy family having densities of the form

$$g_A(x, 1, 0, \lambda^{-1}\gamma, \lambda) = \frac{\lambda}{\pi}[\lambda^2 + (x - \gamma)]^{-1} \tag{1.0.3}$$

play an important role in contemporary physics. Distributions of this type are known under their own names in molecular physics, atomic physics, nuclear physics, and the physics of elementary particles.

For example, the energy distribution for unstable states in the reactions of decay products is called the Lorentz law, while the mass distribution for the particles taking part in the same reactions is called the Breit-Wigner distribution.

The mechanism for the appearance of the Cauchy distribution in quantum mechanics in the description of unstable systems is very specific and is not connected in any apparent way with the common mechanisms leading to stable laws. In our view, it is useful for experts in probability theory to have an idea of how this mechanism acts. We briefly present arguments leading to the appearance of the Lorentz distribution.

In quantum mechanics the state of an unstable physical system is described by a so-called wave function $|\psi(t)\rangle$ that has vector values and is a solution of the time-dependent Cauchy problem for the Schrödinger equation

$$H|\psi(t)\rangle = i\frac{\partial}{\partial t}|\psi(t)\rangle, \qquad |\psi(0)\rangle = |\psi_0\rangle, \tag{1.0.4}$$

where H is the Hamiltonian operator (Hermitian operator) corresponding to the system, and $|\psi_0\rangle$ is a given initial value of the wave function.*

Let $\{|\varphi_E\rangle, |\varphi_k\rangle\}$ be a complete system of eigenvectors of the operator H ($|\varphi_E\rangle$ corresponds to the absolutely continuous component of its spectrum, and $|\varphi_k\rangle$ corresponds to the discrete component), i.e.,

$$H|\varphi_E\rangle = E|\varphi_E\rangle, \qquad \langle\varphi_{E'}|\varphi_E\rangle = \delta(E' - E),$$
$$H|\varphi_k\rangle = E_k|\varphi_k\rangle, \qquad \langle\varphi_k|\varphi_l\rangle = \delta_{kl},$$

*Here we adhere to the notation common in contemporary theoretical physics. The inner product of vectors $|a\rangle$ and $|b\rangle$ is denoted by $\langle a|b\rangle$.

where $\delta(E' - E)$ is the Dirac delta function, and δ_{kl} is the Kronecker symbol.

We shall be interested in the probability $P(t)$ that at time t the system finds itself in the initial state $|\psi_0\rangle$. According to the rules of quantum mechanics [104],

$$P(t) = |\langle\psi_0|\psi(t)\rangle|^2.$$

In solving the Cauchy problem (1.0.4) for the Schrödinger equation, we can assume that $\langle\psi_0|\psi_0\rangle = 1$. In this case the Fock-Krylov theorem [116] gives us that

$$\begin{aligned} f(t) &= \langle\psi_0|\psi(t)\rangle \\ &= \sum_k |c_k|^2 \exp(-iE_k t) + \int_0^\infty |c(E)|^2 \exp(-iEt)\, dE, \end{aligned} \tag{1.0.5}$$

where c_k and $c(E)$ are the Fourier coefficients in the expansion of the vector $|\psi_0\rangle$ in the complete system $\{|\varphi_E\rangle, |\varphi_k\rangle\}$ of eigenvectors:

$$|\psi_0\rangle = \sum_k c_k|\varphi_k\rangle + \int_0^\infty c(E)|\varphi_E\rangle\, dE.$$

Thus, $f(t)$ can be interpreted as the characteristic function of some distribution having discrete components (probabilities of isolated values) $|c_k|^2$ and absolutely continuous component (i.e., density) $|c(E)|^2$. Instability of the system means that the probability $P(t) = |f(t)|^2$ of the system returning to the original state at time t tends to zero as $t \to \infty$.

Since $f(t)$ is a characteristic function, $|f(t)| \to 0$ only if the discrete components of the spectrum of H are missing, i.e., $c_k = 0$. In this case

$$f(t) = \int_0^\infty \omega(E) \exp(-iEt)\, dE, \tag{1.0.6}$$

where $\omega(E) = |c(E)|^2$ signifies the density of the energy distribution of the decaying physical system described by equation (1.0.4).

It turns out that for a very broad class of unstable physical systems the densities $\omega(E)$ are meromorphic functions (see [116]). For a number of reasons the case of a function $\omega(E)$ having only two simple poles (they are complex conjugates in view of the condition $\omega(E) \geq 0$) is of greatest interest. In this case it is obvious that

$$\omega(E) = A[(E - E_0)^2 + \Gamma^2]^{-1}, \qquad E \geq 0,$$

where A is a normalizing constant, and E_0 and Γ are the most probable value and the measure of dispersion (with respect to E_0) of the system's energy. For real unstable systems* the ratio Γ/E_0 is very small as a rule (10^{-15},

*An example of such a system is a neutron with an average lifetime of 18.6 min. decaying at the end of its lifetime into a proton, an electron, and a neutrino ($\mathrm{n} \to \mathrm{p} + \mathrm{e} + \nu$).

or even smaller). Therefore, to compute $P(t)$ we can without appreciable detriment replace the lower limit 0 in the integral (1.0.6) by $-\infty$, after which the density function $\omega(E)$ and the probability $P(t)$ connected with it have the approximate expressions

$$\omega(E) \approx \frac{\Gamma}{\pi}[(E - E_0)^2 + \Gamma^2]^{-1},$$
$$P(t) = |f(t)|^2 \approx \exp(-2\Gamma t).$$

It is clear from the first relation (the Lorentz distribution of the energy of the unstable system) that we are dealing with a Cauchy distribution, and it is clear from the second relation that the lifetime for unstable systems of the type under consideration has an exponential law.

Thus, the Cauchy law appears here only as a more or less good approximation of the real energy distribution for unstable systems. And there are situations when the replacement of 0 by $-\infty$ in (1.0.6) is unacceptably coarse, because the corresponding law $P(t)$ of decay of the system differs essentially from an exponential law (concerning this see [115]).

§1.1. A model of point sources of influence

The mathematical model with which we shall become familiar in this section is of a fairly general character, and certain types of problems from areas of astronomy, physics, etc., turn out to be connected with it. Of course, in the framework of a general mathematical model the solutions of practical problems most likely cannot claim great accuracy in transmitting the properties of the phenomena under study. However, as a first approximation they can nevertheless turn out to be useful for understanding the features of these phenomena and for directing the further efforts of investigators ([1.2]).

We consider some region U of finite or infinite volume in n-dimensional Euclidean space R^n; in concrete problems it can be a region of physical space, space-time, phase space, etc. Dispersed in U is a countable system $\mathfrak{A}$ of point objects which we conditionally call "particles". The particles are characterized by the value of a parameter θ taking values in some set Θ.

Each of the particles in $\mathfrak{A}$ is capable of creating a "field of action" (for brevity, simply a "field") described by means of an "influence function" $v(x, y, z)$; $v(x, y, z)$ takes values in some subset V of m-dimensional Euclidean space R^m and shows the value of the field created at a point $y \in U$ by a particle located at a point $x \in U$ and characterized by the value $\theta = z$. By definition, a finite collection $A = \{X_i\}$ of particles in $\mathfrak{A}$, with the particle indexed by i located at the point X_i and having value Z_i of the characterizing parameter θ, creates a field with influence function equal to

$$\sum\{v(X_i, y, Z_i)\colon X_i \in A\}, \tag{1.1.1}$$

i.e., a field equal to the vector sum of the fields created by the separate particles in the set A.

Both the positions X_i of the particles in $\mathfrak{A}$ and the values Z_i corresponding to them are assumed to be random variables subject to a number of conditions.

Let U_1 and U_2 be some disjoint regions in U of finite volumes u_1 and u_2. A countable system of randomly situated points in U is called a Poisson ensemble (see [22] concerning this concept if it has the following properties.

1°. The numbers N_1 and N_2 of points falling in U_1 and U_2, respectively, are independent random variables.

2°. For any $k \geq 0$ the probability $\mathsf{P}(N_1 = k)$ depends on k and the volume u_1, but not on the shape of U_1.

3°. For small values of the volume u_1

$$\mathsf{P}(N_1 = 1) = \rho u_1 + o(u_1),$$
$$\mathsf{P}(N_1 \geq 2) = o(u_1),$$

where $\rho > 0$ is a constant signifying the average density of the concentration of points of the system in U.

By using these properties it is not hard to show that N_1 has a Poisson distribution with parameter $\lambda = \rho u_1$, i.e.,

$$\mathsf{P}(N_1 = k) = \lambda^k \exp(-\lambda), \qquad k = 0, 1, \dots. \tag{1.1.2}$$

We mention two more properties of Poisson ensembles of points needed in what follows.

4°. Let $X_1, \dots, X_{N_1}$ be the positions of the ensemble points falling in U_1. Then the random variables $X_1, X_2, \dots$ and N_1 are mutually independent.

5°. The conditional distribution of the position of an ensemble point in U_1 under the condition that it falls in this region is uniform, i.e., has distribution density equal to $1/u_1$.

We make the following three assumptions about the system $\mathfrak{A}$ of particles:

a) $\mathfrak{A}$ is a Poisson ensemble (and, consequently, has properties 1°–5°).

b) For any region U_1 of finite volume the number N_1 of particles falling in U_1, the positions $X_1, X_2, \dots$ of these particles, and the values $Z_1, Z_2, \dots$ of the parameter θ characterizing them are independent random variables.

c) The random variables $Z_1, Z_2, \dots$ have the same distribution law P_θ.

For each region U_1 of finite volume the field created by the particles falling in U_1 is random, of course, and, according to the rule (1.1.1), its value at a point $y \in U$ is

$$v_1 = v_1(y) = \sum\{v(X_j, y, Z_j)\colon X_j \in U_1\}. \tag{1.1.3}$$

The field generated by the whole system $\mathfrak{A}$ will be understood as a certain limit. With this goal we consider in R^n the set of spheres $S_T = S_T(y)$ with a common fixed center at a point $y \in U$ and varying radii $T > 0$. Then we form the sets $U_T = U_T(y) = U \cap S_T(y), T > 0$, and denote by u_T their volumes, and by N_T the number of particles of $\mathfrak{A}$ falling in U_T. These particles generate a field which, according to (1.1.3), has at y the value

$$\nu_T = \nu_T(y) = \sum \{v(X_j, y, Z_j) \colon X_j \in U_T\}. \tag{1.1.4}$$

By definition, the value $\nu(y)$ of the field generated at y by all the particles in $\mathfrak{A}$ is equal to the weak limit of $\nu_T(y)$ as $T \to \infty$. To determine conditions for the existence of this limit and to find an analytic description of it we can consider the question of the limit value as $T \to \infty$ of the characteristic function of the variable $\nu_T(y)$:

$$f_T(t) = f_T(t, y) = \mathsf{E} \exp\{i(t, \nu_T)\}, \qquad t \in R^m.$$

With this aim let us compute the function $f_T(t)$ by using our assumptions a)–c). It is clear from (1.1.4) that $\nu_T(y)$ is the sum of a random number N_T of identically distributed random variables $W_j = v(X_j, y, Z_j)$ that are independent of N_T and of one another. Consequently, denoting by $\varphi_T(t)$ the characteristic function of W_1, we can write

$$\begin{aligned} f_T(t) &= \sum_{k=0}^{\infty} \mathsf{P}(N_T = k) \mathsf{E} \exp\{i(t, W_1 + \cdots + W_k)\} \\ &= \sum_{k=0}^{\infty} \exp(-\rho u_T) \frac{(\rho u_T)^k}{k!} \varphi_T^k(t) \\ &= \exp\{\rho u_T(t) - 1)\}. \end{aligned}$$

We transform the expression in the exponent of the last equality:

$$\begin{aligned} \rho u_T(\varphi_T(t) - 1) &= \rho u_T \int_{U_T \times \Theta} (e^{i(t, v(x,y,z))} - 1) \frac{dx}{u_T} P_\theta(dz) \\ &= \rho \int_{U_T \times \Theta} (\exp\{i(t, v(x, y, z))\} - 1) \, dx \, P_\theta(dz). \end{aligned}$$

Thus, the expression

$$\log f_T(t) = \rho \int_{U_T \times \Theta} [\exp\{i(t, v)\} - 1] \, dx \, P_\theta(dz) \tag{1.1.5}$$

is obtained for $\log f_T(t)$; it can be transformed further by taking the function $v = v(x, y, z)$ as a new variable and forming the measure

$$\mu(B) = \rho \int_{B_*} dx \, P_\theta(dz) \tag{1.1.6}$$

on R^m, where B is a Borel subset of $V \subseteq R^m$ and $B_* = \{(x,z)\colon v(x,y,z) \in B\}$. The introduction of this measure clearly presupposes the measurability of the function v with respect to the pair (x,z) for each $y \in U$.

As a consequence of (1.1.5) and (1.1.6),

$$\log f_T(t) = \int_{V_T} [\exp\{i(t,v)\} - 1]\, \mu(dv), \tag{1.1.7}$$

where $V_T = \{(x,z)\colon v(x,y,z) \in U_T \times \Theta\}$.

Thus, the function $f_T(t)$ obtains the canonical Lévy representation for characteristic functions of infinitely divisible distributions on R^m. The question of the convergence of the integral in (1.1.7) does not arise, since $\mu(V_T) = \rho u_T$, where u_T is the volume of U_T. Since the set V_T is clearly nondecreasing as T increases, the question of the existence of the limit of $f_T(t)$ as $T \to \infty$ reduces to the question of the existence of the limit of the integral

$$J_T = \int_{V_T} [\exp\{i(t,v)\} - 1]\, \mu(dv). \tag{1.1.8}$$

Here several essentially different situations can arise. Denote by $I(A)$ the indicator function of an event A, and by L_T the integral (vector)

$$\int_{V_T} vI(|v| < 1)\, \mu(dv),$$

which obviously exists for all $T > 0$. Let us write the integral J_T in the form

$$J_T = i(t, L_T) + \int_{V_T} (e^{i(t,v)} - 1 - i(t,v)I(|v| < 1))\, \mu(dv).$$

In the second term the integration over V_T can be replaced by integration over V if we simultaneously replace the measure μ by the measure μ_T coinciding with μ on V_T and equal to zero outside this set. Furthermore, $\mu_T \to \mu$ as $T \to \infty$, because V_T is nondecreasing as T increases.

Since $\log f_T(t) = J_T$, $f_T(t)$ converges to some characteristic function $f(t)$ as $T \to \infty$ if and only if (this fact follows from the general theory of infinitely divisible laws in R^m)

1) $L_T \to L$ as $T \to \infty$, where L is a vector in R^m, and
2) $D = \int_V \min(1, |v|^2)\, \mu(dv) < \infty$.

Under these conditions the characteristic function $f(t)$ of the limit distribution has the form

$$\log f(t) = i(t, L) + \int_V (e^{i(t,v)} - 1 - i(t,v)I(|v| < 1))\, \mu(dv). \tag{1.1.9}$$

We indicate two cases when (1.1.9) takes a simpler form.

1*. If

$$D_1 = \int_V \min(1, |v|)\mu(dv) < \infty,$$

then conditions 1) and 2) are satisfied and

$$\log f(t) = \int_V (\exp(i(t, v)) - 1)\mu(dv). \tag{1.1.10}$$

2*. If condition 1) holds along with the condition

$$D_2 = \int_V \min(1, |v|)|v|\mu(dv) < \infty,$$

which clearly implies property 2), then

$$\log f(t) = i(t, M) + \int_V (e^{i(t,v)} - 1 - i(t, v))\mu(dv), \tag{1.1.11}$$

where

$$M = L + \int_{|v|\geq 1} v\,\mu(dv). \tag{1.1.12}$$

In the case we are considering the limit value L does not necessarily have an integral form of expression, without which M cannot be interpreted as the mean value of the limit distribution. If, nevertheless, such a representation of L exists, then (1.1.11) is easy to transform.

As a matter of fact, the violation of condition 2) excludes the possibility that the limit distribution of $\nu_T(y)$ exists, even after this variable is also centered. Therefore, we consider the situation when condition 2) holds but condition 1) does not. In this case it makes sense to study not the random variable ν_T itself but its translate $\tilde{\nu}_T = \nu_T - L_T$. The corresponding weak limit $\tilde{\nu}$ of this variable as $T \to \infty$ exists if and only if 2) holds, and the characteristic function $\tilde{f}$ of the random variable $\tilde{\nu}$ under condition 2) has the form

$$\log \tilde{f}(t) = \int_V (e^{i(t,v)} - 1 - i(t, v)I(|v| < 1))\mu(dv). \tag{1.1.13}$$

If the condition $D_2 < \infty$ also holds, then ν_T can be translated by the mathematical expectation $M_T = \mathsf{E}\nu_T$, which is finite in our situation. With this translation the random variable $\bar{\nu}_T = \nu_T - \mathsf{E}\nu_T$ converges weakly to the random variable $\bar{\nu}$ with characteristic function $\bar{f}$ of the form

$$\log \bar{f}(t) = \int (e^{i(t,v)} - 1 - i(t, v))\mu(dv). \tag{1.1.14}$$

The various translations of ν_T are carried out in the cases when ν_T itself does not have a weak limit, i.e., in the cases when the field created by the whole system $\mathfrak{A}$ of particles simply does not exist. Nonetheless, a practical meaning can be given to the above construction of limit approximations of the distributions of the random variables $\tilde{\nu}_T$ and $\bar{\nu}_T$. The fact of the matter is that the appearance of a region U of infinite volume in the model is a mathematical idealization. The region U has in reality a large but finite volume. In this case we study random fluctuations of the field generated by the particles in U, with a certain constant component singled out in it. After singling out the constant component we can simplify the computations by an appropriate expansion of U, provided, of course, that this does not introduce essential distortions into our calculations.

The limit distributions (1.1.9)–(1.1.11) and (1.1.13)–(1.1.14) considered above are infinitely divisible. It is natural to expect that in certain cases we shall encounter m-dimensional stable laws; this is entirely determined by the form of the measure defined by (1.1.6), which, in turn, depends on the properties of the influence functions $v(x, y, z)$ and, to a lesser degree, on the properties of the distribution P_θ.

Below we single out a class of influence functions generating stable distributions. It is connected with transformation of formula (1.1.10), accompanied by the condition $D_1 < \infty$, and of formulas (1.1.11) and (1.1.14) obtained under the condition $D_2 < \infty$.

A semicone (with vertex at the origin) in the subspace R^{n_1} of R^n ($0 < n_1 \leq n$) is defined to be a set U_1 such that if $u_1 \in U_1$ and $c > 0$, then $cu_1 \in U_1$ and $-cu_1 \notin U_1$. With this definition the following assumptions are made about the set U and the points y for which the distributions of the random variables $\nu(y), \tilde{\nu}(y)$, and $\bar{\nu}(y)$ are to be computed.

(i) The set $U_y = \{u\colon u = x - y, x \in U\}$ either is itself a semicone in R^n or is a direct product $U_1 \times U_2$ of semicones U_1 and U_2 in orthogonal subspaces R^{n_1} and $R^{n_2}, n_1 + n_2 = n$.

The first of these two cases obviously reduces to the second if the dimension n_2 is allowed to take the value $n_2 = 0$. This convention will be followed below.

The decomposition of R^n into orthogonal subspaces corresponds to the notation $x = (x_1, x_2)$ for vectors $x \in R^n$, where $x_1 \in R^{n_1}$ and $x_2 \in R^{n_2}$.

(ii) The influence function $v(x, y, z)$ is 0 for all $x \notin U$, is continuous with respect to the variable x at all interior points of U, and has the following structure there:

$$v(x, y, z) = |x_1 - y_1|^{-p} D(|x_2 - y_2|\,|x_1 - y_1|^{-q}, \cdot), \tag{1.1.15}$$

where the symbol "$\cdot$" indicates the dependence of D on y, z and $s_j = (x_j - y_j)/|x_j - y_j|, j = 1, 2$, or on some of these variables.

The condition in (i) that the U_j be semicones could be extended somewhat by allowing U_j to be both a semicone and a cone with vertex at the origin. However, this extension turns out to be connected with only one possible point y if U_j is not the whole subspace R^{n_j}. Anyway, the last case is encompassed by the assumption (i) if R^{n_j} is regarded as a limiting case of an expanded semicone; therefore, the indicated extension involves only a very special situation.

Further, note that in the case when $U_j = R^{n_j}$ the influence function v must be invariant with respect to a translation by the variable x_j, i.e., it must depend only on the difference $x_j - y_j$.

In what follows, the numbers $p > 0$ and $q \geq 0$ will be connected with the dimensions n_1 and n_2 of the subspaces by additional conditions.

We consider (1.1.10) with the condition $D_1 < \infty$, after returning to the original variables x and z:

$$\log f(t) = \rho \int_{U\times\Theta} (e^{i(t,v)} - 1)\, dx\, P_\theta(dz), \tag{1.1.10*}$$

$$D_1 = \rho \int_{U\times\Theta} \min(1, |v|)\, dx\, P_\theta(dz) < \infty.$$

In the integral (1.1.10*) let us pass to a polar system of coordinates in the subspaces R^n by setting ($j = 1, 2$)

$$r_j = |x_j - y_j|, \qquad x_j - y_j = r_j s_j,$$

where s_j is a point on the surface of the unit sphere S_j about the origin. We have that

$$\omega_1 = \int_U (e^{i(t,v)} - 1)\, dx = \int_{S_1}\int_{S_2} \omega_2\, ds_2\, ds_1,$$

where

$$\omega_2 = \iint\limits_0^\infty [\exp\{i(t, r_1^{-p} D(r_2/r_1^q, \cdot))\} - 1] r_2^{n_2-1} r_1^{n_1-1}\, dr_2\, dr_1.$$

We make a change of variable, replacing r_2 by $r_2 r_1^q$, and, after changing the order of integration, get that

$$\omega_2 = \int_0^\infty r_2^{n_2-1}\, dr_2 \int_0^\infty [\exp(i\xi r_1^{-p}) - 1] r_1^{n_1+n_2q-1}\, dr_1,$$

where $\xi = (t, D(r_2, \cdot))$.

Substitution of r_1 for the variable r_1^{-p} gives us that

$$\omega_2 = \frac{1}{p}\int_0^\infty r_2^{n_2-1}\, dr_2 \int_0^\infty (e^{i\xi r_1} - 1) r_1^{-(n_1+qn_2)/p-1}\, dr_1.$$

If

$$\alpha = (n_1 + qn_2)/p < 1, \tag{1.1.16}$$

then the inside integral in the expression for ω_2 converges, and

$$\int_0^\infty (e^{i\xi r_1} - 1) r_1^{-\alpha-1}\, dr_1 = \Gamma(-\alpha)|\xi|^\alpha \exp\left(-i\frac{\pi}{2}\alpha \operatorname{sgn} \xi\right),$$

which implies that

$$\omega_2 = \frac{\Gamma(-\alpha)}{p}\int_0^\infty |(t,s)|^\alpha \exp\left(-i\frac{\pi}{2}\alpha \operatorname{sgn}(t,s)\right) |D|^\alpha r_2^{n_2-1}\, dr_2,$$

where $s = D/|D|$.

Let us introduce a measure χ on the unit sphere S in R^n by setting

$$\chi(B) = \frac{\rho\Gamma(1-\alpha)}{\alpha p}\int_{s\in B} |D(r_2,\cdot)|^\alpha r_2^{n_2-1}\, dr_2\, ds_1\, ds_2\, P_\theta(dz) \tag{1.1.17}$$

for any Borel subset B of S (here the integration is also over all r_2 and z).

Considering the expressions obtained for ω_1 and ω_2, we can transform the right-hand side of (1.1.10*) to the form

$$\log f(t) = -\int_S |(t,s)|^\alpha \exp\left\{-i\frac{\pi}{2}\alpha \operatorname{sgn}(t,s)\right\} \chi(ds), \tag{1.1.18}$$

which corresponds to the canonical form of expression for the characteristic function of an n-dimensional stable law with main parameter $0 < \alpha < 1$. The finiteness of the complete measure $\chi(S)$ is an additional condition present in the description of the n-dimensional stable laws. It is not hard to verify on the basis of (1.1.17) that the condition $\chi(S) < \infty$ is equivalent to the condition $D_1 < \infty$.

The second case is connected with a transformation of the integral

$$\omega = \int_V (e^{i(t,v)} - 1 - i(t,v))\mu(dv)$$

in (1.1.11) and (1.1.14) under the condition $D_2 < \infty$. Repeating the same arguments used above, we have that

$$\begin{aligned}\omega &= \rho\int_\Theta P_\theta(dz)\int_U (e^{i(t,v)} - 1 - i(t,v))\, dx\\ &= \rho\int_\Theta P_\theta(dz)\int_{S_1}\int_{S_2} \omega_2\, ds_2\, ds_1,\end{aligned}$$

where

$$\omega_2 = \iint_0^\infty [\exp\{i(t, r_1^{-p}D)\} - 1 - i(t, r_1^{-p}D)] r_2^{n_2-1} r_1^{n_1-1}\, dr_2\, dr_1.$$

After substitution of $r_2 r_1^q$ for r_2 the quantity D becomes independent of r_1, and after substitution of r_1 for r_1^{-p} the integral ω_2 is transformed to

$$\omega_2 = \frac{1}{p}\int_0^\infty r_2^{n_2-1}\, dr_2 \int_0^\infty (e^{i\xi r_1} - 1 - i\xi r_1) r_1^{-\alpha-1}\, dr_1,$$

where $\xi = (t, D(r_2, \cdot))$ and $\alpha = (n_1 + qn_2)/p$. If $1 < \alpha < 2$, then the inside integral converges and can be computed. It turns out to be equal to

$$\Gamma(-\alpha)|\xi|^\alpha \exp(-i\tfrac{\pi}{2}\alpha \operatorname{sgn} \xi).$$

Consequently, introducing the measure χ just as in (1.1.17) but with the opposite sign, we arrive at the following expression for $\omega = \log f(t) - i(t, M)$ and $\omega = \log \bar{f}(t)$:

$$\omega = \int_S |(t, s)|^\alpha \exp\left\{-i\tfrac{\pi}{2}\alpha \operatorname{sgn}(t, s)\right\} \chi(ds). \tag{1.1.19}$$

Finiteness of the complete measure $\chi(S)$ is equivalent to the condition $D_2 < \infty$.

The cases selected above are among the simplest ones. The stable laws are obtained also on the basis of formulas (1.1.9) and (1.1.13), since in the final analysis the question of whether or not the limit distribution is stable is answered only by the form of the spectral measure $\mu(dv)$, which is directly connected with the form of the function $v(x, y, z)$. Laws with the value $\alpha = (n_1 + qn_2)/p = 1$ also appear in these complicated cases.

REMARK 1. We mention that the field created by a set of particles is not necessarily homogeneous and isotropic. What kind of field this is depends on the structural characteristics of the influence function. Therefore, in the general situation the stable distributions computed for the value of the field $\nu(y)$ at a point y or the value of the previously centered field $\bar{\nu}(y)$ depend on y, even though the main parameter α of these distributions does not change as y changes.

The distributions $\nu(y)$ and $\bar{\nu}(y)$ do not depend on y in the case when the influence function v has the property of homogeneity within the set U:

$$v(x, y, z) = v(x - y, 0, z). \tag{1.1.20}$$

This should be understood in the sense that $v(x + y, y, z)$ does not depend on y within the confines of the semicone $U_1 \times U_2$, which also does not depend on y. In this situation any interior points of the original region U can be chosen in the role of the points $y \in R^n$ satisfying conditions (i) and (ii).

REMARK 2. The assumption (1.1.15) about the structure of the influence function can be generalized as follows.

Let $R^n = R^{n_1} \times \cdots \times R^{n_k}$, where $0 < n_1 \le n$ and $n_j \ge 0$, $j = 2, \dots, k$. In this case each vector $x \in R^n$ can be written in the form $x = (x_1, \dots, x_k), x_j \in R^{n_j}$.

Assume further that $U_y = U_1 \times \cdots \times U_k$, where the U_j are semicones in R^{n_j}, and that the influence function v is equal to zero outside U and can be expressed inside U in the form

$$v(x, y, z) = |x_1 - y_1|^{-p} D(|x_2 - y_2|\,|x_1 - y_1|^{-q_2}, \dots, |x_k - y_k|\,|x_1 - y_1|^{-q_k}, \cdot). \tag{1.1.21}$$

Here $p > 0, q_j \geq 0$, and the symbol "$\cdot$" indicates (as in (1.1.15)) dependence on the collection of variables y, z, and $s_j = (x_j - y_j)/|x_j - y_j|, j = 1, \dots, k$ (or some of them).

Just as in the case $k = 2$ singled out in (1.1.15), the distributions of the random variables $\nu(y), \tilde{\nu}(y)$, and $\bar{\nu}(y)$ in the case (1.1.21) turn out to be stable with the main parameter

$$\alpha = (n_1 + q_2 n_2 + \cdots + q_k n_k)/p.$$

The course of the arguments is somewhat more complicated than in the case of (1.1.15), but is basically the same.

REMARK 3. Assume that the function D in (1.1.15) has the following structure:

$$\begin{aligned} &D(|x_2 - y_2|/|x_1 - y_1|^q, \cdot) \\ &\quad = \varphi(|x_2 - y_2|/|x_1 - y_1|^q, y, z)\psi(s_1, s_2, y), \end{aligned} \tag{1.1.22}$$

where φ is a real function, $s_j = (x_j - y_j)/|x_j - y_j|$, and

$$\psi(-s_1, s_2, y) = -\psi(s_1, s_2, y) \quad \text{or} \quad \psi(s_1, -s_2, y) = -\psi(s_1, s_2, y).$$

Under the corresponding conditions (i.e., the condition $D_1 < \infty$ for ν and the condition $D_2 < \infty$ for $\bar{\nu}$) the variables $\nu(y)$ and $\bar{\nu}(y)$ in this case have spherically symmetric distributions with characteristic functions of the form

$$\exp(-\lambda|t|^\alpha), \qquad 0 < \alpha < 2,$$

where $\alpha = (n_1 + qn_2)/p$, and λ is a constant depending on v and P_θ.

REMARK 4. The question of which influence functions lead to stable distributions in the model of point sources of influence is apparently not exhausted by functions of the types (1.1.15) and (1.1.21). It is therefore of interest to look for other types of functions v which generate stable laws. Another interesting direction for investigation has to do with the desire to weaken the original assumptions a)–c) in the general model. The following special case shows that the search in this direction is promising.

Let

$$v(x, y, z) = |x_1 - y_1|^{-p} D(x_2 - y_2, \cdot), \qquad p > 0, \tag{1.1.23}$$

where the symbol "$\cdot$" denotes dependence on s_1, s_2, y, and z, and let conditions b) and c) be replaced by the following conditions:

b*) For any region U_1 of finite volume the number N_1 of particles falling in U_1 is independent of both the positions $X_1, X_2, \dots$ of these particles and the values $Z_1, Z_2, \dots$ of the parameter θ characterizing them.

c*) The pairs $(X_1, Z_1), (X_2, Z_2), \dots$ of random variables are independent and identically distributed.

In the setting of the new conditions we suppose that the joint distribution $P(dx, dz)$ has the structure

$$P(dx, dz) = dx_1 Q(dx_2, dz). \tag{1.1.24}$$

It turns that in the case when the conditions (1.1.23) and (1.1.24) are in effect the distribution of $\nu(y)$ and $\bar{\nu}(y)$ is stable with parameter $\alpha = n_1/p$. The functions $\log f(t)$ in (1.1.10) and $\log \bar{f}(t)$ in (1.1.14) are computed by the same scheme as used above. The parameter $\alpha = n_1/p$ varies in the interval $(0, 1)$ if the condition $D_1 < \infty$ is used, and in $(1, 2)$ for $D_2 < \infty$.

If, by analogy with (1.1.22),

$$v(x, y, z) = |x_1 - y_1|^{-p} \varphi(|x_2 - y_2|, y, z) \psi(s_1, s_2, y), \tag{1.1.25}$$

where φ is real and ψ is such that

$$\psi(-s_1, s_2, y) = -\psi(s_1, s_2, y) \quad \text{or} \quad \psi(s_1, -s_2, y) = -\psi(s_1, s_2, y),$$

and

$$Q(dx_2, dz) = Q_1(dr_2, dz)\, ds_2,$$

then the distribution generated by v is spherically symmetric and stable, with parameter varying between 0 and 2.

We give some examples of concrete applied problems.

EXAMPLE 1. *The gravitational field of stars.* Holtsmark [114] considered the problem of the distribution of the gravitational effect of stars of various masses uniformly scattered in space with density ρ, on a unit mass at a chosen point (this problem was mentioned in the Introduction). The system of assumptions used by Holtsmark corresponds completely to the conditions in our model. Choose $U = V = R^3$, $\Theta = R^+ = (0, \infty)$ and

$$v(x, y, z) = Gz(x - y)/|x - y|^3. \tag{1.1.26}$$

The function v expresses the familiar Newton law of attraction and gives the force with which a body of mass z located at a point x attracts a unit mass located at a point y; G is the gravitational constant.

The form of v corresponds to the case (1.1.22). The condition $D_2 < \infty$ holds if it is assumed that $h = \mathsf{E}\theta^{3/2} < \infty$. Since the form (1.1.22) of v ensures the equality $M = 0$ in (1.1.11) because of oddness with respect to the variable s_1, we get (according to (1.1.19)) after some computations that the gravitational field is homogeneous and

$$\log f(t) = -\lambda|t|^{3/2} = -\frac{8\sqrt{2}}{15}\rho h G^{3/2}|t|^{3/2},$$

i.e., the total effect ν at any fixed point of all the stars has a spherically symmetric distribution with $\alpha = 3/2$ ($^{1/3}$).

Along with the value of the force ν of the gravitational effect of the system of stars on a unit mass at zero, we can consider also the rate of change of ν with time, which we denote by ν'. The equality

$$\nu = \sum_i v(X_i, 0, Z_i) = G \sum_i Z_i X_i |X_i|^{-3}$$

implies that

$$\nu' = G \sum_i Z_i \{Y_i |X_i|^{-3} - 3X_i(X_i, Y_i)|X_i|^{-5}\},$$

where Y_i is the rate of change of X_i; therefore, making the natural assumptions that the triples (X_i, Y_i, Z_i) of variables are mutually independent and X_i does not depend on the pair Y_i, Z_i, we get the situation described in Remark 3 in the form of the conditions a), b*), and c*).

Let $U = R^6, U_1 = R^3, U_2 = R^3, V = R^3$, and $\Theta = R^+$. Here we choose the influence function

$$v(x, y, z) = |x_1 - y_1|^{-3}\varphi(|x_2 - y_2|, z)\psi(s_1, s_2),$$

where $\varphi(r, z) = Grz$ and $\psi(s_1, s_2) = s_2 - 3s_1(s_1, s_2)$. The joint distribution of Y_1 and Z_1 is denoted by $P(dx_2, dz)$. The form of v shows that it is a special case of (1.1.23).

It follows from physical considerations that this distribution is spherically symmetric in the first argument, i.e., $P(dx_2, dz) = Q(dr_2, dz)\, ds_2$.

The situation we have described is easily seen to be a special case of the situation discussed in Remark 3. We have that $p = 3$ and $n_1 = 3$, whence $\alpha = 1$. Consequently, the rate of change ν' of the force at any point of space has a stable distribution with characteristic function of the form $\exp(-\lambda|t|)$, i.e., ν' has a three-dimensional Cauchy distribution. Simple computations show that $\lambda = -\pi Gkh/2$, where

$$h = \int_0^\infty \int_0^\infty rzQ(dr, dz)$$

and

$$k = \int_{S_1}\int_{S_2} |(e, \psi(s_1, s_2))|\, ds_1\, ds_2, \qquad e = (1, 0, 0).$$

EXAMPLE 2. *Temperature distribution in a nuclear reactor.* The problem is as follows. In a body D with boundary Γ the action of radiation at random times $\tau_1, \tau_2, \dots$ and at random places $Y_1, Y_2, \dots$ causes transient but very intense temperature flashes connected with nuclear transformations. Because of propagation of heat in the body, each flash arising at a time τ_i and at a point Y_i leads to an increase in the temperature at time $r > \tau_i$ at a chosen point $a \in D$ by an amount $L(r - \tau_i, Y_i, a)$, where the function $L(x_1, x_2, a)$ is a solution of the heat conduction equation

$$\frac{\partial L}{\partial x_1} = \Delta L + \delta(a - x_2)\delta(x_1) \tag{1.1.27}$$

with boundary condition $L + b\partial L/\partial n|_\Gamma = 0$. Here Δ is the Laplace operator with respect to the variable a, and $\delta(x)$ is the so-called Dirac δ-function (the generalized function δ can be taken to be the density of the degenerate distribution concentrated at the point x; see [81] concerning solution of this equation).

The temperature $T(r, a)$ at a point a at time r is composed of the contributions by the flashes that have occurred up to this time, i.e.,

$$T(r, a) = \sum\{L(r - \tau_i, Y_i, a) \colon \tau_i < r\}.$$

It is not hard to see that the problem is in complete correspondence to the conditions taken as the basis of the model of point sources of influence if the natural assumptions a)–c) are made about the pairs $X_i = (\tau_i, Y_i)$, with $U_1 = R^+, U_2 = D, U = U_1 \times U_2$, and $V = R^+$. If we assume that the flashes occur because of decay of atoms of a single type, then the set Θ can be taken to be empty, i.e., the function L can be taken to be independent of the parameter θ characterizing the atoms. Otherwise, L must be assumed to depend on this parameter. We analyze the first case, which is the simplest.

The fact that the solution L of (1.1.27) does not include physical characteristics of the material of the body being irradiated such as the coefficient of thermal conductivity, the energy emitted in the process of a single flash, the heat capacity of the substance, etc., is explained by the special choice of the units of measurement: length, temperature, and time; this choice allows us to treat the problem in a dimensionless setting (for details see [117], where this problem was first considered).

Since we are assuming that the collection of pairs $X_i = (\tau_i, Y_i)$ is a Poisson ensemble subject to the conditions b) and c), the answer to the question of the distribution of the random variable $T(r, a)$ connected with a body D of finite volume is simple. According to (1.1.5), the characteristic function $f_D(t)$

of this random variable has the form

$$\log f_D(t) = itM_D + \rho \int_D \int_0^r (e^{itL} - 1 - itL)\, dx_1\, dx_2, \tag{1.1.28}$$

where ρ is the average density of the pairs X_i in the space-time set $D \times R^+$, $L = L(x_1, x_2, a)$, and

$$M_D = \rho \int_D \int_0^r L(x_1, x_2, a)\, dx_1\, dx_2.$$

In studying the stochastic regularities of the variation of the temperature T at various space-time points (a, r) it is natural to exclude from T the constant component M_D and consider the quantity $\nu_D = T - M_D$.

At the point (a, r) we change the time origin, replacing r by $r + \omega$, and let ω go to ∞; this obviously corresponds to a steady-state mode. Let $\bar{\nu}_D$ denote the weak limit of the variables ν_D under this limit. According to (1.1.28), $\bar{\nu}_D$ has characteristic function

$$\log \bar{f}_D(t) = \rho \int_{D \times R^+} (e^{itL} - 1 - itL)\, dx_1\, dx_2.$$

Of course, the distribution of $\bar{\nu}_D$ is now independent of both a and r.

Subsequent success in computing $\bar{f}_D(t)$ is entirely dependent on whether we can find the explicit form of the influence function L.

We consider the case of a body D large enough so that it can be replaced in the computations by the whole space R^3 without great detriment. Here it is natural to take the boundary condition for equation (1.1.27) to be $L(x_1, x_2, a) \to 0$ as $|a| \to \infty$. A simple verification shows that in this case the function

$$L(x_1, x_2, a) = (4\pi x_1)^{-3/2} \exp\{-|x_2 - a|^2 / 2x_1\}$$

is a solution of (1.1.27). We let $x = (x_1, x_2)$ and $v(x) = L(x_1, x_2, 0)$, and compute the characteristic function $f(t)$ corresponding to $D = R^3$. We have that

$$\log \bar{f}(t) = \rho \int_{R^3} \int_{R^+} (e^{itv} - 1 - itv)\, dx_1\, dx_2. \tag{1.1.29}$$

It is not hard to see that this is a situation corresponding to (1.1.14) with influence function of the form (1.1.21). Furthermore, $\rho = 3/2, q = 1/2, n_1 = 1$, and $n_2 = 3$. Consequently, the distribution of ν is stable, with parameter $\alpha = 5/3$.

Let us carry out a more detailed computation of (1.1.29). With this aim we compute the measure $\mu(dv) = dx_1\, dx_2$, taking $\omega = v(u)$ as a new variable. We have that

$$P(\omega) = \int_{v > \omega} dx_1\, dx_2 = 4\pi \int_{v > \omega} r_2^2\, dx_1\, dr_2.$$

A corresponding change of variables gives us that

$$P(\omega) = \omega^{-5/3} \int_{v>1} 4\pi r_2^2 \, dr_2 \, dx_1 = k\omega^{-5/3}.$$

The constant k can be computed because the integration over the region $((r_2, x_1)\colon v > 1)$ reduces to iterated integration. Namely,

$$k = 4\pi \int_0^A dx_1 \int_0^B r_2^2 \, dr_2 = \frac{1}{10\pi} \left(\frac{3}{5}\right)^{3/2},$$

where $A = (4\pi)^{-1}$ and $B = (-6x_1 \log(4\pi x_1))^{-1/2}$. Thus,

$$\mu(d\omega) = \tfrac{5}{3} k\omega^{-8/3} \, d\omega.$$

The subsequent computation of (1.1.29) is not difficult. In summary, $\bar{\nu}$ has a one-dimensional stable distribution with characteristic function

$$\log \bar{f}(t) = \tfrac{3}{2}\Gamma\left(\tfrac{1}{3}\right) \rho k(-it)^{5/3},$$

which corresponds in the form (B) to the values

$$\alpha = \frac{5}{3}, \quad \beta = 1, \quad \gamma = 0, \quad \lambda = \frac{3}{20\pi} \left(\frac{3}{5}\right)^{3/2} \Gamma\left(\frac{1}{3}\right) \rho.$$

EXAMPLE 3. *Distribution of stresses in crystalline lattices.* As is well known, crystalline structures distinguish themselves by a rigid geometric arrangement of their atoms. However, this idealized representation of crystals turns out to be valid only in very small parts of them. Real crystals always have various disturbances in their structure, either because there are sometimes extraneous atoms at a prescribed location, or because there are no atoms at all at these locations.

Such anomalies in crystalline lattices are called *dislocations.* They may be scattered in the body of a crystal, but they may also be concentrated, forming lines and even surfaces with complicated configurations. We consider the case of point dislocations uniformly scattered in the body of a crystal with average density ρ. Each of the point dislocations creates an additional stress at the nodes of the crystalline lattice. This stress is described by a stress tensor $\mathbf{v} = (v_{kl}; k, l = 1, 2, 3;\ k \neq l)$ depending in general on the point y of the dislocation, on the point x at which the stress is being considered, and on the type of dislocation, which is given by the value of some parameter θ. In many concrete cases the tensor $v(x, y, \theta)$, which without damage to the essence of the problem can be regarded as simply a 6-dimensional vector, has the following form (at least for points x and y far from the surface of the crystal):

$$v(x, y, z) = |x - y|^{-3} D(s, z), \qquad s = (x - y)/|x - y|, \tag{1.1.30}$$

where the vector $D(s, z)$ has the property

$$D(s, z) = -D(-s, z). \tag{1.1.31}$$

The total stress at y created by point dislocations located at points X_i and having corresponding types Z_i is equal to the vector sum $\sum_i v(X_i, y, Z_i)$.

It is completely obvious that here there is a basis for using the model of point sources of influence. Indeed, U is the body of the crystal in R^3, and the point dislocations X_j in U form the system $\mathfrak{A}$ of particles that can be assumed to satisfy conditions a)–c) of the general model (familiarity with the specialized literature (see [139] and [108]) corroborates the possibility of this. The "action" of the individual dislocations is described by a single influence function $v(x, y, z)$ with values in some set $V \subset R^6$. The form of v coincides with (1.1.15) ($p = 3, n_1 = 3, n_2 = 0$), and its property of central symmetry allows the use of (1.1.10) with the value $M = 0$ for computing the total effect ν of the system $\mathfrak{A}$.

In summary, ν has a stable centrally symmetric distribution in R^6 with the value $\alpha = 1$:

$$\log f(t) = -\int_S |(t, v)| \chi(dv), \tag{1.1.32}$$

where

$$\chi(B) = \rho\frac{\pi}{6}\int_{D/|D| \in B} |D(s, z)|\, ds\, P_\theta(dz), \qquad B \in S.$$

This distribution is a nonspherical analogue of a six-dimensional Cauchy distribution. Its projections on the coordinate axes are ordinary Cauchy distributions (with different scale parameters λ in general). We remark that [141] contains a solution of this problem on the level of computation of the distributions for the coordinates of the total stresses.

EXAMPLE 4. *Distribution of the magnetic field generated by a network of elementary magnets.** The following scheme is encountered in certain physics problems.

Actually, we consider here not one but two examples. However, they both involve computing the distribution of a magnetic field in the space $U = R^3$. It is generated by the combined and independent action of a system of elementary (point) magnets randomly dispersed in U with average density ρ. This system is assumed to be a Poisson ensemble satisfying conditions b) and c).

---

*My attention was directed to this interesting example of an application of the model of point sources of influence by D. Vandev and N. Trendafilov, who are colleagues at the Institute of Mathematics of the Bulgarian Academy of Sciences, and I am very grateful to them for this.

In the case $U = R^3$ the influence function $v(x, y, z)$ must be invariant with respect to a translation of the variable x; therefore, it depends only on the difference $x - y$, and we assume without loss of generality that $y = 0$.

The following two influence functions are considered ($U = V = \Theta = R^3$):

$$v_1(x, 0, z) = c\frac{|z|}{|x|^2}(s \times e), \tag{1.1.33}$$

$$v_2(x, 0, z) = c\frac{|z|}{|x|^3}(3(s, e)s - e), \tag{1.1.34}$$

where $s = x/|x|, e = z/|z|$, $(s \times e)$ is the vector product, and c is the magnetic constant.

Physically, the characteristic $\theta = z$ signifies the magnetic moment arising at $y = 0$ under the action of an elementary magnet located at the point x. The magnetic moment $\theta = z$ created by the elementary magnet at x can take various values, depending on the individual properties of the magnet. With this characteristic regarded as random it is natural to assume that the distribution $P_\theta(dz)$ is centrally symmetric, and this will be done below.

The first influence function describes a picture arising in the theory of a constant (more precisely, steady-state) magnetic field created by charges moving in the immediate vicinity of their average positions. This case corresponds to the so-called Biot-Savart relation.

The second influence function corresponds to elementary magnets of dipole type. While in the first case we are dealing with a model for a magnetic field in the microcosm, on the level of elementary particles, in the second case the model relates to the macrocosm—the world of planets and stars scattered in space. If a star or planet has a magnetic field, then on the scale of the universe it can be regarded as a point elementary magnet of dipole type.

The influence function $v_1(x, 0, z)$, as well as $v_2(x, 0, z)$, has central symmetry with respect to x, i.e., $v_l(-x, 0, z) = -v_l(x, 0, z)$. They also have the same symmetry with respect to z, i.e., $v_l(x, 0, -z) = -v_l(x, 0, z)$.

These properties, together with the assumption, natural for the problems under consideration, that the distribution $P_\theta(dz)$ of the magnetic moment is centrally symmetric, enable us to consider a centered total magnetic field strength of the type $\bar{\nu}_l = \bar{\nu}_l(0)$ with characteristic function (1.1.14). With the assumption made about $P_\theta(dz)$, the characteristic function of the variable $\bar{\nu}_l$ can be given the form (recall that $U = \Theta = R^3$):

$$\bar{f}_l(t) = \exp\{\rho \int_{U \times \Theta} (\cos(t, v_l) - 1)\, dx\, P_\theta(dz).$$

The form of the influence functions allows us at once to claim that $\bar{\nu}_1$ has a three-dimensional stable distribution with parameter $\alpha = \frac{3}{2}$, while $\bar{\nu}_2$ has a three-dimensional stable distribution with parameter $\alpha = 1$.

More detailed computations show that

$$\log \bar{f}_1(t) = -\tfrac{1}{3}\sqrt{2\pi}\rho C^{3/2}|t|^{3/2}A_1(t),$$

where

$$A_1(t) = \int_\Theta \int_S \left| \left(\frac{t}{|t|}, (s \times e) \right) \right|^{3/2} |z|^{3/2}\, ds\, P_\theta(dz),$$

and

$$\log \bar{f}_2(t) = -\tfrac{\pi}{6}\rho C|t|A_2(t),$$

where

$$A_2(t) = \int_\Theta \int_S \left| \left(\frac{t}{|t|}, 3(s,e)s - e \right) \right| \cdot |z|\, ds\, P_\theta(dz).$$

If the distribution $P_\theta(dz)$ is assumed to be not only centrally symmetric but also spherically symmetric, then the functions $A_l(t)$ are constants, i.e., the distributions of the $\bar{\nu}_l$ are spherically symmetric stable laws; $\bar{\nu}_1$ has the Holtsmark distribution (already encountered in Example 1) ([1.3]), and $\bar{\nu}_2$ has a three-dimensional Cauchy distribution.

§1.2. Stable laws in problems in radio engineering and electronics

We shall become acquainted below with two examples of the occurrence of stable laws in models used in practice. These models are completely different and are combined here for purely formal reasons.

The first example is connected with computing the performance of systems of radio relay stations (in engineering practice they are called radio relay communications lines). In a mathematical setting the part of the general problem to which we give our attention takes its origin from the paper [135], and its solution with the use of stable laws makes up the content of several papers, of which [136] must be regarded as the main one. The transmission of high-quality radio communications over great distances (for example, television transmissions) poses for engineers not only the problem of relaying high-frequency radio signals that can be received only line of sight, but also the problem of combating noise distortions. The following seems to be one of the simplest models in which it is possible to trace both the effects themselves that arise here and the ways of analyzing them quantitatively.

Consider a vector $\mathbf{a} \in R^2$ rotating with a large angular velocity ω. The projection of $\mathbf{a}$ onto the x-axis at time t, under the condition that its motion began from the position defined by the angle φ, is the periodic function

$y(t) = |\mathbf{a}|\cos(\omega t + \varphi)$. The oscillatory excitation at the output of the radio transmitter is described by the function $y(t)$, in which the quantity $|\mathbf{a}|$ is the amplitude of the radio signal (its power is proportional to $|\mathbf{a}|^2$), ω is its frequency, and $\omega t + \varphi$ is its phase (at the time t). The quantities $|\mathbf{a}|$ and ω stay constant at the transmitter output, while transmission of the useful signal is accomplished by modulating the phase of the signal, i.e., by changing φ.

If the radio signal is received by the receiver without change, then its information content—the phase shift φ—could be recovered without difficulties. However, in real circumstances the radio signal comes to the receiver in a distorted and weakened form. This is due to the fact that multiple beam scattering of the signal takes place after output from the transmitter (even if one tries to give the signal a pencil beam direction). As a result, part of the radiation does not hit the receiver antenna because of the restricted size of the latter, and this leads to a drop in the power of the signal, which is naturally greater, the farther the receiver is from the transmitter. The part of the radiation reaching the antenna travels the distance not in a single trajectory, but in a pencil of trajectories. The different trajectories, in turn, cause the components of the pencils to reach the receiver with modified phase. Since the changes are small in themselves, it follows that in combination with the large frequency ω and the fact that there are many such changes we have a distribution that is close to uniform with respect to the phases of the vectors associated with them. As a result, the total two-dimensional vector $\mathbf{X}$ has the nature of a random vector with a circular normal distribution. This implies that the length $|\mathbf{X}|$ of $\mathbf{X}$ has a Rayleigh distribution with density

$$p(x) = D^{-2}x\exp(-x^2/2D^2), \qquad x \geq 0, \tag{1.2.1}$$

where D^2 is the variance of the components of $\mathbf{X}$. The weakening of the power of the signal and the transformation of the determinate (with respect to the value of the amplitude) radio signal into a random quantity does not yet cause serious difficulties for distinguishing the useful signal—the phase shift φ. The complications begin after the incursion of the noise effects created by the receiver apparatus itself.

The influence of the noise can be represented by adding to $\mathbf{X}$ a two-dimensional random vector $\mathbf{Y}$ having circular normal distribution with variance of the components equal to σ^2. The vector $\mathbf{X} + \mathbf{Y}$ has phase differing from that of $\mathbf{X}$ by an angle ψ determined by

$$\tan\psi = \frac{|\mathbf{Y}_{\mathrm{t}}|}{|\mathbf{X} + \mathbf{Y}_{\mathrm{r}}|}\xi(\mathbf{Y}_{\mathrm{t}}), \qquad -\pi < \psi < \pi, \tag{1.2.2}$$

where $\mathbf{Y}_{\mathrm{r}}$ and $\mathbf{Y}_{\mathrm{t}}$ are the radial and tangential components of $\mathbf{Y}$ with respect to $\mathbf{X}$, and $\xi(\mathbf{Y}_{\mathrm{t}})$ is a random variable determined by the direction of $\mathbf{Y}_{\mathrm{t}}$ and taking the values $+1$ and -1 with probabilities $\frac{1}{2}$. Since the whole picture evolves in time, the vectors $\mathbf{X}$ and $\mathbf{Y}$ and the angle ψ determined by them depend on t, i.e., are random processes. The processes $\mathbf{X}(t)$ and $\mathbf{Y}(t)$ (and, consequently, $\psi(t)$), which are connected with different sections of the relay, can be regarded as independent and stationary.

The total distortion $\bar{\psi}$ of the information phase φ at time t after passage of N sections of the relay is determined (if the delay in passing from section to section is ignored) by the equality

$$\tan\bar{\psi}(t) = \tan(\psi_1(t) + \cdots + \psi_N(t)), \qquad -\pi < \bar{\psi} < \pi. \tag{1.2.3}$$

The distribution of $\psi_j(t)$ is symmetric for each t, as is clear from (1.2.2). Therefore, $\mathsf{E}\psi_j(t) = 0$. The quantity $\bar{\psi}(t)$ obviously has the same property. With this property taken into account, the value of the standard deviation of $\bar{\psi}(t)$ from zero on some interval of time with fixed length is usually taken as a measure characterizing the level of noise in the transmission. For example, on an interval of 1 sec,

$$\Psi^2 = \int_0^1 \bar{\psi}^2(t)\,dt.$$

The estimation of Ψ^2 is a fairly complicated problem in the analytic setting. It can be simplified by a certain coarsening. Namely, first, according to (1.2.3),

$$\begin{aligned}\Psi^2 &\le \int_0^1 (\psi_1(t) + \cdots + \psi_N(t))^2\,dt \\ &= \sum_j \int_0^1 \psi_j^2(t)\,dt + \sum_{ij} \int_0^1 \psi_i(t)\psi_j(t)\,dt.\end{aligned} \tag{1.2.4}$$

Second, considering that the stationary processes $\psi_j(t)$ are independent and that $\mathsf{E}\psi_i(t) = 0$, which implies that

$$\int_0^1 \psi_i(t)\psi_j(t)\,dt \approx \mathsf{E}\psi_i(t)\psi_j(t) = 0, \qquad i \ne j,$$

we can ignore the second term in (1.2.4), taking the sum

$$\sum_j \int_0^1 \psi_j^2(t)\,dt \tag{1.2.5}$$

as an estimator of Ψ^2. The terms of this sum, in turn, admit a simplified approximate expression if the quantities $\varepsilon_j = \sigma_j/D_j$ are taken to be small, where σ_j^2 and D_j^2 are the variances of the components of $\mathbf{Y}_j$ and $\mathbf{X}_j$ connected with the jth part of the relay. Indeed, according to (1.2.2),

$$\int_0^1 \psi_j^2(t)\,dt = \int_0^1 \arctan^2\left(\frac{\varepsilon_j|\mathbf{U}_j|}{|\mathbf{V}_j + \varepsilon_j\mathbf{U}_j'|}\right)dt,$$

where $\mathbf{U}_j = \mathbf{Y}_{mj}/\sigma_j, \mathbf{U}'_j = \mathbf{Y}_{pj}/\sigma_j$, and $\mathbf{V}_j = \mathbf{X}_j/D_j$. Consequently,

$$\int_0^1 \psi_j^2(t)\,dt \le \int_0^1 \frac{\varepsilon_j^2|\mathbf{U}_j|^2}{|\mathbf{V}_j+\varepsilon_j\mathbf{U}'_j|^2}\,dt \approx \int_0^1 \varepsilon_j^2\frac{|\mathbf{U}_j|^2}{|\mathbf{V}_j|^2}\,dt.$$

The next step in simplifying the estimator $\tilde{\Psi}_2$ has to do with the circumstance that the vectors $\mathbf{V}_j(t)$ and $\mathbf{U}_j(t)$ vary with sharply different intensity. For example, $\mathbf{V}_j(t)$ is practically constant on a time interval of the length being considered, while $\mathbf{U}_j(t)$ performs an enormous number of rotations on the same interval (of the order of 10^6). For this reason,

$$\begin{aligned}\int_0^1 \varepsilon_j^2\frac{|\mathbf{U}_j(t)|^2}{|\mathbf{V}_j(t)|^2}\,dt &\approx \frac{\varepsilon_j^2}{|\mathbf{V}_j|^2}\int_0^1 |\mathbf{U}_j(t)|^2\,dt \\ &\approx \frac{\varepsilon_j^2}{|\mathbf{V}_j|^2}\mathsf{E}|\mathbf{U}_j(1)|^2 = \varepsilon_j^2|\mathbf{V}_j|^{-2}.\end{aligned}$$

Consequently, a simplified estimator of Ψ^2 can be represented as a sum

$$\tilde{\Psi}^2 = \sum_j \varepsilon_j^2|\mathbf{V}_j|^{-2},$$

where the ε_j are constants small in magnitude, and the $\mathbf{V}_j$ are independent random vectors having normal distribution with covariance matrix with identity. Thus, as an estimator $\tilde{\Psi}^2$ of the random variable Ψ^2 we get a sum of N independent random variables $\varepsilon_j^2|\mathbf{V}_j|^{-2}$ whose distribution functions have the form (as follows from (1.2.1))

$$F_j(x) = \exp(-\varepsilon_j^2/2x), \qquad x > 0.$$

Since the terms $\varepsilon_j^2|\mathbf{V}_j|^{-2}$ are small, while their number is sufficiently large, the distribution of $\tilde{\Psi}^2$ can be sufficiently well approximated by an infinitely divisible law whose spectral function is not difficult to compute with the help of Theorem B in the Introduction. Namely, $H(x) = 0$ if $x < 0$, and

$$H(x) \approx \sum_j (F_j(x) - 1) \approx -\frac{1}{2}\sum_j \varepsilon_j^2 x^{-1}$$

if $x > 0$. This spectral function corresponds, according to Theorem C.1, to a stable distribution with parameters $\alpha = 1$ and $\beta = 1$ (see Remark I.4 about this). We present a more accurate approximation of the distribution of $\tilde{\Psi}_N^2$, clothing this computation as a limit theorem.

Consider a sequence of series of numbers $\varepsilon_{n1}^2, \ldots, \varepsilon_{nn}^2, n = 1, 2, \ldots$, such that $\sum_j \varepsilon_{nj}^2 = 1$ and $\delta_n = \max_j \varepsilon_{nj}^2 \to 0$ as $n \to \infty$, and the sequence of centered sums

$$Z_n = X_{n1} + \cdots + X_{nm} - A_n$$

of independent random variables X_{nj} with distribution functions $F_{nj}(x) = \exp(-\varepsilon_{nj}^2/x), x > 0$. The choice of centering constants A_n will be made later.

The Laplace-Stieltjes transforms of the distributions F_{nj} can be expressed with the help of the Macdonald function $K_1(z)$ as follows (see [27], 3.471(9)):

$$\mathsf{E}\exp(-sX_{nj}) = 2\sqrt{s\varepsilon_{nj}^2}K_1(2\sqrt{s\varepsilon_{nj}^2}), \qquad s > 0.$$

Consequently,

$$\begin{aligned}\log\mathsf{E}\exp(-sZ_n) &= sA_n + \sum_j \log\mathsf{E}\exp(-sX_{nj}) \\ &= sA_n + \sum_j \log\left\{2\sqrt{s\varepsilon_{nj}^2}K_1(2\sqrt{s\varepsilon_{nj}^2})\right\}. \qquad (1.2.6)\end{aligned}$$

If we take into account the fact that

$$2xK_1(2x) = 1 + x^2\log x^2 + (2\mathbb{C} - 1)x^2 + O(x^4\log x^2)$$

as $s \to 0$ (see [27], 8.446; $\mathbb{C}$ is the Euler constant), along with the requirements noted above about the quantities ε_{nj}^2, then the right-hand side of (1.2.6) can be transformed to the form

$$sA_n + s\log s + \left(2\mathbb{C} - 1 + \sum_j \varepsilon_{nj}^2\log\varepsilon_{nj}^2\right)s + O(\delta_n\log(1/\delta_n)).$$

Choose

$$A_n = -\sum_j \varepsilon_{nj}^2\log\varepsilon_{nj}^2 + 1 - 2\mathbb{C}.$$

Then (1.2.6) gives us that

$$\log\mathsf{E}\exp(-sZ_n) = s\log s + O(\delta_n\log(1/\delta_n)).$$

The function $\exp(s\log s)$ is the Laplace-Stieltjes transform of the distribution $G(x,1,1)$ (see, for example, Theorem 2.6.1); hence,

$$\mathsf{P}(Z_n < x) \to G(x,1,1), \qquad n \to \infty.$$

From this, setting $B_N = \sum_1^N \varepsilon_j^2$ in the original problem and letting $\varepsilon_{Nj}^2 = \varepsilon_j^2/B_N$, we can assert that if δ_N is small, then

$$\mathsf{P}(\Psi^2 < x) \approx \mathsf{P}(\tilde{\Psi}^2 < x) \approx G(xB_N^{-1} - A_N, 1, 1).$$

The second example has to do with computing a model for a homogeneous electrical line, which can be, for example, an electrical cable or the circuit of cascaded four-pole networks ([1.4]). A number of properties of such a line can be described with the help of a so-called time function $F(t,\lambda), t \geq 0$, which shows the reaction of a line of length $\lambda > 0$ to a perturbation of "unit shock" type at the initial time. The function $F'_t(t, \lambda)$ is called the pulse reaction of the line, and its Fourier transform $f(\omega, \lambda)$ is called the frequency characteristic of a homogeneous line of length λ.*

It is known from electrical circuit theory that when a homogeneous line of length $\lambda = \lambda_1 + \lambda_2$ is partitioned into sequentially connected sections of lengths λ_1 and λ_2, its time function $F(t, \lambda)$ is formed from the time functions $F(t, \lambda_1)$ and $F(t, \lambda_2)$ of the separate parts via their convolution, i.e.,

$$F(t, \lambda_1 + \lambda_2) = F(t, \lambda_1) * F(t, \lambda_2),$$

which is equivalent to multiplication of the corresponding frequency characteristics:

$$f(\omega, \lambda_1 + \lambda_2) = f(\omega, \lambda_1) f(\omega, \lambda_2).$$

It follows from this that for any $\lambda > 0$

$$f(\omega, \lambda) = f^{\lambda}(\omega, 1). \tag{1.2.7}$$

Connected with the function $f(\omega, \lambda)$ are the quantities

$$a(\omega, \lambda) = -\operatorname{Re} \log f(\omega, \lambda) = \lambda a(\omega, 1),$$
$$b(\omega, \lambda) = \operatorname{Im} \log f(\omega, \lambda) = \lambda b(\omega, 1),$$

called, respectively, the damping and phase of the frequency characteristic.

It turns out that cases when the time function of the line is nondecreasing on the time axis $t > 0$ are not rare. If, in addition, the damping at zero frequency ($\omega = 0$) is equal to zero, then the time function $F(t, \lambda)$ can be regarded as a distribution function concentrated on the half-line $t > 0$. In this case (1.2.7) implies that $F(t, \lambda)$ is an infinitely divisible distribution with characteristic function $f(\omega, \lambda)$ of the form

$$\log f(\omega, \lambda) = \lambda \left(i\omega\gamma + \int_0^\infty (e^{i\omega u} - 1)\, dH(u) \right), \qquad \gamma \geq 0.$$

For frequency characteristics of this form the phase $b(\omega, \lambda)$ can be recovered from the damping $a(\omega, \lambda)$ to within the term $\omega\gamma\lambda$.

Thus, if $a(\omega, \lambda) = \lambda c |\omega|^\alpha$, where $0 < \alpha \leq 2$ and c is a positive constant, then the corresponding damping of the phase has the form

$$b(\omega, \lambda) = \lambda c (\omega\gamma + \tan \tfrac{\pi}{2}\alpha\, |\omega|^\alpha \operatorname{sgn} \omega), \qquad \gamma \geq 0,$$

*Comparing the interest represented by this example with the scope of the exposition, we do not go into details here about the physical content of the functions $F(t, \lambda)$ and $f(\omega, \lambda)$. Those desiring to acquaint themselves more thoroughly with these concepts can turn to [132], where we borrowed the example (the notation was changed) and where there is a list of literature relevant to this problem.

while the time function $F(t, \lambda)$ connected with such a frequency characteristic is a stable distribution, i.e.,

$$F(t, \lambda) = G_A(t, \alpha, 1, \gamma, c\lambda). \tag{1.2.8}$$

Certain forms of cables which have a power character of damping are known in electrical circuit theory. For example, $a(\omega, \lambda) = \lambda c|\omega|^{1/2}$ for the so-called noninductive and coaxial cables. Consequently, according to (1.2.8), the time function of such cables has the form

$$F(t, \lambda) = G_A(t, \tfrac{1}{2}, 1, \gamma, c\lambda) = 2[1 - \Phi(c\lambda(t - c\gamma\lambda)^{-1/2})],$$

where Φ is the distribution function of the standard normal law. The pulse reaction $F'_t(t, \lambda)$ in this case has the simple explicit expression

$$F'_t(t, \lambda) = \frac{c\lambda}{\sqrt{2\pi}}(x - c\gamma\lambda)^{-3/2} \exp\left\{-\frac{c^2\lambda^2}{2}(x - c\gamma\lambda)^{-1}\right\}.$$

§1.3. Stable laws in economics and biology

In the 1960s stable laws began to attract the attention of scholars working in the area of economics, biology, sociology, and mathematical linguistics, due to a series of publications by the American mathematician Mandelbrot and his successors (see [118]–[128]). The fact is that statistical principles described by so-called Zipf-Pareto distributions had been empirically discovered fairly long ago in all these areas of knowledge. Discrete distributions of this type have the form

$$p_k = ck^{-1-\alpha}, \qquad k \geq 1, \ \alpha > 0,$$

while their continuous analogues (densities) have the form

$$p(x) = cx^{-1-\alpha}, \qquad x \geq a > 0.$$

Mandelbrot called attention to the fact that the use of the extremal stable distributions (corresponding to the value $\beta = 1$) for describing empirical principles was preferable to the use of Zipf-Pareto distributions for a number of reasons. It can be seen from many publications, both theoretical and applied, that Mandelbrot's ideas have more and more won the recognition of experts. In this way the hope arose of confirming empirically established principles in the framework of mathematical models and at the same time clearing up the mechanism for the formation of these principles.

Below we present an example of a stochastic model that can be used to explain a so-called law.

In 1922 the English biologist Willis published a book [254] dealing with the study of statistical principles in evolution processes. One of the main results in his investigations was the discovery of the following principle observed in nature.

Biological species are commonly taken as the primary elements in the classification of living organisms. The species are then combined into coarser groups called genera.

We consider a sequence of genera in the animal or plant world, ordering them according to the number of species occurring in them. Then we calculate how many genera in the total number contain a single species, two species, etc.; let these numbers be $M_1, M_2, \ldots$, etc., and let $M = M_1 + M_2 + \cdots$ be the total number of genera involved. Next, form the sequence of frequencies $p_k = M_k/M$ $(k = 1, 2, \ldots)$, which allows us to represent the probability of finding exactly k species in a randomly chosen genus.

In the language of probability theory, the discovery of Willis consisted in seeing that for genera containing sufficiently many species, i.e., for $n \geq n_0$

$$\sum_{k>n} p_k \approx An^{-1/2}, \tag{1.3.1}$$

where A is a constant.

In other words, the probability of finding in a genus at least n species decreases with increasing n at the rate $1/\sqrt{n}$.

In 1924 the English mathematician Yule [253] was able to find a theoretical basis for the relation (1.3.1) in the framework of a stochastic model that can be included in the theory of branching processes.

We demonstrate below (in a way different from that in [253]) how the Willis-Yule law could be explained. Most interesting here is the occurrence in our model of a stable law and its connection with the principle (1.3.1), which is traditionally associated with the Zipf-Pareto distribution.

A model of a random branching process with two types of particles will be taken as a basis for the arguments.

Let us consider the reproduction process as time passes for particles of two types. This goes as follows. At the beginning of the process we have a single particle of type T_1. During a unit of time this particle turns into a collection of μ_{10} particles of type T_0 and μ_{11} particles of type T_1. The particles of type T_0 remain unchanged, but the particles of type T_1 can be further transformed. The numbers μ_{10} and μ_{11} are random variables. We impose some conditions on this random process of reproduction of particles.

1. The transformation of each of the particles of type T_1 takes place independently of its history and independently of what happens with the other particles.

2. The joint probability distribution for the values of the random variables μ_{10} and μ_{11} at the time of transformation of a particle of type T_1 stays the

same for all particles of this type and does not depend on the time at which the transformation takes place. Furthermore, the total number $\mu_{10} + \mu_{11}$ of "descendants" produced in the course of a single transformation cannot exceed some constant number h.

3. The mean number $\delta = \mathsf{E}\mu_{10}$ of particles of type T_0 produced in a single act of transformation is positive, while the mean number of particles of type T_1 is equal to 1. Moreover, $c = \operatorname{Var}\mu_{11} > 0$. The last condition, in particular, does not permit a particle of type T_1 to produce with probability 1 only one particle of type T_1 at the time of transformation. Together with the condition $\mathsf{E}\mu_{11} = 1$, this means that with some nonzero probability a particle of type T_1 does not produce any particles of the same type. Since particles of type T_0 are not transformed, further transformation of the given branch is thereby terminated.

It is known that transformations of a single particle of type T_1 subject to the above conditions ultimately (with probability 1) cease. Its transformation process gives a certain random number U of particles of type T_0, called final particles.

Let us assume that there are initially n particles of type T_1. In the process of their transformations and the transformations of their "descendants" (this process, as mentioned, stops with probability 1) there appear $U_1, \ldots, U_n$ final particles produced by the first, second, etc., initial particle of type T_1.

According to conditions 1 and 2, the random variables U_j are independent and identically distributed. Let us form a normalized sum of these random variables:

$$V_n = (U_1 + \cdots + U_n)(2\delta n^2/c)^{-1}.$$

It can be proved that the distributions of the random variables V_n converge as $n \to \infty$ to the stable distribution with parameters $\alpha = 1/2$, $\beta = 1$, $\gamma = 0$, and $\lambda = 1$ (to the Lévy law).

In the recognized opinion of experts, the diversity of biological species came about as the result of evolution of living beings in the course of strict natural selection. The Earth's climate has changed, both globally and in regional parts, and this has created new requirements on plants and animals—to survive they have had to acquire new qualities.

This has been accomplished due to the variability of characteristics in new generations and due to natural selection of the fittest of their representatives. If we attempt to formalize this selection, then in simplified form the picture will remind us of the random branching process described above. By a *fractional part* we understand some portion of the population united by some characteristic important from the viewpoint of survival, and we understand transformations of parts to be the changes arising in a long series of generations; the fractional parts of type T_0 (the final parts) should be interpreted as the "descendants" that were able to secure qualities needed for stable existence, while the parts of type T_1 should be interpreted as "descendants" lacking such qualities and thus doomed either to extinction or to relatively rapid change.

Of course, the model with final parts is an idealization of the phenomenon. Even well-adapted species are subjected to subsequent changes. But this happens much more rarely and slowly than with the groups which "feel" discomfort in their condition and are pressed toward further variability. Therefore, a model in which fractional parts of both types T_0 and T_1 are transformed is in closer correspondence with the real situation: at the end of its existence a T_0-part turns into μ_{00} T_0-parts and μ_{01} T_1-parts, and a T_1-part turns into μ_{10} T_0-parts and μ_{11} T_1-parts. It is assumed that

$$\mathsf{E}\mu_{00} = 1, \quad \mathsf{E}\mu_{11} = 1, \quad 0 < c_1 < c = \operatorname{Var}\mu_{11} < c_2, \quad \sum \mu_{ij} < h,$$

as in the original model, and that the numbers $\delta = \mathsf{E}\mu_{10}, \varepsilon = \mathsf{E}\mu_{01}$, and $\sigma = \operatorname{Var}\mu_{00}$ are positive but small in absolute value.

In this variant of the process the transformations of the fractional parts do not cease with probability 1, as was the case in the original process. Changes take place over an arbitrarily long period of time with positive probability. This property of the process corresponds more closely to real evolutionary processes. At the same time, if the value of ε is small, then the average lifetime of parts of type T_0 is correspondingly large. The total number of parts in this model increases on the average. However, on any bounded interval of time this growth depends on the value of the sum $\eta = \varepsilon + \delta$ and can be arbitrarily slow if this sum is sufficiently small.

All the foregoing agrees well with our ideas about the flow of evolutionary processes.

Before passing to an explanation of the Willis-Yule law we dwell further on a certain feature of the new variant of the model.

In the new variant there are no final fractional parts. Therefore, one can speak only of the number $V_n(\Delta) = U_1(\Delta) + \cdots + U_n(\Delta)$ of long-lived parts of type T_0 at a time Δ sufficiently long after the initial time.

The principal way in which we see the second model as differing from the first is that T_0-parts are not final but simply long-lived in it. The quantity $\varepsilon = \mathsf{E}\mu_{01}$ will be regarded as the main parameter in the second model, because the first model is obtained by fixing the values h, δ, and c and letting $\varepsilon \to 0$. In other words, in a certain sense the second model exhibits continuity with respect to variation of ε.

Therefore, it seems quite likely (and is actually corroborated by computations) that the distribution of the sum $V_n(\Delta)$ depends continuously on ε as $\varepsilon \to 0$. This can serve as a basis for nevertheless replacing the second model, which better reflects the evolutionary process actually in progress, by the first model in computations, since the latter is more convenient on the analytic level. The analysis of the behavior of the distribution of the sum $V_n(\Delta)$ in the first model can, in turn, be replaced by an analysis of the behavior of this sum corresponding to $\Delta = \infty$. The last condition means that we are considering the distribution of the total number of final fractional parts generated by the original n parts of type T_1.

The validity of this replacement in analyzing the asymptotic behavior of the distribution can be justified. Of course, the error is all the smaller, the greater Δ is. Thus, for our purposes we can use the limit theorem given for the first model with respect to the distributions of the variables V_n. In refined form it asserts that as $n \to \infty$

$$\mathsf{P}(V_n > x) = \mathsf{P}(Y < x)(1 + o(1)),$$

where $Y = Y(\frac{1}{2}, 1, 0, 1)$ and $o(1)$ is a quantity tending to zero uniformly with respect to x.

When considering a set of species (parts of type T_0) within the scope of a single genus it is natural to assume that they all have a common root (i.e., are generated by a single part of type T_1). Between species and genus, of course, there were also intermediate forms not constituting independent units of classification. We think of them as $n-1$ initial parts of type T_1 that in the final analysis produce $U_2 + \cdots + U_n$ parts of type T_0. The initial part of type T_1 also produces U_1 parts of type T_0. Together they produce $W = U_1 + \cdots + U_n$ parts of type T_0. The distribution of this sum is precisely the distribution of the number of species in a genus.

Since δ is small while the quantity c is bounded above and bounded away from zero, the ratio $B^2 = 2\delta n^2/c$ can be regarded as a quantity bounded both above and below (by suitably choosing n). Then the number of species in a genus has a distribution that can be approximately expressed by the stable Lévy law:

$$\mathsf{P}(W > xB^2) = \mathsf{P}(V_n > x) \approx \mathsf{P}(Y > x).$$

If we consider that for large x

$$G(x, \frac{1}{2}, 1) = \mathsf{P}(Y > x) \approx \frac{1}{\sqrt{\pi}} x^{-1/2},$$

then we obtain the Willis-Yule principle (1.3.1):

$$\mathsf{P}(W > x) \approx \mathsf{P}(Y > xB^{-2}) \approx \frac{B}{\sqrt{\pi}} x^{-1/2}.$$

CHAPTER 2

Analytic Properties of Distributions in the Family $\mathfrak{S}$

In contrast to the next chapter, where specific properties of strictly stable laws are considered, the material in the present chapter relates to all the stable laws. The analytic basis for the presentation here is the explicit expression for the characteristic functions of the distributions in the family $\mathfrak{S}$ in one of the two forms (A) or (B). The specific form of parametrization with which the formulation of a particular property is associated is either stated explicitly or indicated by a corresponding subscript. The absence of such a specification means that the formulation is associated with the form (B).

The stable laws (and the random variables having such laws) with parameter values $\gamma = 0$ and $\lambda = 1$ in the form (B) will be called *standard*. The set of standard stable laws is denoted by $\mathfrak{S}_0$.

The concept of a standard law is connected with a definite form (we chose the form (B) as most suitable for this purpose from an analytic point of view) because the sets of standard laws with values $\gamma = 0$ and $\lambda = 1$ do not coincide in the various forms. It is not difficult to establish the values of the parameters of standard laws in the forms (A) and (M) by using the formulas for passing from the form (B) to the forms (A) and (M).

In each of the forms of parametrization for the family $\mathfrak{S}$ or its subfamilies $\mathfrak{W}$ and $\mathfrak{S}_0$ there is an indicated domain of variation of the parameters corresponding to this form, which we shall call the *domain of admissible values of the parameters*.

§2.1. Elementary properties of stable laws

The very simple expression for the characteristic functions of stable laws enables us at once to note a whole series of interrelations between them. With these relations as a basis, it becomes possible, in particular, for us to reduce the study of the analytic properties of the distributions in $\mathfrak{S}$ to the study of the properties of distributions in various subfamilies of $\mathfrak{S}$.

It was already mentioned in the Introduction that in many cases it is convenient and intuitive to assign to relations between stable distributions the same form of relations between random variables having these distributions.* This form is employed for formulating the properties given below (with the exception of the last property). The reverse passage from relations between random variables to relations between the corresponding distributions is sufficiently simple and does not require additional explanations.

PROPERTY 2.1. *Any two admissible parameter quadruples* $(\alpha, \beta, \gamma, \lambda)$ *and* $(\alpha, \beta, \gamma', \lambda')$ *uniquely determine real numbers* $a > 0$ *and* b *such that*

$$Y(\alpha, \beta, \gamma, \lambda) \stackrel{d}{=} aY(\alpha, \beta, \gamma', \lambda') + \lambda b. \tag{2.1.1}$$

In the form (A) *the dependence of* a *and* b *on the parameters is expressed as follows*:

$$a = (\lambda/\lambda')^{1/\alpha},$$

$$b = \begin{cases} \gamma - \gamma'(\lambda/\lambda')^{1/\alpha - 1} & \text{if } \alpha \neq 1, \\ \gamma - \gamma' + \frac{2}{\pi}\beta \log(\lambda/\lambda') & \text{if } \alpha = 1. \end{cases}$$

We single out an important particular case in (2.1.1). Let $\gamma' = 0$ and $\lambda' = 1$. Then

$$Y(\alpha, \beta, \gamma, \lambda) \stackrel{d}{=} \lambda^{1/\alpha} Y(\alpha, \beta) + \lambda(\gamma + b_0), \tag{2.1.2}$$

where $b_0 = 0$ if $\alpha \neq 1$, and $b_0 = \frac{2}{\pi}\beta \log \lambda$ if $\alpha = 1$.

The equality (2.1.2) shows that λ signifies a scale parameter, while γ signifies a translation parameter (more precisely, a pure shift of the distribution is a linear function of γ).

PROPERTY 2.2. *For any admissible parameter quadruple* $(\alpha, \beta, \gamma, \lambda)$

$$Y(\alpha, -\beta, -\gamma, \lambda) \stackrel{d}{=} -Y(\alpha, \beta, \gamma, \lambda). \tag{2.1.3}$$

The useful content of this property consists, in particular, in the fact that it allows us without loss of generality to consider the distribution functions $G(x, \alpha, \beta, \gamma, \lambda)$ with the one (according to our choice) additional condition that the sign of the argument x, the parameter β, or the parameter γ is preserved.

---

*In this section the absence of an index means that the relation is valid for both forms, (B) and (A).

PROPERTY 2.3. *Any admissible parameter quadruples $(\alpha, \beta_k, \gamma_k, \lambda_k)$ and any real numbers h and c_k, $k = 1, \dots, n$, uniquely determine a parameter quadruple $(\alpha, \beta, \gamma, \lambda)$ such that*

$$Y(\alpha, \beta, \gamma, \lambda) = \sum_k c_k Y(\alpha, \beta_k, \gamma_k, \lambda_k) + h. \tag{2.1.4}$$

In the form (A) the dependence of the quadruple $(\alpha, \beta, \gamma, \lambda)$ on the chosen parameters and numbers is as follows:

$$\lambda = \sum_k \lambda_k |c_k|^\alpha,$$

$$\lambda\beta = \sum_k \lambda_k \beta_k |c_k|^\alpha \operatorname{sgn} c_k,$$

$$\lambda\gamma = \sum_k \lambda_k \gamma_k c_k + h_0,$$

where $h_0 = h$ if $\alpha \neq 1$, and $h_0 = h - \frac{2}{\pi} \sum_k \lambda_k \beta_k c_k \log |c_k|$ if $\alpha = 1$.

We mention some special cases of Property 2.3 that are of independent interest.

PROPERTY 2.3.a. *An arbitrary admissible parameter quadruple $(\alpha, \beta, \gamma, \lambda)$ and any β' and β'' with $-1 \leq \beta' \leq \beta \leq \beta'' \leq 1$ determine unique positive numbers c' and c'' and a real number l such that*

$$Y(\alpha, \beta, \gamma, \lambda) \stackrel{d}{=} c' Y(\alpha, \beta') + c'' Y(\alpha, \beta'') + l. \tag{2.15}$$

In the form (A) the dependence of the parameters and the numbers is expressed as follows:

$$c' = \left(\lambda \frac{\beta'' - \beta}{\beta'' - \beta'}\right)^{1/\alpha}, \qquad c'' = \left(\lambda \frac{\beta - \beta'}{\beta'' - \beta'}\right)^{1/\alpha},$$

$$l = \begin{cases} \lambda\gamma & \text{if } \alpha \neq 1, \\ \lambda\gamma + \frac{2}{\pi}(\beta' c' \log c' + \beta'' c'' \log c'') & \text{if } \alpha = 1. \end{cases}$$

Choosing $\beta' = -1$ and $\beta'' = 1$ and using the equality $Y(\alpha, -1) = -Y(\alpha, 1)$, we get that any random variable $Y(\alpha, \beta, \gamma, \lambda)$ can be expressed as a linear combination of two independent random variables of the form $Y(\alpha, 1)$ (in the sense of the equality "$\stackrel{d}{=}$").

PROPERTY 2.3.b. *For any admissible parameter quadruple $(\alpha, \beta, \gamma, \lambda)$*

$$Y(\alpha, \beta, \gamma, \lambda) - Y(\alpha, \beta, \gamma, \lambda) \stackrel{d}{=} Y(\alpha, 0, 0, 2\lambda). \tag{2.1.6}$$

PROPERTY 2.3.c. *Any admissible parameter quadruple $q = (\alpha, \beta, \gamma, \lambda)$ uniquely determines an admissible quadruple $q^* = (\alpha, \beta^*, \gamma^*, \lambda^*)$ such that*

$$Y(\alpha, \beta, \gamma, \lambda) - \tfrac{1}{2} Y(\alpha, \beta, \gamma, \lambda) - \tfrac{1}{2} Y(\alpha, \beta, \gamma, \lambda) \stackrel{d}{=} Y(\alpha, \beta^*, \gamma^*, \lambda^*). \tag{2.1.7}$$

In the form (A) the parameters of q^* are expressed in terms of the parameters of q as follows:

$$\beta^* = \frac{1-2^{1-\alpha}}{1+2^{1-\alpha}}\beta \qquad \left(|\beta^*| \le \frac{|1-2^{1-\alpha}|}{1+2^{1-\alpha}} \le \frac{1}{3}\right),$$

$$\gamma^* = 0 \text{ if } \alpha \ne 1 \quad \text{and} \quad \gamma^* = -(\beta \log 2)/\pi \text{ if } \alpha = 1,$$

$$\lambda^* = (1+2^{1-\alpha})\lambda.$$

It is not hard to see that the random variables on the right-hand sides of (2.1.6) and (2.1.7) have strictly stable distributions. This feature of the transformations of the independent random variables (having an arbitrary stable distribution) on the left-hand sides of (2.1.6) and (2.1.7) turns out to be very useful in the problem of statistical estimation of parameters of stable laws (Chapter 4).

The above properties are all proved according to a single scheme. We consider the characteristic functions of the left and right sides of (2.1.1)–(2.1.7) (for example, in the form (A)) and, writing out their logarithms, compare the coefficients of the linearly independent functions of the variable t. As an example let us prove (2.1.4) in the case when $\alpha \ne 1$. We have that

$$\lambda\left(it\gamma - |t|^\alpha + it|t|^{\alpha-1}\beta \tan \tfrac{\pi}{2}\alpha\right)$$
$$= ith + \sum_k \lambda_k \left(it\gamma_k c_k - |c_k|^\alpha |t|^\alpha + it|t|^{\alpha-1} c_k |c_k|^{\alpha-1} \beta_k \tan \tfrac{\pi}{2}\alpha\right).$$

Comparison of the coefficients of the functions it, $|t|^\alpha$, and $it|t|^{\alpha-1}$ gives relations determining the values of the parameters β, γ, and λ. It is obvious that γ and λ have admissible values. If we write out the parameter β in the final form

$$\beta = \frac{\sum_k \lambda_k |c_k|^\alpha \beta_k \operatorname{sgn} c_k}{\sum_k \lambda_k |c_k|^\alpha},$$

then the satisfaction of the condition $|\beta| \le 1$ also becomes clear, because $|\beta_k| \le 1$.

PROPERTY 2.4. *For any admissible quadruple $q = (\alpha, \beta, \gamma, \lambda)$ of parameter values and any admissible quadruple $q' = (\alpha', \beta', \gamma', \lambda')$ such that $q' \to q$ (i.e., the corresponding parameters converge)*

$$Y_M(\alpha', \beta', \gamma', \lambda') \xrightarrow{d} Y_M(\alpha, \beta, \gamma, \lambda).$$

The convergence $\xrightarrow{d}$ is understood as weak convergence of distributions. However, taking into consideration that stable distributions have densities, we can at once conclude that the convergence $\xrightarrow{d}$ can be replaced by convergence in the sense of nearness of the corresponding distribution functions in the uniform metric.

PROPERTY 2.5. *Let $(\alpha, \beta, \gamma, \lambda)$ be an admissible parameter quadruple, and let* $\operatorname{med} Y$ *be the median of the random variable* $Y = Y(\alpha, \beta, \gamma, \lambda)$. *Then the following inequality holds for any* $0 \leq s < \alpha$:

$$\mathsf{E}|Y - \operatorname{med} Y|^s \leq 4(2\lambda)^{s/\alpha}\Gamma(1 - s/\alpha)\Gamma(s) \sin \tfrac{\pi}{2} s. \tag{2.1.8}$$

REMARK. We need not consider the case $\alpha = 2$ corresponding to a normal law, since (as elementary computations show) we have here that for all $s \geq 0$

$$\begin{aligned}\mathsf{E}|Y - \operatorname{med} Y|^s &= \tfrac{1}{\pi} 2^{1-s/2}(2\lambda)^{s/2}\Gamma(1 - s/2)\Gamma(s) \sin \tfrac{\pi}{2} s \\ &< 4(2\lambda)^{s/2}\Gamma(1 - s/2)\Gamma(s) \sin \tfrac{\pi}{2} s.\end{aligned}$$

An obvious consequence of (2.1.8) is that $\mathsf{E}|Y|^s < \infty$ for any $0 \leq s < \alpha$.

The proof of (2.1.8) is based on the following expression of the absolute moment $\mathsf{E}|X|^s$ of an arbitrary random variable X with the help of the characteristic function $\mathfrak{f}(t)$ of X:

$$\mathsf{E}|X|^s = 2\Gamma(1 + s) \sin \frac{\pi}{2} s \int_0^\infty (1 - \operatorname{Re} \mathfrak{f}(t)) t^{s-1}\, dt. \tag{2.1.9}$$

It is obtained by substituting X for x in the equality

$$|x|^s = 2\Gamma(1 + s) \sin \frac{\pi}{2} s \int_0^\infty (1 - \cos tx) t^{-s-1}\, dt \tag{2.1.10}$$

and taking the expectations of both sides. The equality itself is established by the same device as used to compute the same integrals in Theorem C.2. Namely, for real $p > 0$ we first compute the integral

$$\int_0^\infty (1 - e^{-pt}) t^{-s-1}\, dt = p^s \Gamma(1 - s) s^{-1}; \tag{2.1.11}$$

then we extend it analytically to the half-plane $\operatorname{Re} p > 0$ and use the continuity of this expression on the imaginary axis. After substitution of the value $p = -ix$ the real part of (2.1.11) coincides with the integral in (2.1.10) of interest to us.

Let us consider independent random variables Y' and Y'' with the same distribution as Y. According to (2.1.6), the random variable $Y' - Y''$ has characteristic function $\exp(-2\lambda|t|^\alpha)$. From this, on the basis of (2.1.9), we get

$$\begin{aligned}\mathsf{E}|Y' - Y''|^s &= 2\Gamma(1 + s) \sin \frac{\pi}{2} s \int_0^\infty (1 - \exp(-2\lambda t^\alpha)) t^{-s-1}\, dt \\ &= 2(2\lambda)^{s/\alpha}\Gamma(1 - s/\alpha)\Gamma(s) \sin \frac{\pi}{2} s.\end{aligned} \tag{2.1.12}$$

The right-hand side of (2.1.12), hence also the left, is finite for $0 \leq s < \alpha$. Further, the easily verified inequality

$$\mathsf{P}(|Y' - Y''| \geq x) \geq \tfrac{1}{2}\mathsf{P}(|Y' - \operatorname{med} Y'| \geq x)$$

gives us the inequality

$$\mathsf{E}|Y - \operatorname{med} Y|^s \leq 2\mathsf{E}|Y' - Y''|^s,$$

which together with (2.1.12) leads us to (2.1.8).

We conclude this section with an observation which has to do both with the relations given above among the random variables $Y(\alpha, \beta, \gamma, \lambda)$ and the relations between the random variables $Y(\alpha, \theta)$ and $Z(\alpha, \rho)$ to be introduced ([2.1])* in Chapter 3.

The following theorem is given in [58] after one of the results in the note [17] is generalized.

Let $Y_1 \stackrel{d}{=} Y_2 \stackrel{d}{=} Y(\alpha, 0, 0, \lambda)$ be independent random variables and let $L = (L_1, L_2)$ be an arbitrary two-dimensional vector with $\mathsf{P}(L_1 + L_2 = 0) = 0$ that is independent of Y_1 and Y_2. Then

$$\left(\frac{L_1}{L_1 + L_2}\right)^{1/\alpha} Y_1 - \left(\frac{L_2}{L_1 + L_2}\right)^{1/\alpha} Y_2 \stackrel{d}{=} Y_1.$$

It is not hard to see that this relation extends one variant of (2.1.4) to the case when the coefficients c_k are assumed to be random variables. Arising naturally in this connection is the question of how general the situation is when various parameters in relations between random variables can also be regarded as random if desired, without thereby disturbing the validity of the relations. It turns out that such a transition from nonrandom to random parameters is possible under relatively slight restrictions, and it is based on the following general fact.

LEMMA 2.1.1. *Let $h = h(a)$ be a function defined and measurable on a set $\mathcal{A} \subset R^m$ and taking values in R^n, and let $X(a)$ and $X'(h)$ be random variables whose distributions depend on respective parameter tuples a and h and which are connected for each $a \in \mathcal{A}$ by the relation*

$$X(a) \stackrel{d}{=} X'(h, (a)).$$

Then, for any random variables $V \stackrel{d}{=} V'$ with values in $\mathcal{A}$ and such that V is independent of X while V' is independent of X',

$$X(V) \stackrel{d}{=} X'(h(V')).$$

---

*Raised numbers in this form refer to the corresponding notes in the Comments at the end of the book.

The proof is elementary. We have that

$$\begin{aligned}\mathsf{E}\exp(itX(V)) &= \int \mathsf{E}\exp(itX(a))P_V(da)\\ &= \int \mathsf{E}\exp(itX'(h(a)))P_{V'}(da)\\ &= \mathsf{E}\exp(itX'(h(V'))).\end{aligned}$$

This obviously implies the lemma.

Let us illustrate the foregoing by an example. Let

$$V = (\alpha, \beta, \lambda, c_1, \dots, c_n) \stackrel{d}{=} V' = (\alpha', \beta', \lambda', c_1', \dots, c_n')$$

be random variables such that, with probability 1, $\alpha \neq 1$, β, and λ vary in the domain of admissible values, $\sum_k |c_k|^\alpha = 1$, V is independent of $Y_1, \dots, Y_n$, and V' is independent of Y. Then, according to (2.1.4) and Lemma 2.1.1,

$$Y(\alpha', \beta', 0, \lambda') \stackrel{d}{=} \sum_{k=1}^{n} c_k Y_k(\alpha, \beta, 0, \lambda).$$

§2.2. Representation of stable laws by integrals

Theorem C.2 established an explicit form for the characteristic functions $\mathfrak{g}$ of stable laws. In this section we turn to the question of an expression for the stable distributions themselves, on the basis of the expression for the functions $\mathfrak{g}$ in the form (B). Due to Property 2.1, it is possible without loss of generality to consider only standard stable distributions. What is more, in considering the density $g(x, \alpha, \beta)$ or the distribution function $G(x, \alpha, \beta)$ it suffices to solve the problem of representing these functions in the more restrictive situation when $x \geq 0$ or $\beta \geq 0$, since, according to Property 2.3,

$$\begin{aligned}g(-x, \alpha, \beta) &= g(x, \alpha, -\beta),\\ G(-x, \alpha, \beta) &= 1 - G(x, \alpha, -\beta),\\ \mathfrak{g}(-t, \alpha, \beta) &= \mathfrak{g}(t, \alpha, -\beta).\end{aligned}$$

As already noted in the Introduction, the function $|\mathfrak{g}(t)|$ is integrable on the whole real t-axis, and hence the density g can be expressed with the help of the function $\mathfrak{g}$ by the inversion formula

$$\begin{aligned}g(x, \alpha, \beta) &= \frac{1}{2\pi} \int e^{-itx} \mathfrak{g}(t, \alpha, \beta)\, dt\\ &= \frac{1}{\pi} \operatorname{Re} \int_0^\infty e^{itx} \mathfrak{g}(t, \alpha, -\beta)\, dt. \qquad (2.2.1)\end{aligned}$$

It is natural to begin the search for diverse variants of representations for stable laws by analyzing the inversion formula (2.2.1), which gives a first variant for expression the densities $g(x, \alpha, \beta)$ of standard distributions.

Substituting in (2.2.1) the expression given in the Introduction for the function $\mathfrak{g}(t, \alpha, \beta)$ in the form (B), we get that

$$g(x, \alpha, \beta) = \frac{1}{\pi} \operatorname{Re} \int_0^\infty \exp\left\{-itx - t^\alpha \exp\left(-i\frac{\pi}{2}\beta K(\alpha)\right)\right\} dt \tag{2.2.1a}$$

in the case $\alpha \neq 1$, and

$$g(x, 1, \beta) = \frac{1}{\pi} \operatorname{Re} \int_0^\infty \exp\left\{-itx - \frac{\pi}{2}t - i\beta t \log t\right\} dt \tag{2.2.1b}$$

in the case $\alpha = 1$.

Although, as we see, the functions $\mathfrak{g}$ can be written in a simple form,* there are expressions in elementary functions for the densities corresponding to them in only four cases. That these stable laws actually have the densities given below can be verified by direct computation of the corresponding integrals in (2.2.1a) and (2.2.1b) ([2.2]).

1. The *Lévy distribution*

$$g(x, \tfrac{1}{2}, 1) = \begin{cases} x^{-3/2}e^{-1/4x}/2\sqrt{\pi} & \text{if } x > 0, \\ 0 & \text{if } x < 0. \end{cases}$$

2. The *Cauchy distribution*†

$$g(x, 1, 0) = \tfrac{1}{2}(\pi^2/4 + x^2)^{-1}.$$

3. The *Gauss distribution* (with variance $\sigma^2 = 2$)

$$g(x, 2, \beta) = e^{-x^2/4}/2\sqrt{\pi}.$$

4. The case obtained from the first by symmetric reflection:

$$g(x, \tfrac{1}{2}, -1) = g(-x, \tfrac{1}{2}, 1).$$

For transformations of the integral in (2.2.1) with the purpose of obtaining other expressions for the density it is desirable to have an analytic extension of the function $\mathfrak{g}(t, \alpha, \beta)$ into the complex z-plane with semiaxis $\operatorname{Re} z = t > 0$. When considering analytic extensions of the functions $\mathfrak{g}$, the case $\alpha = 2$ corresponding to a normal law can be skipped as obvious. For it the function $\mathfrak{g}(z, 2, \beta) = \exp(-z^2)$ is entire.

---

*It should be noted that this section does not include all the integral representations of stable distributions which we decided to put in the book. For methodical reasons some of these representations are spread over other sections (see 2.3, 2.5 and 3.4).

†The expression for the density in the form (B) differs from the traditional form of expression, which corresponds to the form (C): $g_C(x, 1, 0) = \frac{1}{\pi}(1 + x^2)^{-1}$.

If $\alpha < 2$, then the function $\mathfrak{g}(t, \alpha, \beta)$ can also be extended analytically from the half-line $t > 0$ (or from $t < 0$), but it then has the branch points $z = 0$ and $z = \infty$. Therefore, to single out a principal branch of it we must make a cut along some contour joining 0 and ∞ without intersecting the real t-axis. In what follows we need an analytic extension of $\mathfrak{g}$ in the half-plane $\operatorname{Re} z \geq 0$, and in this connection we make a cut in the lower half-plane along the ray $\{z: \arg z = -3\pi/4\}$. The analytic extensions of $\mathfrak{g}(t, \alpha, \beta)$ from the semi-axes $t > 0$ and $t < 0$, where it is given by different formulas, are denoted by $\mathfrak{g}^+(z) = \mathfrak{g}^+(z, \alpha, \beta)$ and $\mathfrak{g}^-(z) = \mathfrak{g}^-(z, \alpha, \beta)$. It is not hard to verify by using the expression for $\mathfrak{g}(t, \alpha, \beta)$ in the form (B) that the extensions have the forms

$$\log \mathfrak{g}^+(z) = \begin{cases} -z^\alpha \exp(-i\frac{\pi}{2}\beta K(\alpha)) & \text{if } \alpha \neq 1, \\ -z(\frac{\pi}{2} + i\beta \log z) & \text{if } \alpha = 1 \end{cases} \tag{2.2.2}$$

(where z^α and $\log z$ are understood to be the principal branches of these functions) and

$$\log \mathfrak{g}^-(z) = \begin{cases} -(-z)^\alpha \exp(i\frac{\pi}{2}\beta K(\alpha)) & \text{if } \alpha \neq 1, \\ z(\frac{\pi}{2} - i\beta \log(-z)) & \text{if } \alpha = 1. \end{cases} \tag{2.2.3}$$

An elementary transformation of the functions (2.2.2) and (2.2.3) enables us to give them the desired form ($\varepsilon(\alpha) = \operatorname{sgn}(1 - \alpha)$):

$$\log \mathfrak{g}^+(z) = \begin{cases} -\varepsilon(\alpha)(-iz)^\alpha \exp(-i\frac{\pi}{2}K(\alpha)(\beta - 1)) & \text{if } \alpha \neq 1, \\ (-iz)(i(\beta - 1)\frac{\pi}{2} + \beta \log(-iz)) & \text{if } \alpha = 1, \end{cases} \tag{2.2.4}$$

$$\log \mathfrak{g}^-(z) = \begin{cases} -\varepsilon(\alpha)(-iz)^\alpha \exp(i\frac{\pi}{2}K(\alpha)(\beta - 1)) & \text{if } \alpha \neq 1, \\ (-iz)(i(1 - \beta)\frac{\pi}{2} + \beta \log(-iz)) & \text{if } \alpha = 1. \end{cases} \tag{2.2.5}$$

Comparison of (2.2.4) and (2.2.5) shows that the analytic extensions $\mathfrak{g}^+$ and $\mathfrak{g}^-$ do not coincide, except in the already mentioned simple case when $\alpha = 2$ and the case when $\alpha < 2$ and $\beta = 1$. If we consider the equality $\mathfrak{g}(t, \alpha, \beta) = \mathfrak{g}(-t, \alpha, -\beta)$, then we can add the case when $\alpha < 2$ and $\beta = -1$ to the indicated case. The following assertion summarizes the foregoing.

LEMMA 2.2.1. *Analytic extension of $\mathfrak{g}(t, \alpha, \beta)$ from the whole t-axis to the complex z-plane is possible only in the case $\alpha = 2$, while analytic extension to the complex plane with a cut along the ray $\arg z = -3\pi/4$ is possible only in the cases $\alpha < 2$, $\beta = 1$ and $\alpha < 2$, $\beta = -1$, which correspond to extremal stable distributions. Moreover,*

$$\log \mathfrak{g}^+(z, \alpha, \beta) = \begin{cases} -\varepsilon(\alpha)(-iz)^\alpha & \textit{if } \alpha \neq 1, \\ (-iz)\log(-iz) & \textit{if } \alpha = 1. \end{cases} \tag{2.2.6}$$

REMARK. The function $\psi_\alpha(z) = \log \mathfrak{g}^+(z, \alpha, 1)$ has the following integral representation in the half-plane $\operatorname{Im} z \geq 0$ (see the proof of Theorem C.2, where similar formulas have already been encountered):

$$\psi_\alpha(z) = \begin{cases} \dfrac{\alpha(\alpha-1)}{\Gamma(2-\alpha)} \displaystyle\int_0^\infty (e^{izu} - 1 - izu)u^{-(\alpha+1)}\,du & \text{if } \alpha > 1, \\ \dfrac{\alpha(\alpha-1)}{\Gamma(2-\alpha)} \displaystyle\int_0^\infty (e^{izu} - 1)u^{-(\alpha+1)}\,du & \text{if } \alpha < 1, \\ \displaystyle\int_0^\infty (e^{izu} - 1 - iz\sin u)u^{-2}\,du & \text{if } \alpha = 1. \end{cases}$$

We pass to the construction of the integral representations of interest to us for the density of $\mathfrak{g}$. The transformation of the integral

$$\mathfrak{J} = \int_0^\infty \exp(izx)\mathfrak{g}^+(z, \alpha, -\beta)\,dz$$

in (2.2.1) will be implemented by changing the contour of integration, in such a way that the value of the integral $\mathfrak{J}$, or at least the value of its real part, does not change. The next lemma is needed for justifying the changes of contour.

LEMMA 2.2.2. *Let $0 < \varepsilon < \pi/4$ be a fixed number. Then as $r \to \infty$ or $r \to 0$*

$$Q(C_r) = \int_{C_r} \exp(izx)\mathfrak{g}^+(z, \alpha, -\beta)\,dz \to 0$$

for any sequence of contours of the following form:

1. *If $x > 0$, $\alpha < 1$, and β has any admissible value, or if x is any real number, $\alpha = 1$, and $\beta > 0$, then*

$$C_r = \{z \colon |z| = r,\ 0 \le \arg z \le \pi - \varepsilon\}.$$

2. *If $x > 0$, $\alpha > 1$, and β has any admissible value, then*

$$C_r = \left\{z \colon |z| = r, \frac{\pi}{2\alpha} + \varepsilon \le \arg z - \frac{\pi}{2}\beta\frac{K(\alpha)}{\alpha} \le \frac{\pi}{2\alpha}\right\}.$$

PROOF. In all the cases considered the length of the contour C_r is at most πr. Therefore,

$$\begin{aligned} |Q(C_r)| &\le \pi r \max(|e^{izx}\mathfrak{g}^+(z, \alpha, -\beta)| \colon z \in C_r) \\ &= \pi r \exp\{\max(U \colon z \in C_r)\}, \end{aligned}$$

where $U = -x\operatorname{Im} z + \operatorname{Re}\log\mathfrak{g}^+(z, \alpha, -\beta)$.

In the case when $\alpha < 1$ and $x > 0$ we set $z = re^{i\varphi}$ and get that

$$U = -xr\sin\varphi - r^\alpha\cos\alpha(\varphi + \pi\beta/2),$$

whence

$$\begin{aligned}\max(U: z \in C_r) &= \max(U: 0 \le \varphi \le \pi - \varepsilon)\\ &\le \max(U: 0 \le \varphi < \tfrac{\pi}{2}(1-\alpha))\\ &\quad + \max(U: \tfrac{\pi}{2}(1-\alpha) \le \varphi \le \pi - \varepsilon)\\ &\le r^\alpha(1 - \cos\tfrac{\pi}{2}\alpha(2-\alpha)) - xr\min(\varepsilon, \tfrac{\pi}{2}(1-\alpha)).\end{aligned}$$

For large values of r the last expression can obviously be estimated from above by the quantity $-2\log r$, and this implies that $|Q(C_r)| \le \pi/r \to 0$ as $r \to \infty$.

In the case when $\alpha = 1$ and $\beta > 0$

$$U = -(\beta \log r + x) r \sin\varphi - r(\tfrac{\pi}{2} + \beta\varphi)\cos\varphi.$$

From this we find, as in the preceding case, that

$$\begin{aligned}\max(U: z \in C_r) &= \max(U: 0 \le \varphi \le \pi - \varepsilon)\\ &\le \max(U: 0 \le \varphi \le \pi/6) + \max(U: \pi/6 \le \varphi \le \pi - \varepsilon)\\ &\le 2(4 + |x|)r - \beta r \log r \min(\varepsilon, \tfrac{1}{2}).\end{aligned}$$

Consequently, for large values of r

$$\max(U: z \in C_r) \le -2\log r,$$

and hence $|Q(C_r)| \le \pi/r \to 0$ as $r \to \infty$. From these estimates it is also clear that $|Q(C_r)| = O(r)$ as $r \to 0$.

The case when $\alpha > 1$ and $x > 0$ is handled in exactly the same way.

REMARK. The assertion of the lemma is preserved if the contours C_r are replaced by parts of them.

LEMMA 2.2.3. *In the complex z-plane with a cut along the ray* $\arg z = -3\pi/4$, *consider a family $\{\Gamma\}$ of contours satisfying the following conditions:*

1. *Each contour Γ begins at the point $z = 0$.*
2. *None of the contours Γ intersect the cut.*
3. *By moving from the point $z = 0$ along a contour Γ one goes to infinity in such a way that from some point all the points $z \in \Gamma$ have values of their arguments in the intervals*

$$0 \le \arg z \le \pi - \varepsilon,$$

$$\varepsilon - \pi(1 - \beta K(\alpha))/2\alpha \le \arg z \le \pi(1 + \beta K(\alpha))/2\alpha,$$

where $0 < \alpha < 2$, $|\beta| \le 1$, and $\varepsilon > 0$ is an arbitrarily small number.

Then, for any contour of the indicated form and any pair $(\alpha \ne 1, \beta)$ or $(\alpha = 1, \beta > 0)$ of admissible parameters,

$$\int_0^\infty e^{izx} \mathfrak{g}^+(z, \alpha, -\beta)\, dz = \int_\Gamma e^{izx} \mathfrak{g}^+(z, \alpha, -\beta)\, dz. \tag{2.2.7}$$

PROOF. For a chosen contour Γ and for each $r > 0$ let us also define the following auxiliary contours:

T_r is the part of Γ obtained by moving from the initial point $z = 0$ to the first point z_r where Γ intersects the circle $|z| = r$.

L_r is the part (disjoint from the cut) of the circle $|z| = r$ between the points $z = r$ and z_r, traversed in the direction from $z = r$ to z_r.

We form the closed contour $D_{r,R} = [r, R] \cup L_R \cup (\overline{T}_R \backslash \overline{T}_r) \cup \overline{L}_r$ (an overbar indicates that a contour is traversed in the opposite direction).

Since the function $w(z) = e^{izx}\mathfrak{g}^+(z, \alpha, -\beta)$ is anlaytic in the domain where the contour $D_{r,R}$ lies, Cauchy's theorem gives us that

$$\int_{D_{r,R}} w\,dz = \int_r^R w\,dz + \int_{L_R} w\,dz - \int_{T_R} w\,dz + \int_{T_r} w\,dz - \int_{L_r} w\,dz = 0.$$

Invoking Lemma 2.2.2 (see the remark after it), we find that

$$\int_{L_R} w\,dz + \int_{T_r} w\,dz - \int_{L_r} w\,dz \to 0 \quad \text{as } R \to \infty,\ r \to 0.$$

Consequently, (2.2.7) is valid.

THEOREM 2.2.1. *Let x be an arbitrary real number, and let α and β be admissible parameter values for a standard stable distribution.*

1. *If $\alpha < 1$ and $x > 0$, then*

$$\begin{aligned} g(x, \alpha, \beta) &= \frac{1}{\pi} \operatorname{Im} \int_0^\infty \exp\left\{-xu - u^\alpha \exp\left(-i\frac{\pi}{2}\alpha(1+\beta)\right)\right\} du \\ &= \frac{1}{\pi} \operatorname{Im} \int_0^\infty \exp\{-xu - u^\alpha \exp(-i\pi\rho\alpha)\}\, du. \end{aligned} \tag{2.2.8}$$

2. *If $\alpha = 1$ and $\beta > 0$, then*

$$g(x, 1, \beta) = \frac{1}{\pi} \operatorname{Im} \int_0^\infty \exp\left\{-xu - \beta u \log u + i\frac{\pi}{2}(1+\beta)u\right\} du. \tag{2.2.9}$$

3. *If $\alpha > 1$, then*

$$\begin{aligned} g(x, \alpha, \beta) &= \frac{1}{\pi} \operatorname{Re} \int_0^\infty \exp\left\{-u^\alpha - ixu \exp\left(i\frac{\pi}{2}\theta\right) + i\frac{\pi}{2}\theta\right\} du \\ &= \frac{1}{\pi} \operatorname{Im} \int_0^\infty \exp\{-u^\alpha - xu \exp(i\pi\rho) + i\pi\rho\}\, du. \end{aligned} \tag{2.2.10}$$

The proof of each of the formulas (2.2.8)–(2.2.10) reduces to the following actions:

a) choice of a contour Γ needed for transforming the integral in (2.2.7);

b) justification of the change of the contour of integration in the integral in (2.2.7) with the help of Lemma 2.2.3; and

c) replacement of the variable of integration on the right-hand side of (2.2.7).

The positive part of the imaginary axis is taken as the contour Γ in cases 1 and 2. The use of Lemma 2.2.3 here does not require any additional explanations. For such a contour the change of variable $z = iu$, $u > 0$, leads to the corresponding expression for the density.

In case 3 the ray $\Gamma = \{z\colon \arg z = \pi\theta/2\}$ is chosen for the role of the contour Γ, with the subsequent change of variable $z = u\exp(i\pi\theta/2)$.

The transformation of the integrals are very simple in all the cases considered. For example, if $\alpha < 1$ and $x > 0$, then (2.2.4) gives us that

$$\begin{aligned} g(x, \alpha, \beta) &= \frac{1}{\pi}\operatorname{Re}\int_0^\infty \exp(izx)\mathfrak{g}^+(z, \alpha, -\beta)\,dz \\ &= -\frac{1}{\pi}\operatorname{Im}\int_0^\infty \exp\left\{-xu - u^\alpha \exp\left(i\frac{\pi}{2}\alpha(1+\beta)\right)\right\}du \\ &= \frac{1}{\pi}\operatorname{Im}\int_0^\infty \exp\left\{-xu - u^\alpha \exp\left(-i\frac{\pi}{2}\alpha(1+\beta)\right)\right\}du. \end{aligned}$$

COROLLARY 1. *If $\alpha < 1$, then for any admissible β and any $x > 0$*

$$G(x, \alpha, \beta) = 1 - \frac{1}{\pi}\int_0^\infty e^{-xu}\operatorname{Im}\exp(-u^\alpha e^{-i\pi\rho\alpha})\frac{du}{u}.$$

If $\alpha = 1$, then for any admissible $\beta > 0$ and any real x

$$G(x, 1, \beta) = 1 - \frac{1}{\pi}\int_0^\infty \exp(-xu - \beta u \log u)\sin\left[\frac{\pi}{2}(1+\beta)u\right]\frac{du}{u}.$$

If $\alpha > 1$, then for any admissible β and any $x > 0$

$$G(x, \alpha, \beta) = \frac{1}{2}(1-\theta) - \frac{1}{\pi}\int_0^\infty e^{-u^\alpha}\operatorname{Im}\exp(-xue^{i\pi\rho\alpha})\frac{du}{u}.$$

In the case $\alpha \le 1$ these equalities are obtained by integrating the corresponding expressions for the density. In the case $\alpha > 1$, besides integrating the density from 0 to x, we should use the expression for $G(0, \alpha, \beta)$ given by (2.2.30).

COROLLARY 2. *If $\alpha \ne 1$, then for any admissible values of β*

$$g(0, \alpha, \beta) = \frac{1}{\pi}\Gamma\left(1 + \frac{1}{\alpha}\right)\cos\left(\frac{\pi}{2}\beta\frac{K(\alpha)}{\alpha}\right). \tag{2.2.11}$$

In the case $\alpha > 1$ this equality is obtained by substituting the value $x = 0$ in (2.2.10) and computing the resulting integral.

In the case $\alpha < 1$ we first consider the expression (2.2.8) for the density $g(x, \alpha, -|\beta|)$, $x > 0$, and pass to the limit as $x \to 0$. The integral obtained in the limit is computed. Then we use the fact that

$$g(0, \alpha, \beta) = g(0, \alpha, -\beta) = g(0, \alpha, -|\beta|).$$

In the case when $\alpha = 1$ and $\beta > 0$ there is no elementary expression for the value

$$g(0,1,\beta) = \frac{1}{\pi}\int_0^\infty \exp(-\beta u \log u)\sin\left[\frac{\pi}{2}(1+\beta)u\right]\frac{du}{u}$$

of the density at zero ([2.3]).

The following integral representations of stable laws are connected with the parametrization in the form (A). Let $\mu = \beta\tan(\pi\alpha/2)$ and $y = x + |x|^{1-\alpha}\mu$.

THEOREM 2.2.2. 1. *If $\alpha < 1$, then for any admissible β and $x \neq 0$ such that $y > 0$*

$$\begin{aligned} &g_A(x+\mu|x|^{1-\alpha},\alpha,\beta)\\ &\quad = \frac{1}{\pi|x|}\operatorname{Im}\int_0^\infty \exp\left\{-\frac{xu}{|x|} - |x|^{-\alpha}\right.\\ &\qquad\qquad \left.\times\left[\left(\frac{u}{i}\right)^\alpha - \mu u\left(\left(\frac{u}{i}\right)^{\alpha-1} - 1\right)\right]\right\}du. \qquad (2.2.12)\end{aligned}$$

2. *If $\alpha = 1$, then for any real $x \neq 0$ and any $\beta \neq 0$*

$$\begin{aligned} &g_A\left(x + \frac{2}{\pi}\beta\log|x|, 1, \beta\right)\\ &\quad = \frac{1}{\pi|x|}\operatorname{Im}\int_0^\infty \exp\left\{-\frac{\beta x u}{|\beta x|} - \frac{u}{|x|}\left[\frac{2}{\pi}|\beta|\log u - i(1+|\beta|)\right]\right\}du. \qquad (2.1.13)\end{aligned}$$

PROOF. In the first case we have, according to (2.2.1a),

$$\begin{aligned} &g_A(x+\mu|x|^{1-\alpha},\alpha,\beta)\\ &\quad = \frac{1}{\pi}\operatorname{Re}\int_0^\infty \exp(-izy - z^\alpha + i\mu z^\alpha)\,dz\\ &\quad = \frac{1}{\pi|x|}\operatorname{Re}\int_0^\infty \exp\left\{-i\frac{xu}{|x|} - |x|^{-\alpha}[u^\alpha - i\mu u(u^{\alpha-1} - 1)]\right\}du. \qquad (2.2.14)\end{aligned}$$

Since $\alpha < 1$ and $y > 0$, it is possible to rotate the contour in the integrals (2.2.14) through the angle $-\pi/2$ without changing their real parts. The result is (2.2.12).

The second case is analyzed similarly. Taking (2.2.1b) as a basis, we get, after the corresponding transformations,

$$\begin{aligned} &g_A\left(x + \frac{2}{\pi}\beta\log|x|, 1, \beta\right)\\ &\quad = \frac{1}{\pi|x|}\operatorname{Re}\int_0^\infty \exp\left\{-i\frac{\beta x}{|\beta x|}z - \frac{z}{|x|}\left(1 + i\frac{2}{\pi}|\beta|\log z\right)\right\}dz. \qquad (2.2.15)\end{aligned}$$

Rotation of the contour of integration in (2.2.15) through the angle $-\pi/2$ leads us to (2.2.13) ([2.4]). In both cases the rotation of the contour is justified by standard arguments already used above more than once, with the help of Lemma 2.2.2.

It is interesting to note that (2.2.13) can be obtained from (2.2.12) by passing to the limit as $\alpha \to 1$. The proof of it is based on the following arguments. First of all, we can conclude on the basis of Property 2.2 that in the case $\beta \neq 0$

$$g_A(y, \alpha, \beta) = g_A(y^*, \alpha, |\beta|),$$

where the quantity

$$y^* = y \operatorname{sgn} \beta = \frac{\beta x}{|\beta|} + |\beta| \tan \frac{\pi}{2} \alpha |x|^{1-\alpha}$$

obviously becomes positive for values of α sufficiently close to 1. Therefore, without loss of generality we can consider only the case $\beta > 0$ in (2.2.12). It is not hard to see that in this situation the integral in (2.2.12) is transformed into (2.2.13) as $\alpha \to 1$. It remains to establish that the left-hand sides of these equalities also approach each other. The connection between the parameters of the forms (A) and (M) and the specific joint continuity property of stable distributions in the form (M) with respect to all the parameters noted in the Introduction show that

$$\begin{aligned} &\alpha_A = \alpha_M, \qquad \beta_A = \beta_M, \\ &g_A(x, \alpha, \beta) = g_M(x - \mu, \alpha, \beta), \\ &g_A(x, 1, \beta) = g_M(x, 1, \beta), \\ &g_M(x, \alpha, \beta) \to g_M(x, 1, \beta) \quad \text{as } \alpha \to 1. \end{aligned} \tag{2.2.16}$$

Moreover, it is not hard to verify that

$$\mu(|x|^{1-\alpha} - 1) \to \tfrac{2}{\pi} \beta \log |x| \quad \text{as } \alpha \to 1. \tag{2.2.17}$$

It follows from (2.2.16) and (2.2.17) that

$$\begin{aligned} g_A(x + \mu |x|^{1-\alpha}, \alpha, \beta) &= g_M(x + \mu(|x|^{1-\alpha} - 1), \alpha, \beta) \\ &\to g_M\left(x + \tfrac{2}{\pi} \beta \log |x|, 1, \beta\right) \\ &= g_A\left(x + \tfrac{2}{\pi} \beta \log |x|, 1, \beta\right). \end{aligned}$$

REMARK 1. The case $\alpha = 1$, $\beta = 0$ included in the second part of the theorem corresponds to the well-known Cauchy law, and thus it can be excluded without damaging the formulation.

REMARK 2. The case $\alpha > 1$ was not considered in Theorem 2.2.2. It seems that a direct analogue of (2.2.12) does not exist here. However, there is a formula that can serve as a complement to this theorem in a certain sense. For methodical reasons it is included in the next section (see (2.3.6)).

REMARK 3. There are no representations analogous to (2.2.12) and (2.2.13) for the distribution functions. More precisely, there are such representations, but they are not very simple, because the integrands no longer contain only elementary functions, but also functions such as the incomplete gamma-function and the integral exponential function, which must furthermore be considered in some part of the complex plane.

All the integral representations of stable distributions given above have one feature which sometimes greatly complicates their use, for example, in constructing estimates in tabulating the values of the density and of distribution functions, etc. This feature is that the functions under the integral sign in the corresponding representations oscillate, i.e., change sign infinitely many times. It is natural to ask whether there are representations of stable distributions free from this deficiency. The answer turns out to be positive.

The following notation is used below (some of it was used earlier):

$$\varepsilon(\alpha) = \operatorname{sgn}(1-\alpha), \quad \theta = \beta K(\alpha)/\alpha, \quad \theta^* = \theta \operatorname{sgn} x,$$

$$C(\alpha, \theta) = 1 - \tfrac{1}{4}(1+\theta)(1+\varepsilon(\alpha)),$$

$$U_\alpha(\varphi, \theta) = \left(\frac{\sin \frac{\pi}{2}\alpha(\varphi+\theta)}{\cos \frac{\pi}{2}\varphi}\right)^{\alpha/(1-\alpha)} \frac{\cos \frac{\pi}{2}((\alpha-1)\varphi + \alpha\theta)}{\cos \frac{\pi}{2}\varphi},$$

$$U_1(\varphi, \beta) = \frac{\pi}{2} \frac{(1+\beta\varphi)}{\cos \frac{\pi}{2}\varphi} \exp\left(\frac{\pi}{2}\left(\varphi + \frac{1}{\beta}\right) \tan \frac{\pi}{2}\varphi\right).$$

THEOREM 2.2.3. *The densities of standard stable distributions can be written as follows:*

1. *If $\alpha \neq 1$ and $x \neq 0$, then for any $|\beta| \leq 1$*

$$g(x, \alpha, \beta) = \frac{\alpha |x|^{1/(\alpha-1)}}{2|1-\alpha|} \int_{-\theta^*}^{1} U_\alpha(\varphi, \theta^*) \exp\{-|x|^{\alpha/(\alpha-1)} U_\alpha(\varphi, \theta^*)\}\, d\varphi. \tag{2.2.18}$$

2. *If $\alpha = 1$ and $\beta \neq 0$, then for any x*

$$g(x, 1, \beta) = \frac{1}{2|\beta|} e^{-x/\beta} \int_{-1}^{1} U_1(\varphi, \beta) \exp\{-e^{-x/\beta} U_1(\varphi, \beta)\}\, d\varphi, \tag{2.2.19}$$

We mention that excluding the cases $\alpha \neq 1$, $x = 0$ and $\alpha = 1$, $\beta = 0$ in the formulation of the theorem involves no loss of generality for its assertion, because the first of these cases was consdered in (2.2.11), while the second corresponds to the well-known Cauchy law ($^{2.5}$).

The proof here, as in the preceding theorems on integral representations, has the inversion formula (2.2.1) as a starting point:

$$g(-x, \alpha, -\beta) = \frac{1}{\pi} \operatorname{Re} \int_0^\infty \exp(izx + \psi(z, \alpha, -\beta))\, dz. \tag{2.2.20}$$

The function $\psi(z, \alpha, \beta) = \log \mathfrak{g}^+(z, \alpha, \beta)$ is taken in the form (C), i.e.,

$$\psi(z, \alpha, \beta) = \begin{cases} -z^\alpha \exp(-i\frac{\pi}{2}\theta\alpha) & \text{if } \alpha \neq 1, \\ -\frac{\pi}{2}z - i\beta z \log z & \text{if } \alpha = 1. \end{cases} \tag{2.2.21}$$

Without loss of generality, we assume that $x > 0$ in the case $\alpha \neq 1$ and $\beta > 0$ in the case $\alpha = 1$. In the complex z-plane consider the contour

$$\Gamma = \{z \colon \operatorname{Im}(izx + \psi(z, \alpha - \beta)) = 0, \tfrac{\pi}{2}k \leq \arg z \leq \tfrac{\pi}{2}\},$$

where $k = -\theta$ if $\alpha \neq 1$, and $k = -1$ if $\alpha = 1$. For $\alpha \leq 1$ and$\beta = -1$ we have that $k = 1$. According to (2.2.6), $\psi(z, \alpha, 1) = -(-iz)^\alpha$, and hence the sum $izx + \psi$ takes real values on the positive part I^+ of the imaginary axis, i.e., $\Gamma = I^+$, which explains the meaning of the condition $\pi/2 \leq \arg z \leq \pi/2$ arising in this case.

In the remaining cases the contour Γ is a curve whose equation in polar coordinates $(z = re^{i\varphi}, \pi k/2 < \varphi < \pi/2)$ is obtained by using the form (2.2.21) of the function ψ:

$$\begin{aligned} xr\cos\varphi - r^\alpha \sin\alpha(\varphi + \pi\theta/2) = 0 \quad & \text{if } \alpha \neq 1, \\ xr\cos\varphi + \beta r \log r \cos\varphi - (\pi/2 + \beta\varphi) r \sin\varphi = 0 \quad & \text{if } \alpha = 1. \end{aligned} \tag{2.2.22}$$

The solution $r = r(\varphi)$ is found in the explicit form

$$r(\varphi) = \begin{cases} \left(\dfrac{\sin\alpha(\varphi + \pi\theta/2)}{x\cos q}\right)^{1/(1-\alpha)}, & \alpha \neq 1, \\ \exp\left(-x/\beta + (\varphi + \pi/2\beta)\tan\varphi\right), & \alpha = 1. \end{cases} \tag{2.2.23}$$

The contours described by the functions (2.2.23) differ in their form, depending on the choice of the parameters (α, β) to which they correspond. They can be divided into four groups.

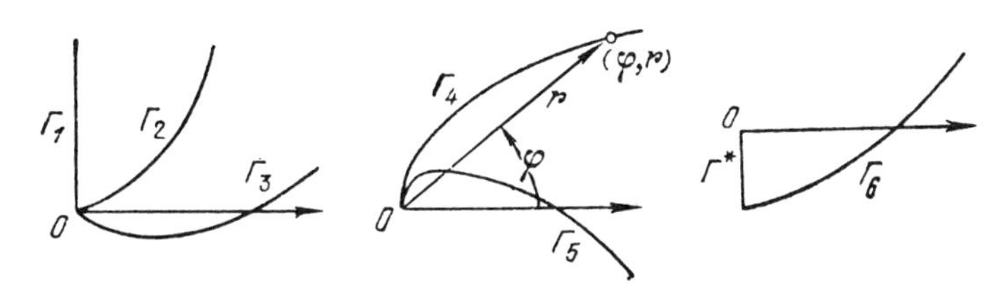

FIGURE 1. Form of the contours Γ_k of integration (in polar coordinates).

1. $(\alpha < 1, \beta = -1)$, with $\Gamma = \Gamma_1 = I^+$.

2. $(\alpha < 1, \beta \neq \pm 1)$ and $(\alpha = 1, 0 < \beta < 1)$. The contours in this group begin at zero and go to infinity as φ approaches $\pi/2$. They approach the point $z = 0$ at different angles, depending on the quantity k, and they have the form Γ_1 if $k = 1$, the form Γ_2 if $0 < k < 1$, and the form Γ_3 if $-1 < k \leq 0$.

3. $\alpha > 1$. Here the contours Γ begin at zero, which they approach at the angle $\pi/2$, and go to infinity as φ approaches $-\pi\theta/2$, i.e., they have the form Γ_4 if $\beta > 0$, or Γ_5 if $\beta \leq 0$.

4. ($\alpha < 1$, $\beta = 1$) and ($\alpha = 1$, $\beta = 1$). The contours Γ_6 in this group go to infinity as φ approaches $\pi/2$, and they begin at the point $-i\tau$, where

$$\tau = \begin{cases} (\alpha/x)^{1/(1-\alpha)} & \text{if } \alpha \neq 1, \\ \exp(-x-1) & \text{if } \alpha = 1. \end{cases}$$

With the help of the contour Γ_6 we form a new contour S by following the rule that $S = \Gamma_6$ if $\beta \neq 1$, and $S = \Gamma_6 \cap \Gamma^*$ if $\beta = 1$, where $\Gamma^* = \{z\colon \operatorname{Re} z = 0, -\tau \leq \operatorname{Im} z \leq 0\}$. A feature of the contour Γ^* is that

$$\operatorname{Im}(izx + \psi(z, \alpha, -1)) = 0 \quad \text{for } z \in \Gamma^*,$$

which clearly implies the equality

$$\frac{1}{\pi} \operatorname{Re} \int_{\Gamma^*} (izx + \psi(z, \alpha, -1))\, dz = 0.$$

Therefore, in the case $\beta = 1$

$$\frac{1}{\pi} \operatorname{Re} \int_S \exp(izx + \psi)\, dz = \frac{1}{\pi} \operatorname{Re} \int_\Gamma \exp(izx + \psi)\, dz.$$

Lemma 2.2.3 makes it possible (by the last equality) to replace the contour of integration in (2.2.20) by Γ, and we see that

$$\begin{aligned} g(-x, \alpha, -\beta) &= \frac{1}{\pi} \operatorname{Re} \int_\Gamma \exp(izx + \psi)\, dz \\ &= \frac{1}{\pi} \int_\Gamma \exp\{\operatorname{Re}(izx + \psi)\}\, d(\operatorname{Re} z) \\ &= \frac{1}{\pi} \int_\Gamma \exp(-W(\varphi))\, d(r \cos\varphi), \end{aligned} \tag{2.2.24}$$

where

$$W(\varphi) = \begin{cases} xr \sin\varphi + r^\alpha \cos\alpha(\varphi + \pi\theta/2), & \alpha \neq 1, \\ xr \sin\varphi + \beta r \log r \sin\varphi + (\pi/2 + \beta\varphi) r \cos\varphi, & \alpha = 1. \end{cases} \tag{2.2.25}$$

We obtain the final form $W(\varphi)$ by substituting in (2.2.25) the expression found for $r(\varphi)$ in (2.2.23).

Suppose that $\alpha \neq 1$. According to (2.2.23),

$$xr = r^\alpha \frac{\sin\alpha(\varphi + \pi\theta/2)}{\cos\varphi},$$

and, therefore,

$$\begin{aligned} W(\varphi) &= r^\alpha \sin\varphi \frac{\sin\alpha(\varphi + \pi\theta/2)}{\cos\varphi} + r^\alpha \cos\alpha(\varphi + \pi\theta/2) \\ &= r^\alpha \frac{\cos[(\alpha - 1)\varphi + \pi\alpha\theta/2]}{\cos\varphi} = x^{\alpha/(\alpha-1)} U_\alpha(2\varphi/\pi, \theta). \end{aligned}$$

If $\alpha = 1$, then the equality $\beta \log r = -x + (\beta\varphi + \pi/2)\tan\varphi$ follows from (2.2.22), and this equality in (2.2.25) gives us finally that

$$\begin{aligned} W(\varphi) &= xr\sin\varphi + (-x + (\beta\varphi + \pi/2)\tan\varphi)r\sin\varphi \\ &\quad + (\beta\varphi + \pi/2)r\cos\varphi \\ &= r(\cos\varphi + \sin^2\varphi/\cos\varphi)(\beta\varphi + \pi/2) \\ &= \exp(-x/\beta)U_1(2\varphi/\pi, \beta). \end{aligned}$$

To conclude the proof we must clear up the form of the differential $d(r\cos\varphi)$. Let us consider the case $\alpha \neq 1$. By (2.2.22), $xr\cos\varphi = r^\alpha \sin\alpha(\varphi + \pi\theta/2)$. Consequently,

$$\begin{aligned} xd(r\cos\varphi) &= \alpha r^\alpha \cos\alpha(\varphi + \pi\theta/2)\,d\varphi + \alpha r^{\alpha-1}\sin\alpha(\varphi + \pi\theta/2)dr \\ &= \alpha r^\alpha \cos\alpha(\varphi + \pi\theta/2)\,d\varphi + \alpha xr\cos\varphi r^{-1}dr \\ &= \alpha xd(r\cos\varphi) + \alpha[xr\sin\varphi + r^\alpha\cos\alpha(\varphi + \pi\theta/2)]d\varphi. \end{aligned}$$

As is clear from (2.2.25), the second term in the last sum is equal to $\alpha W(\varphi)\,d\varphi$.

Thus,

$$xd(r\cos\varphi) = \alpha xd(r\cos\varphi) + \alpha x^{\alpha/(\alpha-1)}U_\alpha(2\varphi/\pi, \theta)d\varphi,$$

which implies the desired form of the differential $d(r\cos\varphi)$:

$$d(r\cos\varphi) = \frac{\alpha}{1-\alpha}x^{1/(\alpha-1)}U_\alpha(2\varphi/\pi, \theta)d\varphi. \tag{2.2.26}$$

The case $\alpha = 1$ is handled similarly. There the result of the computations is the equality

$$d(r\cos\varphi) = W(\varphi)d\varphi/\beta = (1/\beta)\exp(-x/\beta)U_1(2\varphi/\pi, \theta)d\varphi.$$

To conclude the proof it remains for us to substitute the expressions $W(\varphi)$ and $d(r\cos\varphi)$ into (2.2.24) and consider that, as we go along the contour Γ in the direction of increasing r, the angle φ changes from $k\pi/2$ to $\pi/2$ if $\alpha \leq 1$, and from $\pi/2$ to $\pi\theta/2$ if $\alpha > 1$.

Consequently, if $\alpha \neq 1$, then (2.2.24) can be written in the form

$$\begin{aligned} g(-x, \alpha, -\beta) = \frac{\alpha\varepsilon(\alpha)}{\pi(1-\alpha)} \int_{\pi\theta/2}^{\pi/2} & x^{1/(\alpha-1)}U_\alpha(2\varphi/\pi, \theta) \\ & \times \exp\{-x^{\alpha/(\alpha-1)}U_\alpha(2\varphi/\pi, \theta)\}\,d\varphi, \end{aligned}$$

and if $\alpha = 1$, then in the form

$$g(-x, 1, -\beta) = \frac{1}{\pi\beta}e^{-x/\beta}\int_{-\pi/2}^{\pi/2} U_1(2\varphi/\pi, \beta)\exp\{-e^{-x/\beta}U_1(2\varphi, \pi, \beta)\}\,d\varphi.$$

The proof was carried out for densities $g(-x, \alpha, -\beta)$ under the additional conditions $x > 0$ in the case $\alpha \neq 1$ and $\beta < 0$ in the case $\alpha = 1$. The general case can be reduced to the cases stipulated by these conditions. Indeed,

$$g(x, \alpha, \beta) = g(-|x|, \alpha, -\beta^*),$$
$$g(x, 1, \beta) = g(-x^*, 1 - |\beta|),$$

where $\beta^* = \beta \operatorname{sgn} x$ and $x^* = x \operatorname{sgn} \beta$. If we substitute the quantities $|x|$ and β^* (in the case $\alpha \neq 1$) and the quantities x^* and $|\beta|$ (in the case $\alpha = 1$) in the expressions found for the densities, then we get the respective equalities (2.2.18) and (2.2.19) after replacing the variable φ by $\pi\varphi/2$. In the last case it must also be taken into account that

$$-x^*/|\beta| = -x/\beta \quad \text{and} \quad U_1(\varphi, \beta) = U_1(-\varphi, -\beta).$$

We make some remarks about the theorem just proved.

REMARK 1. The distribution functions $G(x, \alpha, \beta)$ of standard stable distributions can be written in the following form:

1.

$$G(x, \alpha, \beta) = C(\alpha, \theta) + \frac{\varepsilon(\alpha)}{2} \int_{-\theta}^{1} \exp\{-x^{\alpha/(\alpha-1)} U_\alpha(\varphi, \theta)\} \, d\varphi \tag{2.2.27}$$

if $\alpha \neq 1$ and $x > 0$.

2.

$$G(x, 1, \beta) = \frac{1}{2} \int_{-1}^{1} \exp(-e^{-x/\beta} U_1(\varphi, \beta)) \, d\varphi \tag{2.2.28}$$

if $\alpha = 1$ and $\beta > 0$.

The cases $\alpha \neq 1$, $x < 0$ and $\alpha = 1$, $\beta < 0$ can be reduced to the corresponding cases $\alpha \neq 1$, $x > 0$ and $\alpha = 1$, $\beta > 0$ with the help of the following equality, which is a consequence of Property 2.2 and is valid for the forms (B) and (A): for any real x and any admissible parameters α and β

$$G(-x, \alpha, \beta) + G(x, \alpha, -\beta) = 1. \tag{2.2.29}$$

The proof of (2.2.27) and (2.2.28) reduces to integration of the corresponding inequalities (2.2.18) and (2.2.19). For example, consider the case $\alpha \neq 1$,

$x > 0$. Then

$$\begin{aligned}
1 - G(x,\alpha,\beta) &= \int_x^\infty g(u,\alpha,\beta)\,du \\
&= \frac{\varepsilon(\alpha)}{2}\int_{-\theta}^1 d\varphi \int_x^\infty \left(-\frac{\alpha}{\alpha-1}u^{\alpha/(\alpha-1)-1}U_\alpha\right) \\
&\qquad\qquad\qquad\qquad \times \exp(-u^{\alpha/(\alpha-1)}U_\alpha)\,du \\
&= \frac{\varepsilon(\alpha)}{2}\int_{-\theta}^1 d\varphi \int_x^\infty d\exp(-u^{\alpha/(\alpha-1)}U_\alpha) \\
&= \frac{\varepsilon(\alpha)}{2}\int_{-\theta}^1 \left[\frac{1}{2}(1+\varepsilon(\alpha)) - \exp(-x^{\alpha/(\alpha-1)}U_\alpha)\right] d\varphi \\
&= \frac{1}{4}(1+\theta)(1+\varepsilon(\alpha)) - \frac{\varepsilon(\alpha)}{2}\int_{-\theta}^1 \exp(-x^{\alpha/(\alpha-1)}U_\alpha)\,d\varphi.
\end{aligned}$$

The case $\alpha = 1$ is analyzed similarly.

REMARK 2. Although (2.2.27) was proved for the values $x > 0$, there is nothing to prevent us from passing to the limit as $x \to 0$ and computing the value of $G(0,\alpha,\beta)$. Passing to the limit under the integral sign does not require any additional justification. As a result, in the case $\alpha \neq 1$

$$G(0,\alpha,\beta) = \tfrac{1}{2}(1 - \beta K(\alpha)/\alpha). \tag{2.2.30}$$

Unfortunately, in the case $\alpha = 1$, $\beta > 0$ no values of a are known such that $G(a,1,\beta)$ can be expressed as a combination of elementary functions of β. However, the expression

$$G(0,1,\beta) = \frac{1}{2|\beta|}\int_{-1}^1 U_1(\varphi,\beta)\exp(-U_1(\varphi,\beta))\,d\varphi \tag{2.2.31}$$

obtained from (2.2.28) also turns out to be useful in the problem of statistical estimation of the parameter β.

REMARK 3. If $\alpha < 1$ and $\beta = 1$, then for all $x < 0$

$$G(x,\alpha,1) = 0. \tag{2.2.32}$$

Since $G(x,\alpha,1)$ is a nondecreasing function tending to zero as $x \to -\infty$, (2.2.32) is a consequence of the equality $G(0,\alpha,1) = 0$ (see (2.2.30)) ([2.6]). In turn, (2.2.29) and (2.2.32) imply that in the case when $\alpha < 1$ and $\beta = -1$

$$G(x,\alpha,-1) = 1 \tag{2.2.33}$$

for all $x > 0$.

REMARK 4. The forms (2.2.18) and (2.2.19) of the densities of standard distributions, together with properties (2.2.32) and (2.2.33), enable us to draw the following conclusion.

The probability measures of standard stable laws are concentrated on the following intervals:

a) the positive semi-axis if $\alpha < 1$ and $\beta = 1$;

b) the negative semi-axis if $\alpha < 1$ and $\beta = -1$;

c) the whole real line in the other cases.

Furthermore, the density $g(x, \alpha, \beta)$ is positive at each interior point x of the interval on which the probability measure is concentrated.

REMARK 5. We get an unusual way of writing a normal distribution in the case $\alpha = 2$ from (2.2.18) and (2.2.27). (Recall that a standard stable distribution with $\alpha = 2$ corresponds to the unbiased normal law with variance $\sigma^2 = 2$; see the beginning of §2.2.) Namely,

$$\frac{1}{\sqrt{2\pi}} \exp\left(-\frac{x^2}{2}\right) = \frac{|x|}{\pi} \int_0^{\pi/2} \exp\left(-\frac{x^2}{2\sin^2\varphi}\right) \frac{d\varphi}{\sin^2\varphi},$$

$$\frac{1}{\sqrt{2\pi}} \int_{-\infty}^{\pi} e^{-t^2/2}\,dt = 1 - \frac{1}{\pi} \int_0^{\pi/2} \exp\left(-\frac{x^2}{2\sin^2\varphi}\right) d\varphi, \qquad x > 0.$$

The equalities (2.2.18) and (2.2.19) obviously make it possible to represent derivatives $g^{(n)}(x, \alpha, \beta)$ of any order n in analogous form. These representations, like the representations of the densities themselves, are far from unique. The fact is that the method used to prove Theorem 2.2.3 can be applied to the derivatives obtained from (2.2.20):

$$g^{(n)}(x, \alpha, \beta) = \frac{1}{\pi} \operatorname{Re} \int_0^{\infty} (iz)^n \exp(izx + \psi(z, \alpha, -\beta))\,dz,$$

i.e., it is possible to replace the integration over the semi-axis $z > 0$ by integration over the contour Γ, and so on. As a result,

$$g^{(n)}(x, \alpha, \beta) = \frac{1}{\pi} \int_{-\pi/2}^{\pi/2} \exp(-W(\varphi)) r^n \left(r' \sin(n+1)\left(\varphi + \frac{\pi}{2}\right)\right.$$
$$\left. + r\cos(n+1)\left(\varphi + \frac{\pi}{2}\right)\right) d\varphi. \quad (2.2.34)$$

It turns out that this representation has a rarely expressed individuality and differs essentially from what is obtained by n-fold differentiation of (2.2.18) and (2.2.19). We illustrate this by an example.

Consider the case when $n = 1$, $0 < \alpha < 2$, $|\beta| = 1$, $x > 0$, and let

$$k(x) = \begin{cases} x^{1/(\alpha-1)} & \text{if } \alpha \neq 1, \\ \exp(-x/\beta) & \text{if } \alpha = 1, \end{cases}$$

$$a(\varphi) = \begin{cases} \left|\dfrac{\sin \alpha\varphi}{\sin\varphi}\right|^{1/(1-\alpha)} \left|\dfrac{\sin(1-\alpha)\varphi}{\sin\alpha\varphi}\right| & \text{if } \alpha \neq 1, \\ \dfrac{\varphi}{\sin\varphi} \exp(\varphi\cot\varphi) & \text{if } \alpha = 1, \end{cases}$$

$$b(\varphi) = \begin{cases} \dfrac{\beta}{1-\alpha}\left|\dfrac{\sin\alpha\varphi}{\sin\varphi}\right|^{2/(1-\alpha)}\left(\dfrac{\alpha\sin(2-\alpha)\varphi}{\sin\alpha\varphi} - 1\right) & \text{if } \alpha \neq 1, \\ \exp(-2\varphi\cot\varphi)(2\varphi\cot\varphi - 1) & \text{if } \alpha = 1, \end{cases}$$

$$u = \begin{cases} -\pi & \text{if } \alpha > 1,\ \beta = 1, \\ 0 & \text{otherwise,} \end{cases}$$

$$v = \begin{cases} 0 & \text{if } \alpha < 1,\ \beta = -1, \\ -\pi\beta/\alpha & \text{if } \beta > 1, \\ \pi & \text{otherwise.} \end{cases}$$

With the help of (2.2.23) the equality (2.2.34) can be transformed into

$$g'(x,\alpha,\beta) = \frac{1}{\pi}k^2(x)\int_u^v b(\varphi)\exp\{-k^\alpha(x)a(\varphi)\}\,d\varphi. \tag{2.2.35}$$

At the same time, differentiation of (2.2.18) and (2.2.19) gives us that, in the same notation,

$$g'(x,\alpha,\beta) = ck'\int_u^v (1-\alpha k^\alpha a)a\exp(-k^\alpha a)\,d\varphi,$$

where $c = \alpha/\{\pi|1-\alpha|\}$ if $\alpha \neq 1$, and $c = 1/\pi$ if $\alpha = 1$.

Even a quick glance shows how different these representations of $g'(x,\alpha,\beta)$ are. In the case $\alpha \neq 1$ the representation (2.2.35) was first considered in [34], where it was used in studying the question of unimodality of stable laws. The main result of that paper is presented below (Theorem 2.7.5).

The equality (2.2.35), in turn, can serve as a starting point for obtaining new expressions for the function $g(x,\alpha,\beta)$. However, as a rule, the formulas thus obtained turn out to be essentially more complicated than those given in Theorem 2.2.3. They have the simplest form in the case when $\alpha = 1/n$, $n = 1, 2, \ldots$. For example,

$$g(x,1,1) = \frac{1}{\pi}\int_0^\pi ba^{-2}(1+ka)\exp(-ka)\,d\varphi,$$

$$g(x,\tfrac{1}{3},1) = \frac{2}{\pi}\int_0^\pi ba^{-4}(a^3x^{-3/2} + 3a^2x^{-1} + 6ax^{-1/2} + 6)\exp(-x^{-1/2}a)\,d\varphi.$$

It is not yet clear how such representations might turn out to be useful in studying the properties of stable distributions.

§2.3. The duality law in the class of strictly stable distributions

This is the name for the relation connecting the distributions with parameter $\alpha \geq 1$ and the distributions with parameters $\alpha' = 1/\alpha$ in the class $\mathfrak{W}$.

It should be said that this relation is not peculiar just to stable distributions. There is an analogue of it (more precisely, of the part of it connecting the extremal stable distributions) in the set of infinitely divisible laws. The duality law proves to be a valuable addition to the elementary properties of stable distributions. Thanks to it, we can better form an idea of the structure of these distributions.

The presentation of specific properties of strictly stable distributions (the duality law is included in these properties) is commited to the next chapter. However, for reasons of a methodical nature we shall become acquainted with the duality law in this chapter. In the course of the present section we do not have to go outside the class $\mathfrak{W}$, so it is natural to parametrize stable laws in the form (C). It will be indicated explicitly when we have to use other forms.

Without loss of generality we can consider only part of the distributions in $\mathfrak{W}$ by fixing the value of the scale parameter: let $\lambda_C = 1$, which corresponds to the densities $g_C(x, \alpha, \theta)$ and distribution functions $G_C(x, \alpha, \theta)$ with characteristic functions of the form

$$\mathfrak{g}_C(t, \alpha, 0) = \exp(-|t|^\alpha \exp(-i\pi\alpha\theta/2)),$$

where $0 < \alpha \leq 2$ and $|\theta| \leq \min(1, 2/\alpha - 1)$.

THEOREM 2.3.1. *For any pairs of admissible parameters $\alpha \geq 1$, θ and any $x > 0$*

$$\begin{aligned} \alpha(1 - G_C(x, \alpha, \theta)) &= G_C(x^{-\alpha}, \alpha', \theta') - \tfrac{1}{2}(1 - \theta') \\ &= G_C(x^{-\alpha}, \alpha', \theta') - G_C(0, \alpha', \theta'), \end{aligned} \tag{2.3.1}$$

where the parameters α', θ' are concerned with α, θ by the equalities

$$\alpha' = 1/\alpha, \qquad 1 + \theta' = \alpha(1 + \theta). \tag{2.3.2}$$

In terms of densities, (2.3.1) *is equivalent to the equality* ([2.7])

$$g_C(x, \alpha, \theta) = x^{-1-\alpha} g_C(x^{-\alpha}, \alpha', \theta'). \tag{2.3.3}$$

The proof of the theorem is based on (2.2.27), which in the form (C) has the same form as in the form (B), because $G_B(x, \alpha, \beta) = G_C(x, \alpha, \theta)$. Consider the case $\alpha > 1$; then

$$\alpha(1 - G_C(x, \alpha, \theta)) = \frac{\alpha}{2} \int_0^1 \exp\{-x^{\alpha/(\alpha-1)} U_\alpha(\varphi, \theta)\} \, d\varphi. \tag{2.3.4}$$

The right-hand side of (2.3.1) can be given an analogous form:

$$\begin{aligned} &G_C(x^{-\alpha}, \alpha', \theta') - G_C(0, \alpha', \theta') \\ &\quad = \frac{1}{2} \int_{-\theta'}^1 \exp\{-(x^{-\alpha})^{\alpha'/(\alpha'-1)} U_{\alpha'}(\varphi, \theta')\} \, d\varphi. \end{aligned} \tag{2.3.5}$$

To establish (2.3.1) it suffices to transform the right-hand side of (2.3.4) in such a way that it takes the form of the right-hand side of of (2.3.5). Since $\alpha' = 1/\alpha$, it follows that

$$x^{\alpha/(\alpha-1)} = (x^{-\alpha})^{\alpha'/(\alpha'-1)}.$$

Further,

$$\begin{aligned} U_\alpha(\varphi,\theta) &= \left(\frac{\sin\frac{\pi}{2}\alpha(\varphi+\theta)}{\cos\frac{\pi}{2}\varphi}\right)^{\alpha/(1-\alpha)} \frac{\cos\frac{\pi}{2}[(\alpha-1)\varphi+\alpha\theta]}{\cos\frac{\pi}{2}\varphi} \\ &= \left(\frac{\sin\frac{\pi}{2}\alpha(\varphi+\theta)}{\cos\frac{\pi}{2}\varphi}\right)^{1/(1-\alpha)} \frac{\cos\frac{\pi}{2}[(\alpha-1)\varphi+\alpha\theta]}{\sin\frac{\pi}{2}\alpha(\varphi+\theta)}. \end{aligned}$$

We perform a change of variable, setting

$$\alpha(\varphi+\theta) = 1-\kappa, \quad \text{i.e.,} \quad \varphi = -\alpha'\kappa + \alpha' - \theta.$$

With this substitution the limits of integration become 1 and $1-\alpha(1+\theta) = -\theta'$. Moreover, we get that

$$\begin{aligned} (\alpha-1)\varphi+\alpha\theta &= (\alpha-1)\varphi + 1 - \kappa - \alpha\varphi = 1-\kappa-\varphi \\ &= 1-\kappa+\alpha'\kappa-\alpha'+\theta = (\alpha'-1)\kappa+\alpha'\theta', \\ \varphi &= -\alpha'\kappa+1-\alpha'\theta' = 1-\alpha'(\kappa+\theta'). \end{aligned}$$

Substituting these expressions in (2.3.4) and considering that $1/(1-\alpha) = -\alpha'/(1-\alpha')$, we obtain the equality

$$U_\alpha(\varphi,\theta) = U_{\alpha'}(\kappa,\theta').$$

As a result, (2.3.4) is transformed into (2.3.5).

The case $\alpha = 1$ need not be considered separately, since the validity of (2.3.1) follows from the joint continuity of the distributions in $\mathfrak{W}$ with respect to α and θ in the whole domain of admissible values of these parameters (a property mentioned in the Introduction). Thus, it suffices for us to pass to the limit as $\alpha \to 1$ in (2.3.1).

In the case $\alpha = 1$, θ we have that $\alpha' = 1$ and $\theta' = \theta$. This case corresponds to the Cauchy distribution with linearly transformed argument. Namely,

$$g_C(x,1,\theta) = \frac{1}{\pi}\cos\frac{\pi}{2}\theta\left(1-2x\sin\frac{\pi}{2}\theta+x^2\right)^{-1},$$

$$G_C(x,1,\theta) = \frac{1}{2}+\frac{1}{\pi}\arctan\left(\frac{x-\sin\frac{\pi}{2}\theta}{\cos\frac{\pi}{2}\theta}\right). \tag{2.3.5a}$$

We now demonstrate the possibilities of using the duality law in the problem of constructing integral representations for densities of complex arguments. The representation to be proved below relates to stable laws with parameters $\alpha > 1$ and serves as a complement to the representations (2.2.12) and (2.2.13) in Theorem 2.2.2.

THEOREM 2.3.2. *Suppose that $\alpha > 1$, β is a pair of admissible parameters corresponding to the form* (A), *and x is a real number. Let*

$$\beta^* = \cot(\pi/2\alpha) \cdot \cot[\pi Q(\alpha,\beta)/2\alpha], \qquad \alpha' = 1/\alpha,$$
$$Q(\alpha,\beta) = 1 - \tfrac{2}{\pi}\arctan(\beta\tan(\pi\alpha/2)),$$
$$D = [1 + (\beta\tan(\pi\alpha/2))^2]^{1/2\alpha}(\sin(\pi Q(\alpha,\beta)/2\alpha))^{-1}.$$

If $y = x + |x|^{1-\alpha'}\beta^*\tan(\pi\alpha'/2)$, then

$$Dy^{-1-1/\alpha} g_A(Dy^{-1/\alpha}, \alpha, \beta) = g_A(y, \alpha', \beta^*). \tag{2.3.6}$$

PROOF. The equality (2.3.6) is verified by successive transformation of its left-hand (or right-hand) side according to the scheme

$$g_A(\cdot,\alpha,\cdot) \leftrightarrow g_B(\cdot,\alpha,\cdot) \leftrightarrow g_C(\cdot,\alpha,\cdot) \leftrightarrow g_C(\cdot,\alpha',\cdot)$$
$$\leftrightarrow g_B(\cdot,\alpha',\cdot) \leftrightarrow g_A(\cdot,\alpha',\cdot). \tag{2.3.7}$$

The middle part of this chain of transformations is the duality law, while the remaining links involve transition from one form to another. Therefore, in content (2.3.6) (if we abstract from the special form of y and regard it as an independent variable) is just an expression of the duality law in the form (A). It will be more convenient for us to go through the chain (2.3.7) of transformations from the end to the beginning.

The first step is a transformation from the form (A) to (B):

$$g_A(y,\alpha',\beta^*) = g_B(y,\alpha',\beta_B^*,0,\lambda_B').$$

Here the parameters β_B^* and λ_B' are connected with the parameters α' and β^* by the equalities (see (I.19))

$$\tan(\pi\alpha'\beta_B^*/2) = \beta^*\tan(\pi\alpha'/2), \quad \lambda_B'\cos(\pi\alpha'\beta_B^*/2) = 1. \tag{2.3.8}$$

Further, by Property 2.1,

$$g_B(y,\alpha',\beta_B^*,0,\lambda_B') = (\lambda_B')^{-1/\alpha} g_B((\lambda_B)^{-1/\alpha}y,\alpha',\beta_B^*)$$
$$= (\lambda_B')^{-\alpha} g_C((\lambda_B')^{-\alpha}y,\alpha',\theta^*),$$

where $\theta^* = \beta_B^*$.

According to the duality law (2.3.3), the last function is equal to

$$(\lambda_B')^{-\alpha}(\lambda_B'y^{-1/\alpha})^{1+\alpha} g_C(\lambda_B'y^{-1/\alpha},\alpha,\theta), \tag{2.3.9}$$

where the parameter θ is connected with θ^* by

$$1 + \theta^* = \alpha(1+\theta). \tag{2.3.10}$$

The next step in (2.3.9)—a transition to the form (B)—does not change the form of the function, since

$$g_C(\lambda_B', y^{-1/\alpha},\alpha,\theta) = g_B(\lambda_B'y^{-1/\alpha},\alpha,\beta_B),$$

where $\theta = K(\alpha)\beta_B/\alpha = (1-2/\alpha)\beta_B$.

It then remains for us to pass to the form (A). We have

$$\begin{aligned}
&\lambda'_B y^{-1-1/\alpha} g_B(\lambda'_B y^{-1/\alpha}, \alpha, \beta_B) \\
&\quad = \lambda'_B y^{-1-1/\alpha} g_A(\lambda'_B y^{-1/\alpha}, \alpha, \beta_B, 0, \lambda_A) \\
&\quad = (\lambda'_B \lambda_A^{-1/\alpha}) y^{-1-1/\alpha} g_A((\lambda'_B \lambda_A^{-1}) y^{-1/\alpha}, \alpha, \beta_A).
\end{aligned}$$

In these equalities the parameters are connected by the relations ($\beta_A = \beta$)

$$\begin{aligned}
\beta \tan(\pi\alpha/2) &= \tan[\pi(\alpha-2)\beta_B/2], \\
\lambda_A &= \cos[\pi(\alpha-2)\beta_B/2].
\end{aligned} \tag{2.3.11}$$

Thus, we have arrived at (2.3.6), in which $D = \lambda'_B \lambda_A^{-1/\alpha}$. But it remains to determine how the quantities β^* and D depend on the original parameters $\alpha > 1$ and β. This is easy to do by using formulas (2.3.8), (2.3.10), and (2.3.11) connecting the parameters. By (2.3.11),

$$\begin{aligned}
\lambda_A &= (1 - \tan^2(\pi(\alpha-2)\beta_B/2))^{-1/2} \\
&= (1 + (\beta \tan(\pi\alpha/2))^2)^{-1/2}.
\end{aligned} \tag{2.3.12}$$

According to (2.3.10),

$$\beta_B^* = \alpha - 1 + (\alpha-2)\beta_B = \alpha - 1 + \tfrac{2}{\pi} \arctan(\beta \tan(\pi\alpha/2)).$$

Therefore,

$$\begin{aligned}
\alpha' \beta_B &= 1 - \alpha' Q(\alpha, \beta), \\
\tan(\pi\alpha'\beta_B^*/2) &= \cot[\pi Q(\alpha,\beta)/2\alpha].
\end{aligned} \tag{2.3.13}$$

From (2.3.13) and (2.3.8) we find expressions for the parameters λ'_B and β^*:

$$\begin{aligned}
\lambda'_B &= (1 + \tan^2(\pi\alpha'\beta_B^*/2))^{1/2} \\
&= [1 + \cot^2(\pi Q(\alpha,\beta)/2\alpha)]^{1/2} = [\sin(\pi Q(\alpha,\beta)/2\alpha)]^{-1}, \\
\beta^* &= \cot(\pi/2\alpha)\cot(\pi Q(\alpha,\beta)/2\alpha).
\end{aligned}$$

REMARK 1. Let $\mu^* = \beta^* \tan \frac{\pi}{2}\alpha'$, $y = x + |x|^{1-\alpha'}\mu^*$, and $l = \operatorname{sgn} y$. The equalities (2.3.6) and (2.2.12) yield an integral expression for the density of the complex function when $\alpha > 1$ in the case not considered in Theorem 2.2.2:

$$\begin{aligned}
&D y^{-1-1/\alpha} g_A(D y^{-1/\alpha}, \alpha, \beta) \\
&\quad = \frac{1}{\pi|x|} \operatorname{Im} \int_0^\infty \exp\left\{ -l \frac{xu}{|x|} - |x|^{-1/\alpha}[(-iu)^{1/\alpha} \right. \\
&\qquad\qquad\qquad \left. - l\mu^* u((-iu)^{1/\alpha - 1} - 1)] \right\} du.
\end{aligned}$$

REMARK 2. We consider a feature of the duality law which is connected with the presence of the condition $x > 0$ in the formulation of the law. On the face of it, this condition is not essential, since the case $x < 0$ can be reduced to the case $x > 0$ by changing the sign of β, i.e., due to the property $g(x, \alpha, \beta) = g(-x, \alpha, -\beta)$, and indeed, the relation

$$g(x, \alpha, \beta) = x^{-1-\alpha} g(x^{-\alpha}, \alpha', \beta'), \qquad \alpha > 1, \ x > 0,$$

is complemented by an analogous relation for $x < 0$:

$$\begin{aligned} g(x, \alpha, \beta) &= g(|x|, \alpha, -\beta) = |x|^{-1-\alpha} g(|x|^{-\alpha}, \alpha', -\beta'') \\ &= |x|^{-1-\alpha} g(-|x|^{-\alpha}, \alpha', \beta''), \qquad \beta'' = \beta' - 2(1-\alpha). \end{aligned}$$

Since $\beta' \neq \beta''$ if $\alpha > 1$, in this case the "intermediaries" of the parts of the distribution $g(\cdot, \alpha', \cdot)$ (i.e., $g_1 = g(\cdot, \alpha, \cdot)$ and $g_2 = g(\cdot, \alpha, \cdot)$, $\alpha' = 1/\alpha$) are not component parts of some single distribution $g(\cdot, \alpha, \cdot)$. Therefore, there is no relation analogous to the duality law between the functions $g(x, \alpha, \beta, \gamma, \lambda)$ and $g(x^{-\alpha}, \alpha', \beta', \gamma', \lambda')$ in the case when $x > 0$, $\alpha > 1$, and $\gamma \neq 0$, i.e., the class $\mathfrak{W}$ is the natural part of the family $\mathfrak{S}$ in which the duality law can act. Furthermore, it is the "halves" $g(x, \alpha, \beta)$, $x > 0$, of the densities of stable laws that appear as independent analytic objects, and not the densities as a whole. More precisely, it is not the "halves" themselves of the densities but the functions

$$\varsigma_C(x, \alpha, \rho) = x g_C(x, \alpha, \theta) = x g(x, \alpha, \beta),$$

if we consider them on the semi-axis $x > 0$. For it is they that are the natural analytic entities for studying the group of properties of strictly stable laws, which properties include the duality law. Let us see how much more symmetric this law appears if it is written in terms of the functions ς_C.

For any $x > 0$ and any admissible parameter values α, ρ and α', ρ' connected by the relations $\alpha\alpha' = 1$ and $\sqrt{\alpha}\rho = \sqrt{\alpha'}\rho'$ we have the equality

$$\varsigma_C\left(x^{\sqrt{\alpha}}, \alpha, \rho\right) = \varsigma_C\left(x^{-\sqrt{\alpha'}}, \alpha', \rho'\right).$$

The group of properties connected with the "halves" of the densities or with the functions ς_C turns out to be so extensive that we allot them the main part of the next chapter.

REMARK 3. The duality property turns out to be a very convenient instrument for solving a number of problems such as, for instance, the problem of representing densities or distribution functions of the class $\mathfrak{S}$ by integrals, by convergent or asymptotic series, and so on. For example, we already know that the natural intervals of the probability distributions for standard stable laws are the half-line $(0, \infty)$ in the case $\alpha < 1$, $\beta = 1$, the half-line $(-\infty, 0)$ in the case $\alpha < 1$, $\beta = -1$, and the whole real axis in the remaining cases.

One of the questions to be discussed later is the asymptotic behavior of the densities $g(x, \alpha, \beta)$ and the distribution functions $G(x, \alpha, \beta)$ on the boundaries of these intervals. The relations (2.3.1) and (2.3.2) shorten the amount of labor involved in studying the asymptotic behavior by almost half, since it is possible to confine oneself to analysis of the cases $\alpha \leq 1$. In the case $\alpha > 1$ the answer is obtained by an uncomplicated recomputation of results connected with the case $\alpha < 1$.

§2.4. The analytic structure of stable distributions and their representation by convergent series

Stable laws have densities with uniformly bounded derivatives of any order. This fact was established in the Introduction. Moreover, the nth derivative of the density of a standard distribution has the estimate

$$|g^{(n)}(x, \alpha, \beta)| \leq \frac{1}{\pi\alpha}\Gamma\left(\frac{n+1}{\alpha}\right)\left(\cos\left[\frac{\pi}{2}K(\alpha)\beta\right]\right)^{-(n+1)/\alpha}.$$

The present section is devoted to explaining the more subtle analytic structure of stable distributions. It is convenient to carry out the analysis within the confines of the class $\mathfrak{S}_0$. Let us consider the set $\mathcal{D}$ of all possible pairs of admissible values of the parameters (α, β) except for the pair $\alpha = 1$, $\beta = 0$. For each pair $(\alpha, \beta) \in \mathcal{D}$ we define on the semi-axis $x > 0$ the functions

$$q(x, \alpha, \beta) = \begin{cases} xg(x, \alpha, \beta) & \text{if } \alpha \geq 1, \\ x^{-1/\alpha}g(x^{-1/\alpha}, \alpha, \beta) & \text{if } \alpha < 1, \end{cases}$$

$$Q(x, \alpha, \beta) = \begin{cases} G(x, \alpha, \beta) & \text{if } \alpha \geq 1, \\ \alpha - \alpha G(x^{-1/\alpha}, \alpha, \beta) & \text{if } \alpha < 1, \end{cases}$$

which are connected by

$$xQ'(x, \alpha, \beta) = q(x, \alpha, \beta). \tag{2.4.1}$$

THEOREM 2.4.1. *For each pair $(\alpha, \beta) \in \mathcal{D}$ the function $Q(x, \alpha, \beta)$ extends analytically from the semi-axis $x > 0$ to the whole complex plane, i.e., is an entire analytic function.*

The derivative of an entire function is itself an entire function. Therefore, the product $q(x, \alpha, \beta) = xQ'(x, \alpha, \beta)$ of two entire functions is also an entire function, and, moreover, it has a zero of first order at the point $z = 0$ ([2.8]).

PROOF. Consider the case $\alpha = 1$, $\beta > 0$. According to Corollary 1 to Theorem 2.2.1, for any real x the function $1 - Q$ can be represented as the following absolutely convergent integral:

$$\begin{aligned} 1 - Q(x, \alpha, \beta) &= 1 - G(x, \alpha, \beta) \\ &= \frac{1}{\pi}\int_0^\infty \exp(-xu - \beta u \log u)\sin\left[(1+\beta)u\frac{\pi}{2}\right]\frac{du}{u}. \end{aligned}$$

If in the entire function $\exp(-xu)$ appearing in the integrand we replace x by an arbitrary complex number z, the integral remains absolutely convergent. This means that the integral converges uniformly in a disk $|z| \leq r$ of arbitrarily large radius r. Consequently, the integral and with it the function Q extend analytically to the whole z-plane.

The case $\alpha < 1$ is handled according to the same scheme. By Corollary 1 to Theorem 2.2.1,

$$\begin{aligned} Q(x, \alpha, \beta) &= \alpha - \alpha G(x^{-1/\alpha}, \alpha, \beta) \\ &= \frac{\alpha}{\pi} \int_0^\infty e^{-u} \operatorname{Im} \exp(-x u^\alpha e^{-i\pi\rho\alpha}) \frac{du}{u} \end{aligned}$$

for any $x > 0$. The integral is absolutely convergent and remains so if x is replaced in the integrand by any complex number z. Consequently, it converges uniformly in a disk $|z| \leq r$ of any radius, i.e., the function Q is entire.

The case $\alpha > 1$ reduces to $\alpha < 1$ with use of (2.3.1):

$$\begin{aligned} Q(x, \alpha, \beta) &= G(x, \alpha, \beta) \\ &= 1 + \tfrac{1}{2}\alpha'(1 - \beta') - \alpha' G(x^{-1/\alpha'}, \alpha', \beta') \\ &= 1 - \tfrac{1}{2}\alpha'(1 + \beta') + Q(x, \alpha', \beta'), \end{aligned} \tag{2.4.2}$$

where $\alpha' = 1/\alpha$ and $\beta' = -1 + \alpha(1 + \beta K(\alpha)/\alpha)$.

We make two remarks about this theorem.

REMARK 1. The case $\alpha = 1$, $\beta = 0$ corresponding to the Cauchy distribution occupies a special position among the standard stable distributions. The distribution function in this case has the form

$$\begin{aligned} G(x, 1, 0) &= \frac{1}{2} + \frac{1}{\pi} \arctan\left(\frac{2}{\pi} x\right) \\ &= \frac{1}{2} + \frac{1}{2\pi i} \log \frac{x - i\pi/2}{x + i\pi/2}, \end{aligned}$$

which shows that although G does extend analytically from the semi-axis $x > 0$, the extension is only to the half-plane with a cut (for example, along the segment $[-i\pi/2, i\pi/2]$) joining the branch points $-i\pi/2$ and $i\pi/2$

REMARK 2. At the end of the preceding section we pointed out a feature of stable distributions consisting in the fact that the "halves" $g(x, \alpha, \beta)$, $x > 0$, and $g(x, \alpha, \beta)$, $x < 0$, of the standard densities in the case $\alpha \neq 1$ appear as independent analytic formations. Theorem 2.4.1 gives another intuitive confirmation. This is especially clear in the case $\alpha < 1$. Indeed, consider the density $g(x, \alpha, \beta)$ with some pair of parameter values $\alpha < 1$ and β. We can find the "half" $g(x, \alpha, \beta)$, $x < 0$, in symmetrically mapped form among the "halves" $g(x, \alpha, \tilde{\beta})$, $x > 0$, for $\tilde{\beta} = -\beta$. Therefore, the functions $q(x, \alpha, \beta)$ and $q(x, \alpha, -\beta)$ form two parts of a single density. However, the analytic extensions of these functions are completely different and cannot be reduced one to the other by means of some simple transformation like a symmetric mapping. For example, if $\alpha < 1$, then $q(x, \alpha, 1)$ has some nonzero extension to the complex plane, while $q(x, \alpha - 1) = 0$, $x > 0$, has as an extension the function identically equal to zero in the whole plane.

An analogous picture is observed in respect to the function Q.

THEOREM 2.4.2. *For any pair $(\alpha, \beta) \in D$ of parameter values the entire function $Q(x, \alpha, \beta)$ can be represented as a power series as follows. If $\alpha < 1$, then*

$$\begin{aligned} Q(x, \alpha, \beta) &= \alpha - \alpha G(x^{-1/\alpha}, \alpha, \beta) \\ &= \frac{1}{\pi} \sum_{n=1}^{\infty} (-1)^{n-1} \frac{\Gamma(n\alpha + 1)}{n\Gamma(n+1)} \sin(\pi n \rho \alpha) x^n. \end{aligned} \tag{2.4.3}$$

If $\alpha > 1$, then, with the same notation α', β' as in (2.4.2),

$$\begin{aligned} Q(x, \alpha, \beta) = G(x, \alpha, \beta) = 1 + \frac{1}{2} \alpha' (1 + \beta') \\ + \frac{1}{\pi} \sum_{n=1}^{\infty} (-1)^{n-1} \frac{\Gamma(n\alpha' + 1)}{n\Gamma(n+1)} \sin(\pi n \rho) x^n. \end{aligned} \tag{2.4.4}$$

In the case when $\alpha = 1$ and $\beta > 0$

$$Q(x, \alpha, \beta) = G(x, \alpha, \beta) = 1 - \frac{1}{\pi} b_0 + \frac{1}{\pi} \sum_{n=1}^{\infty} (-1)^{n-1} b_n x^n, \tag{2.4.5}$$

where

$$b_n = \frac{1}{\Gamma(n+1)} \int_0^{\infty} \exp(-\beta u \log u) u^{n-1} \sin\left[(1+\beta) u \frac{\pi}{2}\right] du.$$

By (2.2.29), the case $\alpha = 1$, $\beta < 0$ can be reduced to the case $\alpha = 1$, $\beta > 0$.

The proof of (2.4.4)–(2.4.5) is based on expansion of the integrands in power series with respect to x and termwise integration of the representations of Q used in Theorem 2.4.1 in the case $\alpha \geq 1$. The expansion (2.4.3) is obtained from (2.4.4) due to (2.4.2).

REMARK. The equalities (2.4.1) and (2.4.3)–(2.4.5) enable us to get (by differentiation of the series) the following representations of the densities $g(x, \alpha, \beta)$ by convergent series ([2.9]):

If $\alpha > 1$, then for any admissible β and any real x

$$g(x, \alpha, \beta) = \frac{1}{\pi} \sum_{n=1}^{\infty} (-1)^{n-1} \frac{\Gamma(n\alpha' + 1)}{\Gamma(n+1)} \sin(\pi n \rho) x^{n-1}. \tag{2.4.6}$$

If $\alpha = 1$ and $\beta > 0$, then for any real x

$$g(x, 1, \beta) = \frac{1}{\pi} \sum_{n=1}^{\infty} (-1)^{n-1} n b_n x^{n-1}. \tag{2.4.7}$$

If $\alpha < 1$, then for any admissible β and any $x > 0$

$$g(x, \alpha, \beta) = \frac{1}{\pi} \sum_{n=1}^{\infty} (-1)^{n-1} \frac{\Gamma(n\alpha + 1)}{\Gamma(n + 1)} \sin(\pi n \rho \alpha) x^{-n\alpha - 1}. \tag{2.4.8}$$

In contrast to (2.4.5), the equalities (2.4.3) and (2.4.4) have the feature that the power series on their right-hand sides contain explicit expressions for the coefficients. This enables us to compute the order and type of the entire function $Q(x, \alpha, \beta)$ by using the following well-known connections between the order σ and type δ of an entire function $\chi(z) = \sum_0^\infty a_n z^n$ and its coefficients (see, for example, [83]):

$$\sigma = \limsup_{n \to \infty} \frac{n \log n}{|\log |a_n||}, \tag{2.4.9}$$

$$\delta = \frac{1}{\sigma e} \limsup_{n \to \infty} n |a_n|^{\sigma/n}. \tag{2.4.10}$$

Let $\tilde{\alpha} = \alpha$ if $\alpha < 1$, and $\tilde{\alpha} = 1/\alpha$ if $\alpha > 1$.

THEOREM 2.4.3. *In the case $\alpha \neq 1$ the entire function $Q(z, \alpha, \beta)$ has order*

$$\sigma = (1 - \tilde{\alpha})^{-1}$$

and type

$$\delta = (1 - \tilde{\alpha}) \tilde{\alpha}^{-\tilde{\alpha}/(1-\tilde{\alpha})}.$$

In the case $\alpha = 1$, $\beta \neq 0$ the entire function $Q(z, 1, \beta) = G(z, 1, \beta)$ has infinite order ([2.10]).

PROOF. For $\alpha \neq 1$ we find with the help of Stirling's formula that

$$\begin{aligned} &\Gamma(n\tilde{\alpha} + 1)/\Gamma(n = 1) \\ &\qquad = \exp\left\{(\tilde{\alpha} - 1) n \log n + n(\tilde{\alpha} \log \tilde{\alpha} - \tilde{\alpha} + 1) + \tfrac{1}{2} \log \tilde{\alpha} + o(1)\right\}, \end{aligned}$$

and the values of σ and δ are found from (2.4.9) and (2.4.10) by using (2.4.3) and (2.4.4).

It can be proved directly that $G(z, 1, 1)$ has infinite order by considering the asymptotic behavior of $G(x, 1, 1)$ on the negative semi-axis. According to Theorem 2.4.3,

$$G(x, 1, 1) \sim \frac{1}{\sqrt{2\pi}} \exp\left\{\frac{1}{2}(1 + x) - \exp(-(1 + x))\right\} \quad \text{as } x \to -\infty.$$

It is known from the theory of analytic functions (see, for example, the well-known book [83]) that if an entire function $w(z)$ has finite order σ, then for any fixed $\varepsilon > 0$ there exist circles $|z| = r$ of arbitrarily large radius r on which

$$\min\{|w(z)|: |z| = r\} > \exp(-r^{\sigma+\varepsilon}).$$

It is easy to see that the asymptotic behavior of $G(x, 1, 1)$ as $x \to -\infty$ is incompatible with the assumption that it has finite order.

We choose another route for proving that the entire function $G(z, 1, \beta)$, $\beta \neq 0$, has infinite order. Note first of all that without loss of generality we can assume that $\beta > 0$. According to Corollary 1 to Theorem 2.2.1, for any real x

$$G(x, 1, \beta) = 1 - \frac{1}{\pi} \int_0^\infty \exp(-xu - \beta u \log u) \sin\left[\frac{\pi}{2}(1+\beta)u\right] \frac{du}{u}.$$

It is not hard to see that the integral remains absolutely convergent if x is replaced by some complex number. This means that the integral gives a representation of $G(z, 1, \beta)$ in the whole complex z-plane. Consider the sequence of complex numbers

$$z_k = -x_k - i\tfrac{\pi}{2}(1+\beta) = -\beta(1 + \log u_k) - i\tfrac{\pi}{2}(1+\beta),$$

where $u_k = (1 + 4k)/(1 + \beta)$, $k = 1, 2, \dots$.

Let $\omega = (2(1+\beta))^{-1}$. Elementary calculations show that

$$\begin{aligned}
\operatorname{Im} G(z_k, 1, \beta)& \\
&= \frac{1}{\pi} \int_0^\infty \exp(x_k u - \beta u \log u) \left(\sin \frac{\pi}{2}(1+\beta)u\right)^2 \frac{du}{u} \\
&\geq \frac{1}{\pi} \int_{|u-u_k|<\omega} \exp(x_k u - \beta u \log u) \left(\sin \frac{\pi}{2}(1+\beta)u\right)^2 \frac{du}{u} \\
&\geq \frac{1}{2\pi} \log\left(\frac{3+8k}{1+8k}\right) \exp\{x_k(u_k - \omega) - \beta(u_k + \omega)\log(u_k + \omega)\}.
\end{aligned}$$

This quantity has the asymptotic expression

$$\begin{aligned}
\frac{1}{8\pi k} &\exp\{\beta u_k - (1+\beta)\omega \log u_k - 2\beta\omega + o(1)\} \\
&= \exp\{\beta \exp(x_k/\beta - 1) + O(x_k)\}
\end{aligned}$$

as $k \to \infty$. Now let

$$M(r) = \max\{|G(z, 1, \beta)|: |z| = r\}$$

and note that

$$\begin{aligned}
M(|z_k|) &\geq \operatorname{Im} G(z_k, 1, \beta) \\
&\geq \exp\{\beta \exp(\tfrac{1}{\beta} \operatorname{Re} z_k - 1) + O(\operatorname{Re} z_k)\}.
\end{aligned}$$

Since $|z_k| = \operatorname{Re} z_k(1+o(1))$, the last inequality gives us

$$\limsup_{r\to\infty} \tfrac{1}{r} \log\log M(r) \geq \tfrac{1}{\beta} > 0,$$

which, in turn, implies that $G(z,1,\beta)$ has infinite order.

REMARK. As is known from the theory of analytic functions, any derivaitve of an entire function has the same order and type as the function itself. In particular, it follows from this that Theorem 2.4.3 is valid also for the function $g(z,\alpha,\beta)$.

In concluding this section we give without proof a certain interesting fact. The reader desiring to see a proof is referred to [7], whence the fact was taken (2.11).

Consider a standard stable distribution with parameters $\alpha < 1$ and $\beta = 1$, and let

$$L_n^{(s)}(x) = \left(\frac{\Gamma(n+1)}{\Gamma(n+1+s)}\right)^{1/2} \sum_{k=0}^{n} (-1)^k \frac{\Gamma(1+s+n)x^k}{\Gamma(k+1)\Gamma(n-k+1)\Gamma(1+s+n)}$$

be the Laguerre polynomials ($s > -1$, $n = 0, 1, \ldots$) which form a complete orthonormal system in the Hilbert space of real functions on $[0,\infty)$ whose squares are integrable with respect to the measure $\mu(dx) = x^s e^{-x} dx$.

THEOREM 2.4.4. *For any $0 < \alpha < 1$ and $x > 0$ the function $q(x,\alpha,1)$ can be represented by the convergent series*

$$q(x,\alpha,1) = x\exp(-x) \sum_{n=0}^{\infty} k_n^{(s)}(\alpha) L_n^{(s)}(x), \tag{2.4.11}$$

where s is any fixed number greater than -1, and

$$k_n^{(s)}(\alpha) = \alpha \left(\frac{\Gamma(n+1)}{\Gamma(n+1+s)}\right)^{1/2} \times \sum_{m=0}^{n} \frac{(-1)^m \Gamma(1+s+n)}{\Gamma(m+1)\Gamma(n-m+1)\Gamma(1+\alpha(s+m))}.$$

In the case $\alpha > 1$, $x > 0$ the expansion in a convergent series of the function

$$q(x,\alpha,1) = \exp(-x) \sum_{n=0}^{\infty} k_n^{(s)}(1/\alpha) L_n^{(s)}(x)$$

is obtained from (2.4.11) with the help of (2.3.3).

§2.5. Asymptotic expansions of stable distributions

As mentioned in Remark 4 after Theorem 2.2.3, the support of the measure of the distribution $G(x,\alpha,\beta)$ is the semi-axis $(0,\infty)$ if $\alpha < 1$ and $\beta = 1$, the semi-axis $(-\infty, 0)$ if $\alpha < 0$ and $\beta = -1$, and the whole real axis otherwise. The material in the preceding section does not enable us to determine the behavior of the entire functions $G(x, \alpha, \beta)$ and $g(x, \alpha, \beta)$ at the point $x = 0$ if $\alpha < 1$, nor at $x = \infty$ if $1 \le \alpha < 2$ (this does not concern the case when $\alpha = 1$ and $\beta = 0$), because these points of the complex plane are singular for them: namely, a branch point in the first case and an essential singularity in the second. Below we correct this deficiency of information by constructing asymptotic expansions of G and g in a neighborhood of the corresponding singular points. It turns out that the forms of the asymptotic expansions differ considerably, depending on whether or not we are dealing with an extremal distribution (i.e., $\beta = \pm 1$) and on whether $\alpha < 1$, $\alpha = 1$, or $\alpha > 1$.

In this connection the following six possible situations distinguish themselves (because of Property 2.2, it suffices to consider the asymptotic expansions only for $x > 0$):

I. $\alpha < 1$, $\beta \neq 1$, $x \to 0$.

II. $\alpha < 1$, $\beta = 1$, $x \to 0$.

III. $\alpha = 1$, $\beta \neq -1$, $\beta \neq 0$, $x \to \infty$.

IV. $\alpha = 1$, $\beta = -1$, $x \to \infty$.

V. $\alpha > 1$, $\beta \neq -1$, $x \to \infty$.

VI. $\alpha > 1$, $\beta = -1$, $x \to \infty$.

The arrangement and grouping of the material in this section does not follow precisely the indicated division into six cases; hence, some explanations are necessary. The asymptotic expansions given below (the concept of an asymptotic series is the one traditionally accepted in mathematical analysis) are divided into two groups. The first (Theorems 2.5.1–2.5.4) contains asymptotic formulas for the densities $g(x, \alpha, \beta)$ and the distribution functions in neighborhoods of the corresponding singular points. The second group (Theorems 2.5.5–2.5.7) combines the asymptotic expansions for $g(r(x), \alpha, \beta)$ and $G(r(x), \alpha, \beta)$, where the $r(x)$ are functions of a special type. Furthermore, in the cases when the expansions of the distribution functions are obtained as consequences of the asymptotic expansions of the densities corresponding to them, the former expansions are not singled out in independent statements. The same principle is followed also when one of the cases mentioned above can be reduced to another by using duality relations (case V reduces to case I, and VI reduces to II). Cases II, IV, and VI can be combined in a single analytic form of expression, so they are considered in the two Theorems 2.5.2 and 2.5.3 combining them, of which one has to do with expansion of the distribution functions.

The case $\alpha = 1$, $\beta = 0$, for which there are simple explicit expressions for the density and distribution function, is not considered at all.

THEOREM 2.5.1. *If $\alpha < 1$ and $\beta \neq 1$, then the asymptotic representation*

$$g(x, \alpha, \beta) \sim \frac{1}{\pi\alpha} \sum_{k=0}^{\infty} \frac{\Gamma((k+1)/\alpha)}{\Gamma(k+1)} \sin \frac{\pi}{2}(k+1)(1-\beta)x^k \tag{2.5.1}$$

holds as $x \to 0$.

PROOF. Let us use the integral expression for the density in the form (2.2.1) and write

$$\begin{aligned} g(x, \alpha, \beta) &= \frac{1}{\pi} \operatorname{Re} \int_0^{\infty} \exp\left(itx - t^{\alpha} \exp\left(i\frac{\pi}{2}\alpha\beta\right)\right) dt \\ &= \frac{1}{\pi} \sum_{k=0}^{n-1} \frac{x^k}{k!} \operatorname{Re} \int_0^{\infty} (it)^k \exp\left(-t^{\alpha} \exp\left(i\frac{\pi}{2}\alpha\beta\right)\right) dt + R_n, \end{aligned}$$

where

$$R_n = \frac{1}{\pi} \operatorname{Re} \int_0^{\infty} \left[e^{itx} - \sum_{k=0}^{n-1} \frac{(itx)^k}{k!}\right] \exp\left(-t^{\alpha} \exp\left(i\frac{\pi}{2}\alpha\beta\right)\right) dt.$$

We have that

$$\begin{aligned} |R_n| &\leq \frac{1}{\pi} \int_0^{\infty} \frac{1}{n!}(tx)^n \exp\left(-t^{\alpha} \cos\frac{\pi}{2}\alpha\beta\right) dt \\ &= \frac{1}{\pi\alpha} \frac{\Gamma((n+1)/\alpha)}{\Gamma(n+1)} \left[\cos\frac{\pi}{2}\alpha\beta\right]^{-(n+1)/\alpha} x^n. \end{aligned} \tag{2.5.2}$$

In the main sum in all the integrals we perform the change of variable $w = t\exp(i\pi\beta/2)$ and a subsequent rotation of the contour of integration in order that the integration again be carried out along the semi-axis $(0, \infty)$. The expansion (2.5.1) is obtained as a result.

COROLLARY 1. *If $\alpha < 1$ and $\beta \neq 1$, then as $x \to 0$*

$$\begin{aligned} G(x, \alpha, \beta) - G(0, \alpha, \beta) &= G(x, \alpha, \beta) - \frac{1}{2}(1-\beta) \\ &\sim \frac{1}{\pi\alpha} \sum_{k=1}^{\infty} \frac{\Gamma(k/\alpha)}{\Gamma(k+1)} \sin\frac{\pi}{2}k(1-\beta)x^k. \end{aligned} \tag{2.5.3}$$

This can be verified with the help of (2.5.1) and L'Hôpital's rule.

COROLLARY 2. *If $\alpha > 1$ and $\beta \neq -1$, then as $x \to 0$*

$$g(x, \alpha, \beta) \sim \frac{\alpha}{\pi} x^{-1} \sum_{k=1}^{\infty} \frac{\Gamma(\alpha k)}{\Gamma(k)} s \in \frac{\pi}{2} k(2-\alpha)(1+\beta)x^{-k\alpha}. \tag{2.5.4}$$

Indeed, according to (2.3.3), for $\alpha > 1$ and $x > 0$

$$g(x, \alpha, \beta) = x^{-1-\alpha} g(x^{-\alpha}, 1/\alpha, \beta'), \tag{2.5.5}$$

where $\beta' = -1 + \alpha(1 + \beta(\alpha-2)/\alpha) = 1 - (2-\alpha)(1+\beta)$.

We have the asymptotic expansion (2.5.1) for the right-hand side of (2.5.5). This implies (2.5.4).

In turn, we find from (2.5.4) that if $\alpha > 1$ and $\beta \neq 1$, then as $x \to \infty$

$$1 - G(x, \alpha, \beta) \sim \frac{1}{\pi} \sum_{k=1}^{\infty} \frac{\Gamma(\alpha k)}{\Gamma(k+1)} \sin \frac{\pi}{2} k(2-\alpha)(1+\beta) x^{-k\alpha}.$$

(This can be established with the help again of L'Hôpital's rule.)

The analysis of the asymptotic behavior of a stable distribution as $x \to 0$ in the case $\alpha < 1$, $\beta = 1$ and as $x \to \infty$ in the case $\alpha = 1$, $\beta = -1$ is based on the well-known method of Laplace for asymptotic representation of integrals. The assertion taken as a basis for the analysis is one of the simplest variants of theorems of this kind. One can become acquainted with such theorems, for example, in the detailed monograph [20].

Let us consider even functions $s(t)$ and $w(t)$ that are analytic on the interval $(-\pi, \pi)$ and have the following additional properties:

1) $w(t)$ is strictly monotone on $(0, \pi)$.
2) $\mu = s(0) > 0$, $\tau = w(0) > 0$, and $\sigma^2 = w''(0) > 0$.
3) $s(t) = O(w(t))$ as $t \to \pi$.

With the help of these functions we form the integral

$$I_N = \frac{1}{2\pi} \int_{-\pi}^{\pi} s(t) \exp\{-N w(t)\}\, dt, \tag{2.5.6}$$

which obviously exists for each $N > 0$.

LEMMA 2.5.1. *The integral I_N has the following representation by an asymptotic series as $N \to \infty$:*

$$I_N \sim \frac{\mu}{\sigma\sqrt{2\pi}} N^{-1/2} \exp(-\tau N) \left(1 + \sum_{n=1}^{\infty} Q_n N^{-n}\right), \tag{2.5.7}$$

where

$$Q_n = \frac{1}{\sqrt{2\pi}} \int_{-\infty}^{\infty} q_n(t) \exp\left(-\frac{t^2}{2}\right) dt \tag{2.5.8}$$

and the functions $q_n(t)$, which are polynomials, are the coefficients of the series expansion, in powers of $h^2 = (\sigma^2 N)^{-1}$, of the even function

$$\begin{aligned} \omega(t,h) &= \frac{1}{\mu} s(th) \exp\left\{-\frac{1}{\sigma^2 h^2}\left[w(th) - \tau - \frac{\sigma^2 t^2 h^2}{2}\right]\right\} \\ &= 1 + \sum_{n=1}^{\infty} q_n(t) h^{2n}. \end{aligned} \tag{2.5.9}$$

PROOF. We break up the integral I_N into two parts, with $\varepsilon = N^{-1/3}$:

$$I_N = \frac{1}{2\pi}\int_{|t|\le\varepsilon} + \frac{1}{2\pi}\int_{\varepsilon<|t|<\pi} = I'_N + I''_N.$$

The second term can be estimated by using properties 1)–3) of the functions $s(t)$ and $w(t)$ and the fact that $x\exp(-x)$ is monotonically decreasing on the semi-axis $x \ge 1$. For sufficiently large N we have (c, c' and c'' are constants)

$$\begin{aligned} I''_N &\le c\int_\varepsilon^\pi w(t)\exp(-Nw(t))\,dt \\ &< \pi c w(\varepsilon)\exp(-Nw(\varepsilon)) < c'\exp(-Nw(\varepsilon)). \end{aligned}$$

Since $w(\varepsilon) = \tau + \sigma^2\varepsilon^2/2 + o(\varepsilon^4)$, it follows that

$$I''_N < c''\exp(-\tau N - \sigma^2 N^{1/3}/2).$$

Consequently, this part of the integral I_N is infinitesimally small as $N \to \infty$ in comparison to any term of the asymptotic expansion (2.5.7).

In the first term, after transforming it to the form

$$I'_N = \frac{N^{-1/2}}{2\pi\sigma}\exp(-\tau N)\int_{-\varepsilon/h}^{\varepsilon/h} e^{-t^2/2}\omega(t,h)\,dt,$$

we expand $\omega(t,h)$ in an asymptotic series (2.5.9) of powers of h^2 and integrate termwise. It is then verified that the increase in the domain of integration of the integrals

$$\int_{-\varepsilon/h}^{\varepsilon/h} q_n(t)\exp(-t^2/2)\,dt$$

to integration over the whole real line changes each of them by no more than $O(\exp(-N^{1/4}))$.

We consider the case $\alpha < 1$, $\beta = 1$ for $x \to 0$ and the case $\alpha \ge 1$, $\beta = -1$ for $x \to \infty$. It turns out that Theorem 2.2.3 and Remark 1 after it make it possible to write $g(x,\alpha,\beta)$ and $G(x,\alpha,\beta)$ in the form of the integral (2.5.6). Let

$$\xi = \xi(x,\alpha) = \begin{cases} |1-\alpha|(x/\alpha)^{\alpha/(\alpha-1)} & \text{if } \alpha \ne 1, \\ \exp(x-1) & \text{if } \alpha = 1, \end{cases}$$

$$w = w(\varphi,\alpha) = \begin{cases} \left(\dfrac{\sin\alpha\varphi}{\alpha\sin\varphi}\right)^{\alpha/(\alpha-1)} \dfrac{\sin(1-\alpha)\varphi}{(1-\alpha)\sin\varphi} & \text{if } \alpha < 1, \\ \dfrac{\varphi}{\sin\varphi}\exp(1-\varphi\cot\varphi) & \text{if } \alpha = 1 \end{cases}$$

(in the case $\alpha > 1$ the function $w = w(\varphi,\alpha)$ is defined with the help of the function w defined forf $\alpha < 1$ by replacing α by $1/\alpha$), and

$$\nu = \nu(\alpha) = \begin{cases} |1-\alpha|^{-1/\alpha} & \text{if } \alpha \ne 1, \\ 1 & \text{if } \alpha = 1. \end{cases}$$

It is not hard to see that with this notation the equalities (2.2.18) and (2.2.19) can be reduced in the cases $\alpha < 1$, $\beta = 1$ and $\alpha = 1$, $\beta = -1$ to the single equality

$$g(x,\alpha,\beta) = \frac{\nu}{2\pi}\xi^{1/\alpha}\int_{-\pi}^{\pi} w\exp(-\xi w)\,d\varphi. \tag{2.5.10}$$

The fact that this formula remains true also when $\alpha > 1$ is quite easy to check by using the daulity relation (2.3.3) and the following properties of the function ξ:

If $\alpha > 1$, then for any $x > 0$

$$\xi(x^{-\alpha}, 1/\alpha) = \xi(x,\alpha),$$
$$\nu(1/\alpha)\xi^{\alpha}(x^{-\alpha}, 1/\alpha)x^{-(1+\alpha)} = \nu(\alpha)\xi^{1/\alpha}(x,\alpha).$$

Similarly, (2.2.27), (2.2.28) and the duality relation (2.3.1) can be used to write the distribution functions in a unified form in the cases $\alpha < 1$, $\beta = 1$ and $\alpha \geq 1$, $\beta = -1$. Namely,

$$\frac{1}{2\pi}\int_{-\pi}^{\pi}\exp(-\xi w)\,d\varphi = \begin{cases} G(x,\alpha,1) & \text{if } \alpha < 1, \\ 1 - G(x,\alpha,-1) & \text{if } \alpha \geq 1. \end{cases} \tag{2.5.11}$$

For convenience in presenting the subsequent material we introduce the quantity

$$\alpha_* = \begin{cases} \alpha & \text{if } \alpha < 1, \\ 1/\alpha & \text{if } \alpha \geq 1. \end{cases} \tag{2.5.12}$$

The function $\xi(x,\alpha)$ converges to ∞ in the case $\alpha < 1$, $\beta = 1$ as $x \to 0$ or in the case $\alpha \geq 1$, $\beta = -1$ as $x \to \infty$, while $w(\varphi,\alpha)$ (as is immediately clear from its definition) is an even analytic function on $(-\pi,\pi)$ for which

$$\tau = w(0,\alpha) = 1, \qquad \sigma^2 = w''(0,\alpha) = \alpha_* > 0.$$

Since $w(\varphi,\alpha)$ differs from the function $a(\varphi)$ in (2.2.35) by only a factor, it follows from Lemma 2.7.5 that w is strictly monotone on $(0,\pi)$.

This means that w satisfies conditions 1) and 2) in Lemma 2.5.1. The fact that 3) also holds is obvious, because the role of $s(t)$ is played in (2.5.10) and (2.5.11) by either $w(t,\alpha)$ or the function $s(t) \equiv 1 \leq w(t,\alpha)$. Thus, Lemma 2.5.1 can be applied to the integrals (2.5.10) and (2.5.11). For the final formation of the asymptotic expansions of interest to us we take a detailed look at the expansion (2.5.9), on which the form of the coefficients Q_n in (2.5.7) depends.

Let us first write out the power series expansion for $\log w(\varphi,\alpha)$ in the case $\alpha \leq 1$ by using the known expansions for the functions $\log(\sin x/x)$ and $x\cot x$ (see 1.411 and 1.518 in [27]). We have that

$$\log w(\varphi,\alpha) = \sum_{n=1}^{\infty} a_n(\alpha)\varphi^{2n}, \tag{2.5.13}$$

where

$$a_n(\alpha) = \frac{2^{2n}|B_{2n}|}{2n(2n)!}\left[\frac{\alpha(1-\alpha^{2n})}{1-\alpha} + 1 - (1-\alpha)^{2n}\right]$$

are polynomials of degree $2n-1$ with constant term equal to zero (the B_n are the Bernoulli numbers). From this we get by the Faa di Bruno formula that if $\alpha \le 1$, then

$$w(\varphi, \alpha) = 1 + \alpha\frac{\varphi^2}{2} + \sum_{n=2}^{\infty} b_n(\alpha)\varphi^{2n}, \tag{2.5.14}$$

where

$$b_n(\alpha) = \frac{1}{n!}C_n(1!a_1, 2!a_2, \dots, n!a_n)$$

and

$$C_n(y_1, \dots, y_n) = \sum\left\{\frac{n!}{k_1!\cdots k_n!}\left(\frac{y_1}{1!}\right)^{k_1}\cdots\left(\frac{y_n}{n!}\right)^{k_n} : \right.$$
$$\left. k_1 + 2k_2 + \dots + nk_n = n,\ k_j \ge 0\right\}$$

are the so-called Bell polynomials (see [70]). It is easy to see that the $b_n(\alpha)$ are polynomials of degree $2n-1$ with constant terms zero.

It follows from (2.5.14) that for $\alpha \ge 1$

$$w(\varphi, \alpha) = 1 + \alpha_*\frac{\varphi^2}{2} + \sum_{n=2}^{\infty} b_n(\alpha_*)\varphi^{2n},$$

where $b_n(\alpha_*)$ is a rational function of α.

Thus, if $s(\varphi) = w(\varphi) = w(w, \alpha)$, then (2.5.9) has the form

$$\omega(\varphi, t) = \exp\left\{\log w(\varphi t, \alpha) - \frac{1}{\alpha_* t^2}\left[w(\varphi t, \alpha) - 1 - \frac{\alpha_*}{2}\varphi^2 t^2\right]\right\}$$
$$= \exp\left(\sum_{n=1}^{\infty} d_n(\varphi, \alpha)t^{2n}\right) = 1 + \sum_{n=1}^{\infty} q_n(\varphi, \alpha)t^{2n}, \tag{2.5.15}$$

where

$$d_n(\varphi, \alpha) = (a_n(\alpha_*) - \varphi^2 b_{n+1}(\alpha_*)/\alpha_*)\varphi^{2n}$$

and

$$q_n(\varphi, \alpha) = C_n(1!d_1, 2!d_2, \dots, n!d_n)/n!$$

are polynomials of degree $2(n+1)$ in the variable φ and polynomials of degree $2n$ in α_*.

But if $s(\varphi) \equiv 1$ and $w(\varphi) = w(\varphi, \alpha)$, then

$$\omega(\varphi, t) = \exp\left\{-\frac{1}{\alpha_* t^2}\left[w(\varphi t, \alpha) - 1 - \frac{\alpha_*\varphi^2 t^2}{2}\right]\right\}$$
$$= 1 + \sum_{n=1}^{\infty} \tilde{q}_n(\varphi, \alpha)t^{2n}, \tag{2.5.16}$$

where $\tilde{q}_n(\varphi, \alpha)$ is obtained from $q_n(\varphi, \alpha)$ by setting $a_1 = a_2 = \cdots = 0$ in (2.5.15). The $\tilde{q}_n(\varphi, \alpha)$ are thus polynomials of degree $2n$ in the variable α_*.

THEOREM 2.5.2. *Suppose that $\alpha < 1$, $\beta = 1$, and $x \to 0$, or that $\alpha \geq 1$, $\beta = -1$, and $x \to \infty$. Then*

$$g(x, \alpha, \beta) \sim \frac{\nu}{\sqrt{2\pi\alpha}} \xi^{(2-\alpha)/2\alpha} \times \exp(-\xi) \left(1 + \sum_{n=1}^{\infty} Q_n(\alpha_*)(\alpha_* \xi)^{-n}\right). \tag{2.5.17}$$

The cofficients Q_n are given by (2.5.8) with the functions q_n in (2.5.15), and are polynomials of degree $2n$ in the variable α_*

The proof of the theorem reduces to the application of Lemma 2.5.1 to the integral expression (2.5.10) for the density. Furthermore, it is clearly necessary to take into account the information contained in (2.5.15).

REMARK. In the formulation of the theorem the variable $\xi(x, \alpha)$ was used both to obtain a concise expression and to demonstrate a single form of the asymptotic expansion in quite strongly divergent situations. For convenience in using (2.5.17) we write out its principal term ([2.12]):

$$\frac{\nu}{\sqrt{2\pi\alpha}} \xi^{(2-\alpha)/2\alpha} e^{-\xi} = \frac{(x/\alpha)^{(2-\alpha)/2(\alpha-1)}}{\sqrt{2\pi\alpha|1-\alpha|}} \times \exp\left\{-|1-\alpha| \left(\frac{x}{\alpha}\right)^{\alpha/(\alpha-1)}\right\} \tag{2.5.18}$$

in the case $\alpha \neq 1$, and

$$\frac{\nu}{\sqrt{2\pi\alpha}} \xi^{(2-\alpha)/2\alpha} e^{-\xi} = \frac{1}{\sqrt{2\pi}} \exp\left(\frac{x-1}{2} - e^{x-1}\right) \tag{2.5.19}$$

in the case $\alpha = 1$.

If as a starting point we now take the representation (2.5.11) for the distribution functions of the stable laws under consideration, then Lemma 2.5.1 together with the expansion (2.5.16) leads us to the following assertion.

THEOREM 2.5.3. *Suppose that $\alpha < 1$, $\beta = 1$, and $x \to 0$, or that $\alpha \geq 1$, $\beta = -1$, and $x \to \infty$. Then the asymptotic expansion of $G(x, \alpha, \beta)$ in the case $\alpha < 1$ or of $1 - G(x, \alpha, \beta)$ in the case $\alpha \geq 1$ has the form*

$$\frac{1}{\sqrt{2\pi\alpha\xi}} e^{-\xi} \left(1 + \sum_{n=1}^{\infty} \tilde{Q}_n(\alpha_*)(\alpha_* \xi)^{-n}\right), \tag{2.5.20}$$

where the polynomials $\tilde{Q}_n$ of degree $2n$ are given by (2.5.8), in which the functions $\tilde{q}_n(\varphi, \alpha)$ are determined by (2.5.16).

The principal term in (2.5.20) is transformed as follows if the function $\xi(x,\alpha)$ is replaced by its expression:

$$\frac{1}{\sqrt{2\pi\alpha\xi}}e^{-\xi} = \frac{(x/\alpha)^{-\alpha/2(\alpha-1)}}{\sqrt{2\pi\alpha|1-\alpha|}}\exp\left(-|1-\alpha|\left(\frac{x}{\alpha}\right)^{\alpha/(\alpha-1)}\right) \tag{2.5.21}$$

if $\alpha \neq 1$, and

$$\frac{1}{\sqrt{2\pi\alpha\xi}}e^{-\xi} = \frac{1}{\sqrt{2\pi}}\exp\left(-\frac{x-1}{2} - e^{x-1}\right) \tag{2.5.22}$$

if $\alpha = 1$.

REMARK 1. The asymptotic expansion (2.5.20) can be differentiated with respect to x. We first obtain (2.5.17), and then asymptotic expansions for the derivatives of the density (with a corresponding sign change). The admissibility of this formal procedure follows from the form of the representation (2.5.11), which remains an integral of the type I_N in Lemma 2.5.1 after differentiation. It is useful to keep in view that (differentiate (2.5.20))

$$\partial\xi/\partial x = -\varepsilon(\alpha)\nu(\alpha)\xi.$$

In this way it is not hard to establish a connection between the polynomials Q_n and $\tilde{Q}_n$. Since the form of the polynomials does not change for any $0 < \alpha \leq 2$, it follows that without loss of generality we can assume that $\alpha < 1$. Differentiating (2.5.20), we have

$$\begin{aligned} G'(x,\alpha,1) &\sim -\frac{\xi'}{\sqrt{2\pi\alpha}}\xi^{-1/2}e^{-\xi}\left\{\xi^{-1}\sum_{n\geq 1}\tilde{Q}_n n(\alpha\xi)^{-n}\right. \\ &\qquad\left. + \left(1+\frac{1}{2}\xi^{-1}\right)\left(1+\sum_{n\geq 1}\tilde{Q}_n(\alpha\xi)^{-n}\right)\right\} \\ &= \frac{\nu}{\sqrt{2\pi\alpha}}\xi^{(2-\alpha)/2\alpha}e^{-\xi}\left\{1+\sum_{n\geq 1}\left(\tilde{Q}_n + \frac{\alpha}{2}(2n+1)\tilde{Q}_{n-1}\right)(\alpha\xi)^{-n}\right\}. \end{aligned}$$

From this we get

$$Q_n(\alpha) = \frac{\alpha}{2}(2n+1)\tilde{Q}_{n-1}(\alpha) + \tilde{Q}_n(\alpha).$$

Direct computation of the polynomials $\tilde{Q}_n$ can be realized somewhat easier than for Q_n, and thus the recursion relation between them proves to be useful. For example, a computation shows that

$$\tilde{Q}_1(\alpha) = -\tfrac{1}{24}(2+7\alpha+2\alpha^2).$$

From this, considering that $\tilde{Q}_0 = 1$, we have

$$Q_1(\alpha) = -\tfrac{1}{24}(2-29\alpha+2\alpha^2).$$

REMARK 2. The explicit expression for the polynomials a_n shows that they have rational coefficients. The mechanism for forming the polynomials b_n, Q_n, and $\tilde{Q}_n$ preserves this property for them. It is interesting to note that the polynomials q_n, b_n, Q_n, and $\tilde{Q}_n$ have symmetric coefficients. Indeed, simple transformations show that in a sufficiently small neighborhood of zero

$$w(\varphi, 1/\alpha) = w(\varphi/\alpha, \alpha), \qquad 0 < \alpha \leq 2.$$

Expanding both sides of this equality in series of powers of φ, we get

$$b_n(1/\alpha) = \alpha^{-2n} b_n(\alpha),$$

and, expanding the logarithms of both sides in series, we get

$$a_n(1/\alpha) = \alpha^{-2n} a_n(\alpha).$$

Consequently, the polynomials Q_n and $\tilde{Q}_n$ formed from them have the same property. But this means that all these polynomials have symmetric coefficients.

The last of the cases being considered, namely, $\alpha = 1$ and $\beta \neq -1$, is very distinctive.

THEOREM 2.5.4. *If $\alpha = 1$ and $-1 < \beta \leq 1$, then the following asymptotic expansions are valid as $x \to \infty$:*

$$g(x, 1, \beta) \sim \frac{1}{\pi} \sum_{n=1}^{\infty} \frac{1}{n!} P_n(\log x) x^{-n-1}, \tag{2.5.23}$$

where $P_n(y) = \sum_{l=0}^{n} r_{ln} y^l$, and

$$r_{ln} = \sum_{m=l}^{n} \binom{n}{m} \binom{m}{l} (-1)^{m-l} \Gamma^{(m-l)}(1+n)$$
$$\times \beta^m \left(\frac{\pi}{2}(1+\beta)\right)^{n-m} \sin \frac{\pi}{2}(n-m);$$

and

$$1 - G(x, 1, \beta) \sim \frac{1}{\pi} \sum_{n=1}^{\infty} \frac{1}{n!} P_n^*(\log x) x^{-n}, \tag{2.5.24}$$

where $P_n^*(y) = \sum_{l=0}^{n} r_{ln}^* y^l$, and the coefficients r_{ln}^* differ from r_{ln} only in that the derivatives of the Γ-function at the point $1+n$ appearing in them are replaced by the same derivatives at the point n.

PROOF.* In the representation (2.2.1b) of the density $g(x,1,\beta)$ we change the contour of integration, replacing the nonnegative semi-axis by the compound contour $L = L_1 \cup L_2$, where

$$L_1 = \{x = u + iv : u = 0,\ -A \le v \le 0\},$$
$$L_2 = \{z = u + iv : u \ge 0,\ v = -A\}, \qquad A = x^{3/4} > 0.$$

The justification for replacing the contour of integration in (2.2.1b) by the contour L is elementary and reduces to a proof that the integral along the contour $L_N = \{z = u + iv : u = N,\ -A \le v \le 0\}$ tends to zero as $N \to \infty$. In summary

$$\begin{aligned} g(x,1,\beta) &= \frac{1}{\pi} \operatorname{Re} \int_0^\infty \exp\Big(-itx - \frac{\pi}{2} t - i\beta t \log t\Big)\, dt \\ &= \frac{1}{\pi} \operatorname{Re} \int_L = \frac{1}{\pi} \operatorname{Re} \int_{L_1} + \frac{1}{\pi} \int_{L_2} = \frac{1}{\pi} \operatorname{Re}(I_1 + I_2). \end{aligned}$$

Let us estimate the integral I_2. After the change $t = u - iA$ of the variable of integration we find that

$$|I_2| \le \int_0^\infty \exp\left(-A - \frac{\pi}{2} u - \beta d(x,u)\right) du,$$

where $d(x,u) = \operatorname{Im}[(u - ix^{-1/4}) \log(u - ix^{-1/4})]$. Since $|\beta| \le 1$ and $|d(x,u)| \le x^{-1/4} + \frac{1}{2} x^{-1/4} \log(u^2 + x^{-1/2})$, it follows that for $x \ge 16$

$$\begin{aligned} |I_2| &\le \exp(-x^{3/4} + 1) \int_0^\infty \exp\left(\frac{\pi}{2} u + \frac{1}{2} \log(u^2 + 1)\right) du \\ &= C_1 \exp(-x^{3/4}), \end{aligned}$$

where C_1 is a constant. Let us now estimate I_1. In this integral we make the change of variable $t = -iv$ and get that

$$\begin{aligned} ixI_1 &= \int_0^A \exp\left(-v + i\frac{\pi}{2}(1+\beta)\frac{v}{x} - \beta \frac{v}{x} \log \frac{v}{x}\right) dv \\ &= \int_0^A e^{-v} \sum_{k=1}^N \frac{1}{k!} \left(\frac{v}{x}\right)^k \left[i\frac{\pi}{2}(1+\beta) - \beta \log \frac{v}{x}\right]^k dv + R_N, \end{aligned}$$

**Added in translation.* There are some gaps in the proof of Theorem 2.5.4 in the Russian edition. The proof below is free of these defects. It was proposed to the author by A. Lisitskiĭ. He pointed out that the asymptotic series (2.5.32) is actually a convergent series (at least for sufficiently large values of x). Note the great similarity between the proofs of Theorems 2.5.4 and 2.5.5, although they treat different asymptotic expansions.

where

$$R_N = \int_0^A e^{-v}\left(\frac{v}{x}\right)^{N+1}\left[i\frac{\pi}{2}(1+\beta) - \beta\log\frac{v}{x}\right]^{N+1} dv$$
$$\times \frac{1}{N!}\int_0^1 (1-\xi)^N \exp\left\{\xi\frac{v}{x}\left(i\frac{\pi}{2}(1+\beta) - \beta\log\frac{v}{x}\right)\right\} d\xi.$$

We estimate R_N:

$$|R_N| \le \frac{1}{N!}\int_0^A \left(\frac{v}{x}\right)^{N+1}\left(\frac{\pi}{2}(1+\beta) + \left|\log\frac{v}{x}\right|\right)^{N+1} \exp\left(\frac{v}{x}\left|\log\frac{v}{x}\right| - v\right) dv$$
$$\le x^{-(N+1)}e^{1/e}\frac{1}{N!}\int_0^A (\pi + \log x + |\log v|)^{N+1}e^{-v}\, dv$$
$$< C_2(N)(x^{-1}\log x)^{N+1},$$

where $C_2(N)$ is a constant depending only on N.

Next, we consider the integral

$$I_3^k = \int_A^\infty \left(\frac{v}{x}\right)^k \left[i\frac{\pi}{2}(1+\beta) - \beta\log\frac{v}{x}\right]^k e^{-v}\, dv$$

and estimate it for $x > k+2$:

$$|I_3| \le \int_A^\infty \left(\frac{v}{x}\right)^k (\pi + \log v + \log x)^k e^{-v}\, dv$$
$$\le (\pi+2)^k(\log x)^k x^{-k}\int_A^\infty v^{k+2}e^{-v}\frac{dv}{v^2}$$
$$\le [(\pi+2)x^{-1}\log x]^k A^{k+1}e^{-A}$$
$$\le C_3(k)\exp(x^{3/4}/2).$$

Combination of the estimates obtained for the integrals I_2, I_3, and R_N now gives us that for any integer $N \ge 1$

$$g(x,1,\beta) = \sum_{k=0}^N \frac{x^{k-1}}{k!}\operatorname{Im}\int_0^\infty \left[i\frac{\pi}{2}(1+\beta) - \beta\log\frac{v}{x}\right]^k v^k e^{-v}\, dv$$
$$+ O(x^{N-2}\log^{N+1} x)$$

as $x \to \infty$.

The last step of the proof consists in a transformation of the integral which is the coefficient of the power x^{-k-1}. We have that

$$\operatorname{Im}\int_0^\infty \left[i\frac{\pi}{2}(1+\beta) + \beta\log\frac{x}{v}\right]^k v^k e^{-v}\, dv$$
$$= \sum_{l=0}^k \binom{k}{l}\int_0^\infty \left(\beta\log\frac{x}{v}\right)^l \left(\frac{\pi}{2}(1+\beta)\right)^{k-l} \sin\frac{\pi}{2}(k-l)v^k e^{-v}\, dv.$$

A further simple transformation of this sum, with the equality

$$\int_0^\infty e^{-v} v^k (\log v)^m \, dv = \Gamma^{(m)}(1+k)$$

taken into account, leads us to the conclusion that it is a polynomial $P_k(\log x)$ of the form indicated in the theorem.

It is not hard to see that we are justified in integrating the asymptotic representation for the density from x to ∞. This gives an asymptotic series for the function $1 - G(x, 1, \beta)$ as $x \to \infty$.

A sequence of simple transformations shows that

$$\begin{aligned}\int_x^\infty & P_k(\log y) y^{-k-1} \, dy \\ &= \int_x^\infty y^{-k-1} \, dy \operatorname{Im} \int_0^\infty \left[i\frac{\pi}{2}(1+\beta) - \beta \log \frac{v}{y} \right]^k v^k e^{-v} \, dv \\ &= x^{-k} P_k^*(\log x).\end{aligned}$$

This concludes the proof of the theorem.

THEOREM 2.5.5. *If $\alpha = 1$ and $-1 < \beta \leq 1$, then the following asymptotic expansions are valid as $x \to \infty$* (2.13):

$$g(x, \beta \log x, 1, \beta) \sim \frac{1}{\pi} \sum_{n=1}^\infty d_n x^{-n-1}, \tag{2.5.25}$$

where

$$\begin{aligned} d_n = \frac{1}{n!} \sum_{m=0}^{[(n-1)/2]} & (-1)^{n+m-1} \binom{n}{2m+1} \left(\frac{\pi}{2}(1+\beta) \right)^{2m+1} \\ & \times \beta^{n-2m-1} \Gamma^{(n-2m-1)}(n+1); \end{aligned}$$

$$1 - G(x + \beta \log x, 1, \beta) \sim \frac{1}{\pi} \sum_{n=1}^\infty \frac{1}{n} (d_n + \beta d_{n-1}) x^{-n}, \tag{2.5.26}$$

where it is assumed that $d_0 = 0$.

PROOF. By the inversion formula (2.2.1b), the density $g(x, 1, \beta)$, with the variable x replaced by $x + \beta \log x$, satisifes

$$\begin{aligned} g(x + \beta \log x, 1, \beta) &= \frac{1}{\pi x} \operatorname{Re} I \\ &= \frac{1}{\pi x} \operatorname{Re} \int_0^\infty \exp\left(it - \frac{t}{x} \left(\frac{\pi}{2} - i\beta \log t \right) \right) dt. \end{aligned} \tag{2.5.27}$$

We replace the integration along the semi-axis $t > 0$ in this integral by integration along the contour $L = L_1 \cup L_2$ in the complex plane of $z = \sigma + i\xi$, where

$$L_1 = \{z\colon \sigma = 0,\ 0 \le \xi \le x\}, \quad L_2 = \{z\colon \sigma \ge 0,\ \xi = x\}.$$

The justification for this change is elementary, and we omit it. The integral I in (2.5.27) after the sign for the real part can be writen as a sum $I_1 + I_2$ of integrals, the first along L_1 and the second along L_2. After the change of variable $t = i\xi$ in the integral I_1 we get that

$$\begin{aligned} I_1 &= i\int_0^x \exp\left\{-\xi - \frac{\xi}{x}\left[i\frac{\pi}{2}(1+\beta) + \beta\log\xi\right]\right\} d\xi \\ &= i\int_0^x \exp\left(-\xi - \frac{Q}{x}\right) d\xi \\ &= i\sum_{n=0}^{N-1} \frac{(-x)^{-n}}{n!}\int_0^x e^{-\xi}Q^n\,d\xi + R_N. \end{aligned}$$

For any $\xi > 0$

$$|Q| \le \xi(\pi + |\log\xi|), \qquad |\operatorname{Re} Q| \le \xi|\log\xi|,$$

and so, for any $N \ge 1$ and $0 < \xi \le x$,

$$\begin{aligned} &\left|\exp\left(-\frac{Q}{x}\right) - \sum_{n=0}^{N-1}\frac{1}{n!}\left(-\frac{1}{x}\right)^n Q^n\right| \\ &\quad \le \frac{1}{N!}\left(\frac{|Q|^N}{x}\right)^N \exp\left(\frac{1}{x}|\operatorname{Re} Q|\right) \le \left(\frac{|Q|}{x}\right)^N \max\left(\xi, \frac{1}{\xi}\right), \end{aligned}$$

which implies the estimate

$$|R_N| \le \int_0^x e^{-\xi}\left(\frac{|Q|}{x}\right)^N \max\left(\xi, \frac{1}{\xi}\right) d\xi \le cx^{-N},$$

where c is a numerical constant. Further, let

$$\begin{aligned} I_n &= i\int_0^\infty e^{-\xi}Q^n\,d\xi \\ &= i\sum_{k=0}^n \binom{n}{k}\left(i\frac{\pi}{2}(1+\beta)\right)^k \int_0^\infty e^{-\xi}(\beta\log\xi)^{n-k}\xi^n\,d\xi \\ &= \sum_{k=0}^n \binom{n}{k} i^{k+1}\left(\frac{\pi}{2}(1+\beta)\right)^k \beta^{n-k}\Gamma^{(n-k)}(n+1). \end{aligned}$$

An elementary computation shows that as $x \to \infty$

$$i\int_0^x e^{-\xi}Q^n\,d\xi = l_n + O(x^n\log^n x\exp(-x)).$$

Consequently, as $x \to \infty$

$$\operatorname{Re} I_1 = \sum_{n=0}^{N-1} \frac{1}{n!} \operatorname{Re} l_n \left(-\frac{1}{x}\right)^n + O(x^{-N}). \tag{2.5.28}$$

Let us estimate the second integral. Since

$$\begin{aligned} I_2 &= \int_{L_2} \exp\left(iz - \frac{z}{x}\left(\frac{\pi}{2} - i\beta \log z\right)\right) dz \\ &= \int_0^\infty \exp\left\{-x + i\sigma - \frac{\sigma + ix}{2}\left(\frac{\pi}{2} - i\beta \log(\sigma + ix)\right)\right\} d\sigma \\ &= xe^{-x} \int_0^\infty \exp\left\{ix\sigma - (\sigma + i)\left(\frac{\pi}{2} - i\beta \log[(\sigma + i)x]\right)\right\} d\sigma, \end{aligned}$$

it follows that

$$|I_2| \leq x^{1-\beta} e^{-x} \int_0^\infty \exp\left\{-\frac{\pi}{2}\sigma - \beta(\log \sigma + \sigma \arg(\sigma + i))\right\} d\sigma. \tag{2.5.29}$$

Combining (2.5.28) and (2.5.29), we get for any fixed integer $N \geq 1$

$$g(x, \beta \log x, 1, \beta) = \frac{1}{\pi x} \sum_{n=0}^{N=1} \frac{1}{n!} \operatorname{Re} l_n \left(-\frac{1}{x}\right)^n + O(x^{-N}) \tag{2.5.30}$$

as $x \to \infty$, and it remains only to check that $(-1)^n \operatorname{Re} l_n = d_n$.

The asymptotic expansion (2.5.26) is obtained as a consequence of (2.5.25). Indeed, since

$$1 - G(x + \beta \log x, 1, \beta) = \int_x^\infty (1 + \beta/u) g(u + \beta \log u, 1, \beta)\, du, \tag{2.5.31}$$

substitution of the expression (2.5.30) for the density into (2.5.31) gives us the expansion (2.5.26) after the corresponding transformation.

REMARK. The case $\beta = -1$ could have been formally included in the statement of the theorem, since all the arguments remain in force. However, in this case all the d_n are 0, and the substance of the theorem's assertion reduces to the claim that the functions $g(x - \log x, 1, -1)$ and $1 - G(x - \log x, 1, -1)$ decrease more rapidly than any power function.

Theorems 2.2.2 and 2.3.2 can be used to obtain another type of asymptotic expansion of stable distributions related to (2.5.25) and (2.5.26).

THEOREM 2.5.6. *The following asymptotic expansions are valid as* $x \to \infty$ $(\mu = \beta \tan(\pi\alpha/2))$:

If $\alpha < 1$, *then for any admissible* β

$$g_A(x + \mu x^{1-\alpha}, \alpha, \beta) \sim \frac{1}{\pi} \sum_{n=1}^{\infty} A_n(\alpha, \beta) x^{-\alpha n - 1}, \tag{2.5.32}$$

where

$$A_n(\alpha,\beta)=\operatorname{Im}\sum_{k=1}^{n}\frac{\Gamma(\alpha k+n-k+1)}{\Gamma(k+1)\Gamma(n-k+1)}(-\mu)^{n-k}e^{-i\pi\alpha k/2}(\mu e^{i\pi/2}-1)^k;$$

$$1-G_A(x+\mu x^{1-\alpha},\alpha,\beta)$$
$$\sim\frac{1}{\pi}\sum_{n=1}^{\infty}\frac{1}{\alpha n}[A_n(\alpha,\beta)+(1-\alpha)\mu A_{n-1}(\alpha,\beta)]x^{-\alpha n},\qquad(2.5.33)$$

where $A_0=0$ *and* A_n, $n\geq 1$, *are the same as in* (2.5.32).

The proof of (2.5.32) is carried out by expanding the integrand of the corresponding representation of the density (2.2.12) in an asymptotic series. The asymptotic expansion (2.5.33) is obtained from the latter by multiplying by $1+(1-\alpha)\mu x^{-\alpha}$ and integrating from x to ∞. The computation of the form of the coefficients $A_n(\alpha,\beta)$ in these expansions is elementary and will not be given.

REMARK. The series (2.5.32) is interpreted as asymptotic in the theorem. In fact, it can be shown to be convergent (what is more, absolutely convergent) for all sufficiently large x.

Another expansion, though no longer at infinity but in a neighborhod of zero, is obtained with the help of (2.3.6) and (2.5.32). Let us use the notation in Theorem 2.3.2 ($\alpha'=1/\alpha$, $y=x+\mu' x^{1-\alpha'}$, $\beta=\beta^*$, $\mu'=\beta'\tan(\pi\alpha'/2)$, etc.).

THEOREM 2.5.7. *If* $\alpha>1$, *then the following asymptotic expansions hold as* $x\to\infty$ *for any admissible* β:

$$Dy^{-1-1/\alpha}g_A(Ky^{-1/\alpha},\alpha,\beta)\sim\frac{1}{\pi}\sum_{n=1}^{\infty}A_n(\alpha',\beta')x^{-1-n/\alpha},\qquad(2.5.34)$$

$$G_A(0,\alpha,\beta)-G_A(Dy^{-1/\alpha},\alpha,\beta)$$
$$\sim\frac{1}{\pi\alpha}\sum_{n=1}^{\infty}[A_n(\alpha',\beta')+(1-\alpha')\mu' A_{n-1}(\alpha',\beta')]x^{-n/\alpha},$$

where the $A_n(\alpha,\beta)$ *are the same as in* (2.5.33).

These relations are proved by direct comparison of (2.3.6) and the expansions (2.5.32) and (2.5.33).

§2.6. Integral transformations of stable distributions

For the information on the properties of the distributions in $\mathfrak{S}$ gathered in the preceding sections of this chapter we are indebted first and foremost to the fact that the characteristic functions of these distributions have a simple form. Thus, the explicit form for the Fourier transforms of the densities of stable laws can to some extent compensate us for the absence of explicit expressions for the densities themselves. At the same time, the Fourier transformation is only one of many integral transformations known in function theory. And it is natural to expect that each integral transformation of the densities of stable distributions (or of some functions connected with them) that is obtainable in a sufficiently simple form can act as a source of interesting information about their properties (see Chapter 3).

The material in the present section is divided into two parts. One contains several assertions to be proved. It precedes another part which carries the main idea of an algorithm for simplifying the analytic computations in a number of cases. The arguments connected with this algorithm bear a formal character because of the desire to isolate it in pure form, and further refinement and substantiation is required in each concrete case. (For example, it must be specified in what sense the integral defining the integral transform converges, or what the conditions are which allow changing the order of integration, and so on.)

Denote by D some subset of the real axis R, by I a subset of the complex plane C, and by $f(x)$ and $h(s,x)$ functions with domains D and $I \times D$, respectively. Let us consider the integral

$$(\mathcal{H} f)(s) = \int_D h(s,x) f(x)\, dx, \qquad s \in I, \tag{2.6.1}$$

where the sets D and I and the function h called the kernel of the integral transformation $\mathcal{H}$, are fixed, while f, which is chosen from some set $\mathfrak{F}$, is a function variable. The kernel h is chosen so that functions $f \in \mathfrak{F}$ can be reproduced in their domain D from the corresponding transforms $(\mathcal{H} f)(s)$, considered in the domain I.*

We give some examples of well-known integral transformations with which we have been or shall be concerned.

(a) The Fourier transformation ($\mathcal{H} = \mathcal{F}$) corresponds in (2.6.1) to the kernel $h(s,x) = \exp(isx)$ and the sets $D = I = R$.

(b) The two-sided Laplace transformation ($\mathcal{H} = \mathcal{N}$) is connected with the kernel of form $h(s,x) = \exp(-sx)$ and with the sets $D = R$ and $I = \{z\colon -c_1 \le \operatorname{Re} z \le c_2\}$, where c_1 and c_2 are nonnegative numbers. It can be said that this transformation is related to the transformation $\mathcal{F}$, as is clear from the relation connecting them:

$$(\mathcal{N} f)(-is) = (\mathcal{F} f)(s), \qquad s \in R.$$

*The very detailed monograph [82] can be recommended for learning about the well-developed theory of integral transformations with various types of kernels.

(c) The one-sided Laplace transformation ($\mathcal{H} = \mathcal{L}$). It corresponds to the kernel $h(s,x) = \exp(-sx)$ and the sets $D = R^+ = [0,\infty)$ and $I = \{z: -c_1 \le \operatorname{Re} z \le c_2\}$, where c_1 and c_2 are nonnegative numbers. Obviously, $(\mathcal{H}f)(s) = (\mathcal{L}f)(s)$ on the set of functions f equal to zero on the semi-axis $x < 0$.

(d) The Mellin transformation ($\mathcal{H} = \mathcal{M}$). The sets D and I here are of the same form as in the one-sided Laplace transformation, but the kernel has the form $h(s,x) = x^s$.

Instead of the kernel x^{s-1} usually taken for the Mellin transformation (see [82]), we use the kernel x^s. This inessential change is due to the desire to simplify the analytic expressions for the transforms $(\mathcal{M}g)(s)$ of densities of stable laws. We could have kept the traditional choice of the kernel but then considered transformations of the functions xg.

To these known transformations we add one more, which has so far found application only in probability theory and probabilistic number theory.

(e) The characteristic transformation ($\mathcal{H} = \mathcal{W}$) has as kernel the function $h(s,x)$ taking values in the set of 2×2 diagonal matrices

$$h(s,x) = \begin{pmatrix} h_0(s,x) & 0 \\ 0 & h_1(s,x) \end{pmatrix},$$

where $h_k(s,x) = |x|^{si}(\operatorname{sgn} x)^k, k = 0,1$, and $0^{si} = 0$ for any complex numbers s, while the sets are $D = R$ and $I = \{z: -c_1 \le \operatorname{Re} z \le c_2\}$, where c_1 and c_2 are nonnegative numbers.

If the function f in (2.6.1) is written in the form of the integral

$$f(x) = \int_{D_*} h_*(x,u) f_*(u)\, du, \qquad x \in D, \tag{2.6.2}$$

then the transformation $(\mathcal{H}f)(s)$ gets an expression as a double integral. This kind of situation is encountered in considering integral transforms of functions of the form

$$f(x) = l(x) g(r(x), \alpha, \beta, \gamma, \lambda).$$

We have observed individual cases of such representations of functions by integrals in (2.2.8)–(2.2.10), (2.2.12), (2.2.13), (2.2.17), and (2.3.14).

The original transform $(\mathcal{H}f)(s)$ gets a simplified representation by single integrals in the cases when the integral $\int_D h(s,x)h_*(x,u)\,dx$ can be computed in explicit form by regarding it as an integral transform with kernel h or h_*.

We present some formal equalities that can serve as a starting point in the indicated direction for finding explicit or simplified expressions for transformations $\mathcal{H}$ of functions of the form

$$f(x) = l(x)g(r(x), \alpha, \beta, \gamma, \lambda).$$

Let $D = R$ and $I = R$. Using the inversion formulas (2.2.1a) and (2.2.1b) for expressing the density $g(x, \alpha, \beta)$, we find for functions of the form $f(x) = lg = l(x)g(x, \alpha, \beta)$ that

$$(\mathcal{H}lg)(s) = \frac{1}{\pi} \operatorname{Re} \int_0^\infty \mathfrak{g}(t, \alpha, \beta)(\mathcal{F}lh)(-t)\, dt. \tag{2.6.3}$$

The representations (2.2.8) and (2.2.10) of the densities $g(x, \alpha, \beta)$ can also be used, but for certain values of β they prove to be inconvenient for our purposes (complications automatically arise in justifying a change in the order of integration), and thus they require a preliminary change. Modifications of the equalities are obtained by rotating the contour of integration in (2.2.8) by the angle $\pi\rho/2$ and in (2.2.10) by the angle $-\pi\rho/2$. In all the cases with $\alpha \neq 1$ the new variant of the expression for the density $g(x, \alpha, \beta)$ is the same (however, in the case $\alpha < 1$ it is nevertheless assumed that $x > 0$):

$$g(x, \alpha, \beta) = \frac{1}{\pi} \operatorname{Im} \int_0^\infty \exp\left(-xue^{i\pi\rho/2} - u^\alpha e^{-i\pi\rho\alpha/2} + \frac{i\pi\rho}{2}\right) du. \tag{2.6.4}$$

For $D = R^+$ we get from this the following formula for the transformation $\mathcal{H}$ of the function $f(x) = l(x)g(x, \alpha, \beta)$, $\alpha \neq 1$:

$$(\mathcal{H}lg)(s) = \frac{1}{\pi} \operatorname{Im} \int_0^\infty \exp\left(-u^\alpha e^{-i\pi\rho\alpha/2} + \frac{i\pi\rho}{2}\right) (\mathcal{L}lh)(ue^{i\pi\rho/2})\, du. \tag{2.6.5}$$

But if $\alpha = 1$ and $\beta > 0$, then to get an analogous formula we must use the representation (2.2.9) of $g(x, 1, \beta)$, and we obtain

$$(\mathcal{H}lg)(s) = \frac{1}{\pi} \int_0^\infty \exp(-\beta u \log u) \sin[\pi(1+\beta)u/2](\mathcal{L}lh)(u)\, du. \tag{2.6.6}$$

In the case when $D = R$, $\alpha = 1$, and $\beta > 0$, the same representation for the density leads to the formula

$$(\mathcal{H}lg)(s) = \frac{1}{\pi} \int_0^\infty \exp(-\beta u \log u) \sin[\pi(1+\beta)u/2](\mathcal{N}lh)(u)\, du. \tag{2.6.7}$$

Let us consider the case $D = R^+$ and the functions $f(x) = l(x)g(x, \alpha, \beta)$ with densities g in the class $\mathfrak{W}$. Using the expression (2.2.18) for the density and writing

$$w(s, x) = l(x^{1-1/\alpha})h(s, x^{1-1/\alpha}),$$

we get that

$$(\mathcal{H}lg)(s) = \frac{1}{2}\int_0^1 U_\alpha(\varphi,\theta)(\mathcal{L}w)(U_\alpha(\varphi,\theta))\,d\varphi. \tag{2.6.8}$$

In the case of a compound argument $r(x)$ of special form

$$r(x) = \begin{cases} x^{-1/\alpha} + x^{1-1/\alpha}\beta\tan\frac{\pi}{2}\alpha & \text{if } \alpha < 1,\\ x^{-1} - \frac{2}{\pi}\beta\log x & \text{if } \alpha = 1,\end{cases}$$

the representations (2.2.12) and (2.2.13) for $lg_A = l(x)g_A(r(x),\alpha,\beta)$ lead to the following formula for a transformation $\mathcal{H}$ with the set $D = R^+$ ($\alpha \le 1$, $\beta > 0$):

$$(\mathcal{H}lg_A)(s) = \frac{1}{\pi}\operatorname{Im}\int_0^\infty e^{-u}(\mathcal{L}x^{1/\alpha}lh)(T)\,du, \tag{2.6.9}$$

where

$$T(u) = \begin{cases} -(iu)^\alpha - u\beta\tan\frac{\pi}{2}\alpha[(-iu)^{\alpha-1} - 1] & \text{if } \alpha < 1,\\ u[\frac{2}{\pi}\beta\log u - i(1+\beta)] & \text{if } \alpha = 1.\end{cases}$$

In the case $\alpha > 1$ an analogous formula is obtained with the help of (2.3.14). We do not give it because of its unwieldiness, but it is not difficult to reproduce after the remarks above.

Formulas (2.6.5)–(2.6.9) serve as a source of many interesting analytic relations. However, very little has as yet been done in this direction, and the subject of integral transforms of stable distribution still awaits its enthusiasts.

In the final analysis, the variable s in (2.6.5)–(2.6.9) is taken from some subset of the complex plane, but the initial computations should be done under the assumption that s is real, with passage to the complex values of s only in determining an explicit form of the transform or a simplified expression for it.

The form of expression for a transformation $\mathcal{H}$ of the functions $f = lg$ can depend essentially on what representation is used for the density. In order not to make this section excessively longer, we do not consider all possible variants of the integral transform formulas (b)–(d) obtained by using the density representations in §2.2, but give preference to those formulas leading to the expressions that are simplest in our view. The form of the Fourier transforms of stable laws (i.e., the characteristic functions) is known, so we begin with the two-sided Laplace transformation.

(b) For a density $g = g(x,\alpha,\beta,\gamma,\lambda)$ (with the system of parameters in the form (B)) let

$$\Lambda(s,\alpha,\beta,\gamma,\lambda) = (\mathcal{N}g)(s).$$

It is clear from the asymptotic formulas (2.4.8) and (2.5.4) that in the case $\beta \ne \pm 1$ the density $g(x,\alpha,\beta,\gamma,\lambda)$ decreases polynomially as $x \to \infty$ and as

$x \to -\infty$. Consequently, the two-sided Laplace transform of the density g exists only if $s = -it$, $t \in R$, i.e., only if it coincides with the characteristic function corresponding to this density. For the same reason, it does not exist in the half-plane $\operatorname{Re} s > 0$ when $\beta = -1$. The transform Λ need not be considered in the half-plane $\operatorname{Re} s < 0$, since, according to Property 2.2,

$$\Lambda(s, \alpha, \beta, \gamma, \lambda) = \Lambda(-s, \alpha, -\beta, -\gamma, \lambda),$$

i.e., it suffices to confine oneself to analysis of the case $\operatorname{Re} s > 0$, $\beta = 1$.

THEOREM 2.6.1. *The following equality holds in the half-plane* $\operatorname{Re} s \geq 0$ ([2.14]):

$$\log \Lambda(s, \alpha, 1, \gamma, \lambda) = \begin{cases} \lambda(-s\gamma - \varepsilon(\alpha) s^{\alpha}) & \text{if } \alpha \neq 1, \\ \lambda(-s\gamma + s \log s) & \text{if } \alpha = 1. \end{cases} \tag{2.6.10}$$

PROOF. As is clear from the asymptotic formulas (2.5.17) and Remark 3 after Theorem 2.2.3, the density $g(x, \alpha, 1, \gamma, \lambda)$ decreases as $x \to -\infty$ more rapidly than $\exp(cx)$ for any $c > 0$. Therefore, the transform Λ exists for any values of s in the half-plane $\operatorname{Re} s \geq 0$, and is analytic at interior points and continuous on the axis $\operatorname{Re} s = 0$.

In this case the characteristic function $\mathfrak{g}(t, \alpha, 1)$ in §2.2 was given the special form (2.2.6) facilitating its analytic extension from the real t-axis to the half-plane $\operatorname{Im} z \geq 0$. If we make the substitution $z = is$, $\operatorname{Re} s \geq 0$, in this extension, we get the function $\Lambda(s, \alpha, 1, \gamma, \lambda)$. The equality (2.6.10) follows from this.

(c) The one-sided Laplace transform of the density $g = g(x, \alpha, \beta, \gamma, \lambda)$, denoted by

$$L(s, \alpha, \beta, \gamma, \lambda) = (\mathcal{L}g)(s), \quad L(s, \alpha, \beta) = L(s, \alpha, \beta, 0, 1),$$

exists in the half-plane $\operatorname{Re} s \geq 0$ for any admissible values of the parameters. It suffices to consider only the case $\operatorname{Re} s \geq 0$, since, according to Corollary 2.2,

$$L(s, \alpha, \beta, \gamma, \lambda) = L(-s, \alpha, -\beta, -\gamma, \lambda).$$

As already mentioned above, it suffices to get all the formulas for the function $L(s, \alpha, \beta, \gamma, \lambda)$ under the assumption that $s > 0$. They can be generalized to the case of complex s by analytic extension. According to the property

$$g(x, \alpha, \beta, \gamma, \lambda) = \lambda^{-1/\alpha} g(\lambda^{-1/\alpha}(x - l), \alpha, \beta),$$

where $l = \gamma\lambda$ if $\alpha \neq 1$ and $l = \gamma\lambda + \beta\lambda \log \lambda$ if $\alpha = 1$, the transform L can be written in the form

$$\begin{aligned} L(s, \alpha, \beta, \gamma, \lambda) &= \exp(-ls) L(\lambda^{1/\alpha} s, \alpha, \beta) \\ &\quad + \exp(-ls) \int_{-l\lambda^{-1/\alpha}}^{0} \exp(-s\lambda^{1/\alpha} x) g(x, \alpha, \beta)\, dx. \end{aligned}$$

Both terms can be simplified, and in a common way. Thus, to avoid repetition we confine ourselves to the analysis of the case $l = 0$. However, the conclusion of these arguments nevertheless leads to an expression for $L(s, \alpha, \beta, \gamma, \lambda)$ in the case when $\alpha \geq 1$.

THEOREM 2.6.2. *For any s in the half-plane* $\operatorname{Re} s \geq 0$ *the following expressions are valid for the one-sided Laplace transforms $L(s, \alpha, \beta)$ of standard stable distributions* ([2.15]):

If $\alpha \neq 1$, then for any admissible β

$$L(s, \alpha, \beta) = \frac{1}{\pi} \int_0^\infty \exp(-(su)^\alpha) \frac{\sin \pi\rho}{u^2 + 2u\cos \pi\rho + 1}\, du. \tag{2.6.11}$$

If $\alpha = 1$ and $\beta > 0$, then

$$L(s, 1, \beta) = \frac{1}{\pi} \int_0^\infty \exp(-\beta u \log u) \frac{\sin \frac{\pi}{2}(1+\beta)u}{s + u}\, du. \tag{2.6.12}$$

If $\alpha = 1$, then for any admissible β

$$L(s, 1, \beta) = \frac{1}{\pi} \int_0^\infty [s\cos(\beta u \log u) - u \sin(\beta u \log u)] \times \frac{\exp(-\pi u/2)}{\sqrt{s^2 + u^2}}\, du. \tag{2.6.13.}$$

PROOF. For $\alpha \neq 1$ let us substitute in $(\mathcal{L}g)(s)$ the modified expression (2.6.4) for the density and change the order of integration, which is allowed because the double integral converges absolutely. We have that

$$\begin{aligned} L(s, \alpha, \beta) &= \frac{1}{\pi} \operatorname{Im} \int_0^\infty \exp(-u^\alpha e^{-i\pi\rho\alpha/2} + i\pi\rho/2)\, du \\ &\quad \times \int_0^\infty \exp(-sx - xue^{i\pi\rho/2})\, dx \\ &= \frac{1}{\pi} \operatorname{Im} \int_0^\infty \exp[-(ue^{-i\pi\rho/2})^\alpha + i\pi\rho/2](s + ue^{i\pi\rho/2})^{-1}\, du. \end{aligned}$$

Rotation of the contour of integration through the angle $-\pi\rho/2$ and a change of variable give us that

$$L(s, \alpha, \beta) = \frac{1}{\pi} \operatorname{Im} \int_0^\infty \exp(-u^\alpha) \frac{e^{i\pi\rho} + u}{|s + ue^{i\pi\rho}|^2}\, du.$$

Subsequent replacement of u by su brings us to (2.6.11).

In the case $\alpha = 1$ the arguments are mostly the same.The difference reduces only to what expressions we use for the density $g(x, 1, \beta)$. To get (2.6.12) it is necessary to use (2.2.9), and to get (2.6.13) the density is taken in the form (2.2.1b).

We make some remarks about Theorem 2.6.2.

REMARK 1. The rational function under the integral sign in (2.6.11) is equal to $\pi g_C(x, 1, \theta)$, as shown by a comparison with (2.3.5a). Since $L(s, \alpha, \beta) = L_C(s, \alpha, \theta)$, (2.6.11) can be given the form

$$L_C(s, \alpha, \theta) = \int_0^\infty \exp(-(us)^\alpha) g_C(u, 1, \theta)\, du. \tag{2.6.14}$$

If $\alpha < 1$, then, by (2.2.6) and (2.2.32),

$$\exp(-s^\alpha) = (\mathcal{L}g(x, \alpha, 1))(s) = L(s, \alpha, 1).$$

Substitution of this expression into (2.6.14) gives us

$$\begin{aligned} L_C(s, \alpha, \theta) &= \int_0^\infty \int_0^\infty \exp(-suv) g_C(v, \alpha, 1) g_C(u, 1, \theta)\, du\, dv \\ &= \int_0^\infty \exp(-sv)\, dv \int_0^\infty g_C\left(\frac{v}{u}, \alpha, 1\right) g_C(u, 1, \theta) \frac{du}{u}. \end{aligned}$$

Comparison of the left and right sides of this equality (as one-sided Laplace transforms) leads to the following relation between the densities of stable laws for the values $\alpha < 1$ and $x > 0$ ([2.16]):

$$g_C(x, \alpha, \theta) = \int_0^\infty g_C\left(\frac{x}{u}, \alpha, 1\right) g_C(u, 1, \theta) \frac{du}{u}. \tag{2.6.15}$$

REMARK 2. The only case when $L(s, \alpha, \beta, \gamma, \lambda)$, $\operatorname{Re} s \geq 0$, can be expressed in elementary functions corresponds to the values $\alpha < 1$, $\beta = 1$, and $\gamma = 0$. For such values of the parameters, $g(x, \alpha, 1, \gamma, \lambda) = 0$ on the semi-axis $x < 0$, i.e.,

$$L(s, \alpha, 1, \gamma, \lambda) = \Lambda(s, \alpha, 1, \gamma, \lambda) = \exp(-\lambda(\gamma s + s^\alpha)).$$

REMARK 3. In the case $\alpha > 1$ the transform $L(s, \alpha, 1)$ can be expressed with the help of the Mittag-Leffler functions (see §2.10 for more details).

REMARK 4. The transforms $L(s, \alpha, \beta, \gamma, \lambda)$ with any admissible values of the parameters have integral expressions analogous to those given in Theorem 2.6.2. For example, in the case $\alpha > 1$

$$L(s, \alpha, \beta, \gamma, \lambda) = \frac{1}{\pi} \operatorname{Im} \int_0^\infty \exp\{-\lambda[(su)^\alpha - s\gamma(ue^{i\pi\rho} + 1)]\} \frac{e^{i\pi\rho}}{1 + ue^{i\pi\rho}}\, du,$$

and in the case $\alpha = 1$, $\beta > 0$

$$\begin{aligned} L(s, 1, \beta, \gamma, \lambda) = \frac{1}{\pi} \int_0^\infty & \exp\{-\lambda[\beta u \log u - \gamma(s + u) - s\beta \log \lambda]\} \\ & \times \frac{\sin \frac{\pi}{2} \lambda (1 + \beta)^u}{s + u}\, du. \end{aligned}$$

In the case $\alpha < 1$ the expression $L(s, \alpha, \beta, \gamma, \lambda)$ turns out to be precisely the same as in the case $\alpha > 1$, provided that $\gamma \cos \pi\rho \leq 0$. For $\gamma \cos \pi\rho > 0$ the form of the expression for $L(s, \alpha, \beta, \gamma, \lambda)$ changes, but the difference from the case $\alpha > 1$ reduces only to a rotation of the contour of integration by an angle ω such that $\gamma \cos(\pi\rho - \omega) < 0$.

The transform $(\mathcal{L}f)(s)$ of the functions $f(x) = l(x) g_A(r(x), \alpha, \beta)$ are analyzed according to the indicated method with the use of the integral expressions (2.2.12)–(2.2.15) for the functions $g_A(r(x), \alpha, \beta)$. We do not give a general analysis of such transforms, but confine ourselves to two examples which are interesting in themselves and are good illustrations of the possibilities of the method as a whole. Both examples are connected with the densities $g(x, 1, \beta)$, $\beta > 0$.

Consider on the semi-axis $x > 0$ the function

$$f(x) = x^{-1} g_A(r(x), 1, \beta),$$

which, according to (2.2.13), has the integral expression

$$f(x) = \frac{1}{\pi} \operatorname{Im} \int_0^\infty \exp(-u - xT(u))\, du$$

(the functions r and T are the same as in the general formula (2.6.9)). We have

$$(\mathcal{L}f)(s) = \frac{1}{\pi} \operatorname{Im} \int_0^\infty \int_0^\infty \exp(-sx - u - xT(u))\, du\, dx.$$

Since

$$\operatorname{Re} T(u) \geq -\frac{2}{\pi e}\beta + \frac{2}{\pi}\beta u |\log u|,$$

the double integral converges absolutely, and we may change the order of integration. Consequently,

$$\begin{aligned}
&\int_0^\infty e^{-sx} x^{-1} g_A\left(x^{-1} - \frac{2}{\pi}\beta \log x, 1, \beta\right) dx \\
&\quad = \frac{1}{\pi} \operatorname{Im} \int_0^\infty e^{-u}\, du \int_0^\infty \exp\{-(s+T)x\}\, du \\
&\quad = \frac{1}{\pi} \operatorname{Im} \int_0^\infty e^{-u} \frac{du}{s + T(u)} \\
&\quad = \frac{1}{\pi}(1+\beta) \int_0^\infty e^{-u} \left[\left(s + \frac{2}{\pi}\beta u \log u\right)^2 + (1+\beta)^2 u^2\right]^{-1} u\, du.
\end{aligned}$$

In the second example we consider the function

$$f(x) = x^{-1} g(-\beta \log x, 1, \beta), \qquad \beta > 0,$$

which, according to (2.2.19), can be written in the form

$$f(x) = \frac{1}{2\beta}\int_{-1}^{1} U_1(\varphi,\beta)\exp\{-xU_1(\varphi,\beta)\}\,d\varphi, \tag{2.6.16}$$

where

$$U_1(\varphi,\beta) = \frac{\pi(1+\beta\varphi)}{2\cos(\pi\varphi/2)}\exp\left[\frac{\pi}{2}\left(\varphi+\frac{1}{\varphi}\right)\tan\frac{\pi}{2}\varphi\right].$$

Using (2.6.9), we get (changing the order of integration is obviously allowed in this case, because all the functions after the double integral sign are non-negative) that

$$\begin{aligned}(\mathcal{L}f)(s) &= \int_0^\infty e^{-sx}x^{-1}g(-\beta\log x,1,\beta)\,dx\\ &= \frac{1}{2\beta}\int_{-1}^{1} U_1(\varphi,\beta)\,d\varphi\int_0^\infty \exp\{-(s+U_1(\varphi,\beta))x\}\,dx\\ &= \frac{1}{2\beta}\int_{-1}^{1}\left[1+\frac{s}{U_1(\varphi,\beta)}\right]^{-1}d\varphi. \end{aligned}\tag{2.6.17}$$

The equality (2.6.17) has one interesting analytic corollary. Consider the case $\beta=1$. According to (2.5.17), $g(-\log x,1,1)$ decreases as $x\to\infty$ more rapidly than $\exp(-x/e)$. This means that the transform $(\mathcal{L}f)(s)$, as a function of s, admits an analytic extension to the half-plane $\operatorname{Re}s > -1/e$, i.e., it is an analytic function in the disk $|s|<1/e$.

We expand both sides of (2.6.17) in series of powers of s and equate the coefficients of the same powers of s. This gives us the system of equalities $(k=0,1,\ldots)$

$$\int_0^\infty x^{k-1}g(-\log x,1,1)\,dx = \frac{1}{2}\int_{-1}^{1} U_1^{-k}(\varphi,1)\,d\varphi, \tag{2.6.18}$$

which are nontrivial for $k\ge 1$. A change of the variable in the integral on the left-hand side gives us that

$$\int_{-\infty}^{\infty} e^{-kx}g(x,1,1)\,dx = \frac{1}{2}\int_{-1}^{1} U_1^{-k}(\varphi,1)\,d\varphi, \qquad k=1,2,\ldots.$$

The left-hand side of this equality coincides with $\Lambda(k,1,1)$ and, according to (2.6.10), is equal to $\exp(k\log k)=k^k$. Consequently, for any integers $k\ge 1$

$$\frac{1}{2}\int_{-1}^{1}\left(\frac{\cos(\pi\varphi/2)}{1+\varphi}\right)^k \exp\left\{-\frac{\pi}{2}k(1+\varphi)\tan\frac{\pi}{2}\varphi\right\}d\varphi = \left(\frac{\pi}{2}k\right)^k.$$

(d) Consider the Mellin transform of the density $g=g(x,\alpha,\beta,\gamma,\lambda)$. Let

$$m(s,\alpha,\beta,\gamma,\lambda) = (\mathcal{M}g)(s).$$

For $0 < s < 1$ we consider the function

$$m(-s,\alpha,\beta,\gamma,\lambda) = \int_0^\infty x^{-s} g(x,\alpha,\beta,\gamma,\lambda)\,dx.$$

Using the inversion formula for the density, we find that

$$\begin{aligned} m(-s,\alpha,\beta,\gamma,\lambda) &= \frac{1}{\pi}\operatorname{Re}\int_0^\infty\int_0^\infty x^{-s}e^{-itx}\mathfrak{g}(t,\alpha,\beta,\gamma,\lambda)\,dt \\ &= \frac{1}{\pi}\operatorname{Re}\int_0^\infty \mathfrak{g}(t,\alpha,\beta,\gamma,\lambda)\,dt\int_0^\infty x^{-s}e^{-itx}\,dx \\ &= \frac{1}{\pi}\Gamma(1-s)\operatorname{Re}\int_0^\infty (it)^{s-1}\mathfrak{g}(t,\alpha,\beta,\gamma,\lambda)\,dt. \quad (2.6.19) \end{aligned}$$

Of course, these formal transformations of integrals require proofs, but we shall not give them here. On the one hand, they are very unwieldy, and, on the other hand, formula (2.6.19) will not be used below in so general a form. The solution of the same problem turns out to be of considerable interest for stable distributions in the class $\mathfrak{W}$, since in this case it is possible to find an explicit expression for the function $m(s,\alpha,\beta,\gamma,\lambda)$.

We mention some easily verified facts needed in the course of the subsequent arguments.

1) The transform $m(s,\alpha,\beta,\gamma,\lambda)$ is finite for any s in the strip $-1 < \operatorname{Re} s < \alpha$ (a consequence of the finiteness of the density g and of the existence of moments of order less than α).

2) Within the class $\mathfrak{W}$

$$\begin{aligned} m(s,\alpha,\beta,\gamma,\lambda) &= \lambda^{s/\alpha} m(s,\alpha,\beta,\gamma,1) \\ &= \lambda^{s/\alpha} m_C(s,\alpha,\theta) = \lambda^{s/\alpha} m_C(s,\alpha,\rho) \end{aligned}$$

(a consequence of Property 2.2 and the definition of the class $\mathfrak{W}$; $\rho = (1+\theta)/2$ is a modification of the parameter θ).

3) If $\alpha \neq 1$, then

$$m_C(s,\alpha,\rho) = m(s,\alpha,\beta,0,1) = m(s,\alpha,\beta)$$

(see the connection between the parametrization systems (B) and (C)).

4) The function $m_C(s,\alpha,\rho)$ is jointly continuous in the parameters α and ρ in their domain of admissible values (this fact was mentioned in the Introduction).

THEOREM 2.6.3. *For any strictly stable law the following equality holds in the strip* $-1 < \operatorname{Re} s < \alpha$ ([2.17]):

$$m_C(s,\alpha,\rho) = \frac{\sin \pi\rho s}{\sin \pi s}\frac{\Gamma(1-s/\alpha)}{\Gamma(1-s)}. \qquad (2.6.20)$$

PROOF. Let $\alpha \neq 1$ and let β have any admissible value. Let the expression (2.6.4) for the density be substituted in the integral expression for $m(s, \alpha, \beta)$. For values $-1 < s < 0$

$$\begin{aligned} m(s,\alpha,\beta) &= \int_0^\infty x^s g(x,\alpha,\beta)\,dx \\ &= \frac{1}{\pi}\operatorname{Im}\int_0^\infty x^s\,dx \int_0^\infty \exp\left(-xue^{i\pi\rho/2} - u^\alpha e^{-i\pi\rho\alpha/2} + \frac{i\pi\rho}{2}\right)du. \end{aligned}$$

It is not hard to verify that the conditions for changing the order of integration are satisfied here. We have that

$$\begin{aligned} m(s,\alpha,\beta) &= \frac{1}{\pi}\operatorname{Im}\int_0^\infty \exp\left(-u^\alpha e^{-i\pi\rho\alpha/2} + \frac{i\pi\rho}{2}\right)du \\ &\quad\times \int_0^\infty x^s \exp(-xue^{i\pi\rho/2})\,dx \\ &= \Gamma(1+s)\frac{1}{\pi}\operatorname{Im}\int_0^\infty (ue^{i\pi\rho/2})^{-s}\exp(-u^\alpha e^{-i\pi\rho\alpha/2})\frac{du}{u}. \end{aligned}$$

Rotation of the contour of integration in the last integral through the angle $\pi\rho/2$ gives us, after a corresponding change of variable, that

$$\begin{aligned} m(s,\alpha,\beta) &= \frac{1}{\varphi}\Gamma(1+s)\sin\pi\rho s\int_0^\infty u^{-s-1}\exp(-u^\alpha)\,du \\ &= \frac{1}{\pi\alpha}\Gamma(1+s)\Gamma\left(-\frac{s}{\alpha}\right)\sin\pi\rho s \\ &= \frac{\sin\pi\rho s}{\sin\pi s}\frac{\Gamma(1-s/\alpha)}{\Gamma(1-s)}. \end{aligned}$$

Both the left-hand side and the right-hand side of the last equality contain functions analytic in the strip $-1 < \operatorname{Re} s < \alpha$. Therefore, the equality proved for $-1 < s < 0$ is preserved for values of s in the whole strip. Thus, (2.6.20) is proved in the case $\alpha \neq 1$. The fact that this equality is preserved for $\alpha = 1$ follows from the continuity property mentioned above for the function $m_C(s, \alpha, \rho)$ at $\alpha = 1$.

(e) The characteristic transformation will be the last of the integral transformations we consider. Since this transformation cannot be said to be well known, we devote somewhat more attention to it than to the others. First of all we acquaint ourselves with the main properties of characteristic transforms of densities $p_X(x)$ of arbitrary random variables X:

$$W_X(t) = \begin{pmatrix} w_0(t)_X & 0 \\ 0 & w_1(t)_X \end{pmatrix} = (\mathcal{W}p_X)(t).$$

Although this definition assumes the existence of a density for the distribution of X, it is easily freed of this restriction by setting

$$\omega_k(t)_x = \mathsf{E}|X|^{it}(\operatorname{sgn} X)^k, \qquad t \in R,\ k = 0, 1. \tag{2.6.21}$$

The transform $W_X(t)$ was first introduced in [101] and later found an application in multiplicative problems in number theory. Characteristic transforms of random variables play the same role in the scheme of multiplication of random variables as characteristic functions play in the scheme of summation. This analogue is not hard to see in the light of the following properties of characteristic transforms.

1) The characteristic transform exists for any random variable X, as is clear from the very definition (2.6.21) of the functions $w_k(t)_X$.

2) The distribution F_X is uniquely determined by the characteristic transformation W_X.

Indeed, let

$$c^+ = \mathsf{P}(X > 0), \qquad c^- = \mathsf{P}(X < 0),$$
$$c^+\mathfrak{f}^+ = \mathsf{E}|X|^{it}I(X > 0), \qquad c^-\mathfrak{f}^-(t) = \mathsf{E}|X|^{it}I(X < 0),$$

where $\mathfrak{f}^+$ and $\mathfrak{f}^-$ are certain characteristic functions uniquely connected with the parts of the distribution $F_X(x)$ on the semi-axes $x > 0$ and $x < 0$ when the corresponding coefficients c^+ and c^- are nonzero. We have that $c^+ + c^- = 1 - \mathsf{P}(X = 0)$, and

$$\begin{aligned} w_k(t)_X &= c^+\mathfrak{f}^+(t) + (-1)^k c^-\mathfrak{f}^-(t), \qquad k = 0, 1, \\ c^+\mathfrak{f}^+(t) &= \tfrac{1}{2}(w_0(t)_X + w_1(t)_X), \\ c^-\mathfrak{f}^-(t) &= \tfrac{1}{2}(w_0(t)_X - w_1(t)_X). \end{aligned} \tag{2.6.22}$$

Consequently, the distribution F_X can really be recovered if we know the functions w_0 and w_1.

3) If U and V are independent random variables, then, for their product $X = UV$,

$$W_X(t) = W_U(t)W_V(t), \qquad t \in R. \tag{2.6.23}$$

We have

$$w_k(t)_X = \mathsf{E}|UV|^{it}(\operatorname{sgn} UV)^k = w_k(t)_U w_k(t)_V, \qquad k = 0, 1,$$

which implies (2.6.23).

4) For any sequence of random variables $X, X_1, X_2, \dots$ the relation

$$L(F_{X_n}, F_X) + |\mathsf{P}(X_n = 0) - \mathsf{P}(X = 0)| \to 0 \quad \text{as } n \to \infty$$

(where L denotes the Lévy metric) holds if and only if

$$W_{X_n}(t) \to W_X(t) \quad \text{as } n \to \infty$$

in any finite interval of values t.

This assertion is a simple consequence of (2.6.22) and known facts on weak convergence of distributions and the characteristic functions corresponding to them.

Suppose that $g = g(x, \alpha, \beta, \gamma, \lambda)$ and s is a complex number in the strip $-1 < \operatorname{Re} < \alpha$. Let

$$W(s, \alpha, \beta, \gamma, \lambda) = (\mathcal{W}g)(-is) = \begin{pmatrix} w_0(s, \alpha, \beta, \gamma\lambda) & 0 \\ 0 & w_1(s, \alpha, \beta, \gamma, \lambda) \end{pmatrix}.$$

This function exists for any values of s in the indicated strip and any admissible values of the parameters, and is closely connected with the Mellin transform. Namely,

$$w_0(s, \alpha, \beta, \gamma, \lambda) = m(s, \alpha, \beta, \gamma, \lambda) + m(s, \alpha, -\beta, -\gamma, \lambda), \tag{2.6.24}$$

$$w_1(s, \alpha, \beta, \gamma, \lambda) = m(s, \alpha, \beta, \gamma, \lambda) - m(s, \alpha, -\beta, -\gamma, \lambda). \tag{2.6.25}$$

Therefore, what was noted with regard to the transforms m remains valid also for the elements w_k of the transform W. Theorem 2.6.3 enables us easily to obtain an explicit expression for the functions $w_k(s, \alpha, \beta, \gamma, \lambda)$ corresponding to the distributions in the class $\mathfrak{W}$ (in the same form (C)).

THEOREM 2.6.4. *The characteristic transforms of the densities of strictly stable distributions have the following form in the strip* $-1 < \operatorname{Re} s < \alpha$ [2.18]:

$$w_k(s, \alpha, \theta, \lambda)_C = \lambda^{s/\alpha} \frac{\cos \frac{\pi}{2}(k - \theta s)}{\cos \frac{\pi}{2}(k - s)} \frac{\Gamma(1 - s/\alpha)}{\Gamma(1 - s)}, \qquad k = 0, 1. \tag{2.6.26}$$

PROOF. On the basis of (2.6.20)–(2.6.22),

$$w_k(s, \alpha, \beta, \gamma, \lambda) = \lambda^{s/\alpha} w_k(s, \alpha, \beta, \gamma, 1), \qquad k = 0, 1,$$

$$\begin{aligned} w_0(s, \alpha, \theta)_C &= m_C(s, \alpha, \theta) + m_C(s, \alpha, -\theta) \\ &= \frac{\sin \frac{\pi}{2}(1 + \theta)s}{\sin \pi s} \frac{\Gamma(1 - s/\alpha)}{\Gamma(1 - s)} \\ &\quad + \frac{\sin \frac{\pi}{2}(1 - \theta)s}{\sin \pi s} \frac{\Gamma(1 - s/\alpha)}{\Gamma(1 - s)} \\ &= \frac{\cos \frac{\pi}{2}\theta s}{\cos \frac{\pi}{2} s} \frac{\Gamma(1 - s/\alpha)}{\Gamma(1 - s)}. \end{aligned}$$

In a completely analogous way,

$$\begin{aligned} w_1(s, \alpha, \theta)_C &= m_C(s, \alpha, \theta) - m_C(s, \alpha, -\theta) \\ &= \frac{\sin \frac{\pi}{2}\theta s}{\sin \frac{\pi}{2} s} \frac{\Gamma(1 - s/\alpha)}{\Gamma(1 - s)}. \end{aligned}$$

REMARK. In the general case neither the Melling transforms $m(s, \alpha, \beta, \gamma, \lambda)$ nor the characteristic transforms $W(s, \alpha, \beta, \gamma, \lambda)$ have explicit expressions. This is because the cases of simple expressions for the transforms m and W are organically connected with the natural division of the densities of stable distributions into two parts. For example, the point $x = 0$ makes such a division in the class $\mathfrak{W}$ (if we use the parametrization system (C)).

§2.7. Unimodality of stable distributions. The form of the densities

Among the diverse properties of probability measures regarded as worthy of attention in probability theory is one called unimodality. Encountering this property in the course of solving various problems usually gladdens statisticians, and does not leave probabilists indifferent. The property is enjoyed by many known distributions, including all stable distributions, as will be shown below.

The term "unimodality" was apparently introduced in mathematical statistics by the light hand of Pearson. In statistics a "mode" of an absolutely continuous distribution is defined to be a point of the real axis at which the density attains, roughly speaking, a maximal value. Of course, a distribution can have not one but a whole set of modes. However, when statisticians say that a distribution is unimodal, they mean that it has precisely one mode (see [12]). It is not hard to see that the unimodality of a distribution, understood in this way, is inconvenient for its formal definition. The definition of unimodality commonly used at present was proposed in 1938 by Khintchine [44].

DEFINITION. A distribution function $F(x)$ is said to be *unimodal* if there exists at least one value $x = a$ such that $F(x)$ is convex for $x < a$ and concave for $x > a$. In this case one says that the distribution $F(x)$ has a mode at the point $x = a$ (or, more briefly, that $x = a$ is a mode of $F(x)$). Here the convexity and concavity of F are understood in the broad sense, i.e., for example, F is convex on the semi-axis $x < a$ if for any points $x_1, x_2 < a$

$$2F((x_1 + x_2)/2) \leq F(x_1) + F(x_2).$$

It follows from the theory of convex functions (see, for example, [30]) that a unimodal distribution function $F(x)$ with mode at a point $x = a$ has the following properties:

1) $F(x)$ has left-sided and right-sided derivatives at each point $x \neq a$, and the one-sided derivatives coincide everywhere except for points forming a set which is at most countable.

2) The derivative $F'(x)$ (where F' is chosen at each point as one of the one-sided derivatives) is nondecreasing on the semi-axis $x < a$ and is nonincreasing on the semi-axis $x > a$.

Thus, the distribution $F(x)$ is absolutely continuous. As will become clear below, the most convenient variant of the density $F'(x)$ is the one which is a left-continuous function (except perhaps for a single point). The class of distribution functions with such densities will be denoted by $\mathcal{F}_0$. We need it in formulating general criteria for distributions to be unimodal.

This definition of unimodality generalizes the original conception of the property of unimodality, since a distribution that is unimodal in the sense of this definition can have not one but a whole set of modes. This is shown by the simple example of the uniform distribution on the interval $(0,1)$, for which each point of $[0,1]$ is obviously a mode.

It is clear from the definition of unimodality that if points $a_1 < a_2$ can be chosen as modes of a distribution $F(x)$, then each point of $[a_1, a_2]$ is also a mode of $F(x)$.

The problem of unimodality of stable laws stems from a 1936 paper of Wintner [87], where it is proved that all symmetric stable laws are unimodal. It should be mentioned that in his consideration of symmetric distributions he defined unimodality in the same way as Khintchine, by means of the property of convexity of the distribution function on the negative semi-axis. Instead of the term “unimodality” he used “convexity”.

Wintner's paper could serve as a good basis both for finding a natural definition of unimodality of distributions and for posing the general problem of unimodality of stable laws. It is known that this problem was discussed in 1939 at a seminar led by Gnedenko at Moscow University. However, it turned out to be very complicated, and the complete solution of it was stretched over many years, involving in the process such a number of tragicomic situations as perhaps no other problem in probability theory can boast of. There is not yet an independent analytic proof of unimodality for stable laws. The property was proved in 1978 by Yamazato [94] as a corollary of a more general fact: the unimodality of laws in the class L ([2.19]).

The main content of this section is a presentation of Yamazato's theorem in a somewhat simplified version connected with the subclass $\mathfrak{S}$ of L. But together with it we give proofs of particular results obtained earlier by Yamazato and by other methods. On the one hand, such information will be useful to anyone who wants to look for an independent analytic proof of the unimodality of stable distributions. On the other hand, familiarity with the methods used may serve well in solving the more complex problem of unimodality for multidimensional stable laws.

We begin by giving several general criteria for unimodality of distributions. Consider first of all that properties 1) and 2), which follow from the definition of unimodality, can be regarded as necessary conditions. And conversely, by starting from properties 1) and 2) it is not hard to see that a distribution function $F(x)$ having these properties is unimodal with mode at the point $x = a$. We thereby obtain the following criterion.

THEOREM 2.7.1. *A distribution function $F(x)$ is unimodal with mode at the point $x = a$ if and only if it has properties* 1) *and* 2).

For each distribution function $F \in \mathcal{F}_0$ we define on the real x-axis the left-continuous (except perhaps for a single point) function

$$V(x) = F(x) - xF'(x).$$

THEOREM 2.7.2. *A distribution function $F \in \mathcal{F}_0$ is unimodal with mode at the point $x = 0$ if and only if the function V corresponding to it is a distribution function* ([2.20]).

PROOF. *Necessity.* Consider the differential of $V(x)$, understood as the value of the jump at points of discontinuity. Then at all points $x \neq 0$

$$dV(x) = dF(x) - F'(x)\,dx - x dF'(x) = -x\,dF'(x). \qquad (2.7.1)$$

Since F is unimodal with mode at $x = 0$, the right-hand side of (2.7.1) is nonnegative. Consequently, $V(x)$ is nondecreasing on the semi-axes $x < 0$ and $x > 0$.

If $x > 0$, then

$$\begin{aligned} 0 \leq xF'(x) \leq 2\int_{x/2}^{x} F'(u)\,du &= 2[F(x) - F(x/2)] \\ &\leq 2\min\{1 - F(x/2), F(x) - F(+0)\}, \end{aligned}$$

i.e., $xF'(x) \to 0$ as $x \to 0$ or $x \to \infty$. Analogous arguments in the case $x < 0$ lead us to a conclusion which, combined with the treated case $x > 0$, can be given the form

$$xF'(x) \to 0 \quad \text{if } |x| \to 0,\ x \neq 0,\ \text{or } |x| \to \infty. \qquad (2.7.2)$$

Using the expression for $V(x)$ and (2.7.2), we find that

$$V(+0) = F(+0) > F(-0) = V(-0),$$
$$1 - V(\infty) = 1 - F(\infty) = V(-\infty) = F(-\infty) = 0.$$

Together with the property $dV(x) \geq 0$ $(x \neq 0)$ established above, these equalities give us that $V(x)$ is a distribution function.

Sufficiency. Let $F \in \mathfrak{F}_0$ and suppose that the function $V(x)$ associated with it is a distribution function. Then (2.7.2) obviously holds, and a particular case of it is the relation $F'(x) \to 0$ as $|x| \to \infty$. Using this property and integrating the transformed equality (2.7.1)

$$dF'(x) = -\frac{1}{x}dV(x), \qquad x \neq 0,$$

from x to $+\infty \operatorname{sgn} x$, we find that

$$F'(x) = \int_x^{+\infty \operatorname{sgn} x} \frac{1}{v}dV(v), \qquad x \neq 0.$$

This makes it clear that $F'(x)$ is nondecreasing on the semi-axis $x < 0$ and nonincreasing on the semi-axis $x > 0$, i.e., $F(x)$ is unimodal with mode at $x = 0$.

The following general criterion for unimodality of distributions is formulated in terms of random variables (recall that the symbol $\stackrel{d}{=}$ means equality of distributions) ($^{2.2.1}$).

THEOREM 2.7.3. *A distribution F_X of a random variable X is unimodal with mode at the point $x = 0$ if and only if X can be represented in the form*

$$X \stackrel{d}{=} X_0 Z, \tag{2.7.3}$$

where X_0 and Z are independent random variables, with X_0 uniformly distributed on the interval $(0, 1)$.

REMARK. The condition that X can be represented as a product

$$X \stackrel{d}{=} X'Z' \tag{2.7.4}$$

of independent random variables, one of which (say X') has a unimodal distribution with mode at zero, is also a criterion for the unimodality of F_X.

Indeed, the fact that (2.7.4) is a necessary condition follows from the necessity of (2.7.3). Let us verify the sufficiency of (2.7.4). If X' has a unimodal distribution, then $X' \stackrel{d}{=} X_0 Z''$, according to (2.7.3). Consequently,

$$X \stackrel{d}{=} X'Z' \stackrel{d}{=} X_0(Z'Z'') \stackrel{d}{=} X_0 Z.$$

This and the sufficiency of (2.7.3) give us that F_X is unimodal with mode at zero. The criterion (2.7.4) is more attractive than (2.7.3) only as a sufficient condition for unimodality.

PROOF OF THE THEOREM. According to the condition of Theorem 2.7.2, the distribution function F_X of X is unimodal with mode at zero if and only if there is a random variable Z such that $F_Z(x) = F_X(x) - xF'_X(x)$.

Let $F_X(x)$ be unimodal with mode at $x = 0$. We compute the characteristic transform $W_Z(t)$. By (2.7.1),

$$\begin{aligned} w_k(t)_Z &= \mathsf{E}|Z|^{it}(\operatorname{sgn} Z)^k \\ &= \int_{x\neq 0} |x|^{it} x(\operatorname{sgn} x)^k \, dF'_X(x), \qquad k = 0, 1. \end{aligned}$$

Integration by parts and property (2.7.2) give us that

$$\begin{aligned} w_k(t)_Z &= -\int_{x\neq 0} |x|^{1+it}(\operatorname{sgn} x)^{1+k} \, dF'_X(x) \\ &= (1+it)\int_{x\neq 0} |x|^{it}(\operatorname{sgn} x)^k F'_X(x)\, dx \\ &= (1+it) w_k(t)_X, \qquad k = 0, 1. \end{aligned}$$

Consequently, for all $t \in R$

$$W_Z(t) = (1+it)W_X(t). \tag{2.7.5}$$

Simple computations show that the characteristic transform of a random variable X_0 uniformly distributed in $(0, 1)$ has the form

$$W_{X_0}(t) = \begin{pmatrix} (1+it)^{-1} & 0 \\ 0 & (1+it)^{-1} \end{pmatrix}, \tag{2.7.6}$$

and so it follows from (2.7.5) and (2.7.6) that

$$W_X(t) = W_{X_0}(t) W_Z(t). \tag{2.7.7}$$

The representation (2.7.1) is now obtained as a consequence of (2.7.7) and property 2 given at the end of the last equation for characteristic transforms.

Assume the opposite, i.e., that X can be represented in the form (2.7.1). Then

$$\begin{aligned} F_X(x) &= \int_0^1 F_Z\left(\frac{x}{u}\right) du = -\int_1^\infty F_Z(xu)\, d\frac{1}{u} \\ &= F_Z(x) + x\int_x^{\infty \operatorname{sgn} x} \frac{1}{u}\, dF_Z(u). \end{aligned}$$

If $dF_Z(x)$ is understood as the value of the jump of $F_Z(x)$ at points of discontinuity, then the last equality gives us by differentiation that

$$dF_X(x) = \int_x^{+\infty \operatorname{sgn} x} \frac{1}{u}\, dF_Z(u)\, dx.$$

This means that $F'_X(x)$ is nonincreasing on the semi-axis $x > 0$ and nondecreasing on the semi-axis $x < 0$, i.e., F_X is unimodal with mode at zero.

COROLLARY 1. *The distribution F_X is unimodal with mode at the point $x = 0$ if and only if the product $(1 + it)W_X(t)$ is a characteristic transform.*

COROLLARY 2. *Suppose that $p(x)$ is the density of some unimodal distribution F with mode at the point $x = 0$. Then the distribution F_X with characteristic function $\mathfrak{f}_X(t)$ is unimodal with mode at zero if and only if*

$$\mathfrak{f}_X(t) = \int p(u)\mathfrak{f}_Z(tu)\,du, \tag{2.7.8}$$

where $\mathfrak{f}_Z(t)$ is a characteristic function.

In particular, if F is the uniform distribution on the interval $(0, 1)$, then condition (2.7.8) has the form

$$\mathfrak{f}_X(t) = \frac{1}{t}\int_0^t \mathfrak{f}_Z(u)\,du.$$

(This criterion was found by Khintchine [26].)

LEMMA 2.7.1. *Let $F_1, F_2, \dots$ be a sequence of unimodal distribution functions with modes at the respective points $a_1, a_2, \dots$. If F_n converges weakly to the distribution function F as $n \to \infty$, then F is unimodal with mode at the point $a = \lim_{n\to\infty}\sup a_n$.*

PROOF. Unimodality of F_n at a_n means that for any x_1 and x_2 to the left of a_n

$$2F_n((x_1 + x_2)/2) \leq F_n(x_1) + F_n(x_2)$$

and that the reverse inequality holds for points to the right of a_n. We choose a sequence of positive integers n_k such that $a_{n_k} \to a$ as $k \to \infty$. Then any two points x_1 and x_2 to the left of a are to the left of the points a_{n_k} for all sufficiently large k. Passage to the limit $F_{n_k} \to F$ as $k \to \infty$ shows that the liimt function F has analogous properties of convexity and convexity, i.e.,

$$2F((x_1 + x_2)/2) \leq F(x_1) + F(x_2)$$

for points to the left of a, and the opposity inequality for points to the right. But this means that F is unimodal with mode at a.

LEMMA 2.7.2. *If F_1 and F_2 are symmetric unimodal distribution functions, then their convolution $F = F_1 * F_2$ is also symmetric and unimodal* (2.22).

PROOF. Since the functions F_j are symmetric, i.e., $1 - F_j(x) = F_j(-x)$ at all points of continuity, unimodality is equivalent to the inequality

$$2F_j((x_1 + x_2)/2) \geq F_j(x_1) + F_j(x_2)$$

for any points $x_1 \geq x_2 > 0$. We form the function

$$G_h(y) = F_1(y) - \tfrac{1}{2}[F_1(y+h) - F_1(y-h)].$$

It is not hard to verify that

$$G_h(y) = -G_h(-y), \qquad G_h(y) \operatorname{sgn} y \geq 0,$$

and that $d|F_2(y-x) - F_2(y+x)| \geq 0$ on the semi-axis $y > 0$ for any $x > 0$. Let $x_1 = x + h$ and $x_2 = x - h$, $x > h > 0$. Then

$$\int_0^\infty G_h(y)\, d[F_2(y-x) - F_2(y+x)] \geq 0.$$

Using the definition of the function $G_h(y)$, we can transform this inequality to the form

$$\begin{aligned} F((x_1+x_2)/2) &= \int F_1((x_1+x_2)/2 - y)\, dF_2(y) \\ &\geq \int \frac{1}{2}[F_1(x_1 - y) + F_1(x_2 - y)]\, dF_2(y) \\ &= \frac{1}{2}[F(x_1) + F(x_2)]. \end{aligned}$$

Since a convolution of symmetric distributions is symmetric, this inequality means (as remarked above) that F is unimodal.

REMARK. Without the condition of symmetry for the distributions F_j the statement of the lemma is false, as shown by examples (the first such example was given by Chung [10]). Here is one of the simplest examples:

Let $p_a(x)$ be the density of the uniform distribution on $(0, a)$. For $a < b$ let

$$q(x) = \varepsilon p_\alpha(x) + (1-\varepsilon) p_b(x).$$

It is easy to see that this is the density of a unimodal distribution with mode at zero (each point in $[0, a]$ is a mode of q). We consider the density

$$\begin{aligned} \omega(x) &= (q * q)(x) \\ &= \varepsilon^2 p_a^{2*}(x) + 2\varepsilon(1-\varepsilon)(p_a * p_b)(x) + (1-\varepsilon)^2 p_b^{2*}(x). \end{aligned}$$

The density $p_a^{2*}(x)$ has the form of an isosceles triangle with base $(0, 2a)$ and altitude $1/a$. Consequently,

$$\omega(a) > \frac{\varepsilon^2}{a}, \quad \omega(b) > \frac{(1-\varepsilon)^2}{b}, \quad \int_b^{2a} \omega(x)\, dx > \frac{(1-\varepsilon)^2}{2}.$$

Assume that $\omega(x) > \omega(a)$ in the interval $a < x < b$. Then

$$\int_0^b \omega(x)\, dx \geq (b-a)\omega(a) > \frac{\varepsilon^2 b}{a} - \varepsilon^2,$$

whence

$$\int_a^{2b} \omega(x)\, dx > \frac{\varepsilon^2 b}{a} - \varepsilon^2 + \frac{(1-\varepsilon)^2}{2}.$$

If now $\varepsilon = \frac{1}{4}$, $a = \frac{1}{16}$ and $b = 1$, then we see that the lower bound is greater than 1. This means that the assumption that $\omega(x) > \omega(a)$ is not justified for this choice of parameters, i.e., between a and b the density $\omega(x)$ takes values less than $\omega(a)$ and $\omega(b)$. The distribution for $\omega(x)$ is thus not unimodal.

THEOREM 2.7.4. *Symmetric stable distributions are unimodal.*

PROOF. The case $\alpha = 2$ corresponding to a normal distribution is obvious, and without loss of generality we can confine ourselves to an analysis of the case $0 < \alpha < 2$.

According to Theorem C.1, the canonical form of the expression (I.17) for the characteristic function $\mathfrak{g}(t, \alpha, 0, 0, \lambda) = \exp(-\lambda |t|^\alpha)$ of a symmetric stable law has the form

$$\log \mathfrak{g}(t, \alpha, 0, 0, \lambda) = \int_{x \neq 0} (\cos tx - 1) H'(x)\, dx,$$

where

$$H(x) = c|x|^{-1-\alpha} = \frac{\lambda}{\pi} \Gamma(1+\alpha) \sin \frac{\pi}{2} \alpha |x|^{-1-\alpha}.$$

We consider a sequence of series of mutually independent (in each series) random variables X_{nk}, $1 \le k \le n$, $n = 1, 2, \ldots$, having the symmetric distribution function

$$\begin{aligned} F_{nk}(x) &= 1 - F_{nk}(-x) \\ &= \begin{cases} \frac{1}{n} H(x), & x \le -\frac{1}{n} \\ \frac{1}{n}[H(-\frac{1}{n}) + H'(-\frac{1}{n})(x + \frac{1}{n})], & -\frac{1}{n} < x < 0. \end{cases} \end{aligned}$$

For $n\varepsilon > 1$

$$\mathsf{P}(|X_{nk}| \ge \varepsilon) = 2H(-\varepsilon)/n \to 0 \quad \text{as } n \to \infty$$

uniformly with respect to k, i.e., the random variables X_{nk} satisfy the condition (I.2) of limiting uniform smallness; hence the limit distribution of the sum $X_n = X_{n1} + \cdots + X_{nn}$ can be determined by using Theorem B. For $x < 0$

$$nF_{nk}(x) = n[1 - F_{nk}(-x)] \to H(x) \quad \text{as } n \to \infty.$$

Moreover,

$$\begin{aligned}\sigma_n^\varepsilon &= n\operatorname{Var}(X_{nj}I(|X_{nj}|<\varepsilon))\\ &= 2\int_0^{1/n} x^2 H'\left(\frac{1}{n}\right)dx + 2\int_{1/n}^{\varepsilon} x^2 H'(x)\,dx\\ &= \frac{2}{3}\frac{1}{n^3}H'\left(\frac{1}{n}\right) + 2c\int_{1/n}^{\varepsilon} x^{1-\alpha}\,dx\\ &= \frac{2}{3}cn^{\alpha-2} + \frac{2c}{2-\alpha}(\varepsilon^{2-\alpha} - n^{\alpha-2}).\end{aligned}$$

Consequently,

$$\lim_{\varepsilon\to 0}\lim_{n\to\infty}\sup \sigma_n^\varepsilon = 0.$$

Therefore, on the basis of Theorem B,

$$F_n(x) = \mathsf{P}(X_n < x) \to G(x,\alpha,0,0,\lambda)$$

as $n\to\infty$.

The distributions F_{nk} are symmetric and unimodal by construction. Consequently, by Lemma 2.7.2, their convolution $F_n(x)$ is also symmetric and unimodal. To obtain the assertion it now remains to apply Lemma 2.7.1 to the sequence $\{F_n\}$ of distributions.

We now use a purely analytic method to prove unimodality of the extremal stable laws (i.e., those corresponding to the values $\beta=\pm1$). By Properties 2.1 and 2.2, the proof of unimodality for the distribution functions $G(x,\alpha,\beta,\gamma,\lambda)$ reduces to an analysis of the extremal standard distributions $G(x,\alpha,1)$. Furthermore, it is clear that the case $\alpha=2$ need not be considered, and it can be assumed that $0<\alpha<2$.

THEOREM 2.7.5 ([2.23]). *Extremal stable distributions are unimodal and have exactly one mode. Moreover, in the case of a standard distribution $G(x,\alpha,1)$ the corresponding mode is positive if $\alpha<1$ and negative if $1<\alpha<2$.*

The proof is based on the representation (2.2.35) of the function $g'(x,\alpha,\beta)$, $0<\alpha<2$, $\beta=\pm1$, $x>0$. In the course of the proof we use some facts which we isolate as individual lemmas, with the notaiton in (2.2.35).

LEMMA 2.7.3. *If $1<\alpha<2$, then $g'(x,\alpha,1)<0$ for all $x>0$.*

PROOF. According to (2.2.35),

$$g'(x,\alpha,1) = \frac{1}{\pi}k^2\int_{-\pi}^{-\pi/\alpha} b\exp(-k^\alpha a)\,d\varphi, \tag{2.7.9}$$

where the function

$$b(\varphi) = (\alpha-1)^{-1}\left|\frac{\sin\varphi}{\sin\alpha\varphi}\right|^{2/(\alpha-1)}\left(\frac{\alpha\sin(2-\alpha)\varphi}{\sin\alpha\varphi}-1\right)$$

is easily seen to be negative on the interval $(-\pi, -\pi/\alpha)$. Therefore, the integral (2.7.9) is also negative.

LEMMA 2.7.4. *If $1 < \alpha < 2$, then*

$$g'(0,\alpha,1) = -g'(0,\alpha,-1) = \tfrac{1}{2\pi}\Gamma(1+2/\alpha)\sin(2\pi/\alpha) < 0.$$

PROOF. Using the representation (2.2.10) for the density $g(x,\alpha,\beta)$, we find that

$$g'(x,\alpha,\beta) = -\frac{1}{\pi}\operatorname{Im}\int_0^\infty y\exp(-y^\alpha - xye^{i\pi\rho} + 2\pi i\rho)\,dy.$$

This gives us that for $\beta = \pm 1$

$$\begin{aligned} g'(0,\alpha,\beta) &= -\frac{1}{\pi}\sin 2\pi\rho\int_0^\infty u\exp(-u^\alpha)\,du \\ &= \frac{1}{\alpha\pi}\sin\pi\theta\Gamma\left(\frac{2}{\alpha}\right) = \frac{\beta}{2\pi}\Gamma\left(1+\frac{2}{\alpha}\right)\sin\frac{2\pi}{\alpha} < 0. \end{aligned}$$

As is clear from its definition, the function $b(\varphi)$ has as factors a function that is trivially positive in the corresponding interval (u,v), and the function

$$w(\varphi) = \begin{cases} \operatorname{sgn}(1-\alpha)\left(\dfrac{\alpha\sin(2-\alpha)\varphi}{\sin\alpha\varphi}-1\right) & \text{if } \alpha\neq 1, \\ 2\varphi\cot\varphi - 1 & \text{if } \alpha = 1, \end{cases}$$

whose behavior is not obvious.

LEMMA 2.7.5. *On the corresponding interval (u,v)*

1) *the function $a(\varphi)$ is positive and monotonically increasing, and*

2) *the function $w(\varphi)$ is monotonically decreasing if $\alpha \le 1$ and monotonically increasing if $\alpha > 1$.*

PROOF. Consider the function $h(\alpha,\varphi) = \alpha\cot\alpha\varphi - \cot\varphi$ in the domain $0 < \varphi < \min(\pi, \pi/\alpha)$, $0 < \alpha < 2$. In this domain

$$\frac{\partial h}{\partial\alpha} = (2\sin^2\alpha\varphi)^{-1}(\sin 2\alpha\varphi - 2\alpha\varphi) < 0;$$

therefore, for fixed φ the function $h(\alpha,\varphi)$ decreases as α increases. Since $h(1,\varphi) = 0$,

$$\begin{aligned} h(\alpha,\varphi) &> 0 \quad \text{if } 0<\alpha<1, \\ h(\alpha,\varphi) &< 0 \quad \text{if } 1<\alpha<2. \end{aligned} \tag{2.7.10}$$

The case when $\alpha < 1$ and $\beta = -1$, for which $u = v$, corresponds to the trivial situation ($g(x, \alpha, \beta) = 0$ on the semi-axis $x > 0$) and thus does not contradict the assertion of the lemma.

In the case when $\alpha < 1$ and $\beta = 1$ we have that $u = 0$ and $v = \pi$, and

$$a(\varphi) = \left(\frac{\sin \alpha\varphi}{\sin \varphi}\right)^{1/(1-\alpha)} \frac{\sin(1-\alpha)\varphi}{\sin \alpha\varphi},$$

$$w(\varphi) = \frac{\alpha \sin(2-\alpha)\varphi}{\sin \alpha\varphi} - 1.$$

(2.7.10) gives us that

$$a'(\varphi)/a(\varphi) = [\alpha h(\alpha, \varphi) + (1-\alpha)h(1-\alpha, \varphi)]/(1-\alpha) > 0.$$

Since $a(\varphi)$ is obviously positive in $(0, \pi)$, it follows that $a'(\varphi) > 0$.

Further, by (2.7.10),

$$w'(\varphi) = \alpha[h(2-\alpha, \varphi) - h(\alpha, \varphi)] < 0.$$

If $\alpha > 1$ and $\beta = -1$, then $u = 0$ and $v = \pi/\alpha$, and on this interval

$$\frac{a'(\varphi)}{a(\varphi)} = -\frac{\alpha}{\alpha - 1} h(\alpha, \varphi) + h(\alpha - 1, \varphi) > 0$$

by (2.7.10). On $(0, \pi/\alpha)$ the function $w(\varphi) = -1 + \alpha \sin(2-\alpha)\varphi / \sin \alpha\varphi$ has a derivative whose sign coincides with that of the function

$$\alpha[-h(\alpha, \varphi) + h(2-\alpha, \varphi)] > 0.$$

Suppose now that $\alpha > 1$ and $\beta = 1$. In this case $u = -\pi$ and $v = -\pi/\alpha$, and

$$a(\varphi) = \left(-\frac{\sin \varphi}{\sin \alpha\varphi}\right)^{1/(\alpha-1)} \frac{\sin(1-\alpha)\varphi}{\sin \alpha\varphi},$$

$$w(\varphi) = -\frac{\alpha \sin(2-\alpha)\varphi}{\sin \alpha\varphi} + 1.$$

The function $a(\varphi)$ is even. Therefore, it suffices for us to prove that it is decreasing on $(\pi/\alpha, \pi)$. We have that $a'(\varphi)/a(\varphi) = -q(\varphi)/(\alpha - 1)$, where

$$\begin{aligned} q(\varphi) &= \alpha^2 \cot \alpha\varphi - \cot \varphi - (\alpha-1)^2 \cot(\alpha-1)\varphi \\ &= \frac{\alpha^2 \sin^2 \varphi + 2\alpha \cos(\alpha-1)\varphi \sin \varphi \sin(-\alpha\varphi) + \sin^2 \alpha\varphi}{\sin(\alpha-1)\varphi \sin \varphi \sin(-\alpha\varphi)}. \end{aligned}$$

The numerator of the fraction is positive on $(\pi/\alpha, \pi)$, and the denominator can be estimated from below by the quantity

$$\alpha^2 \sin^2 \varphi - 2\alpha \sin \varphi \sin(-\alpha\varphi) + \sin^2 \alpha\varphi = (\alpha \sin \varphi + \sin \alpha\varphi)^2.$$

This implies that $a(\varphi)$ is decreasing on $(\pi/\alpha, \pi)$. The function $w(\varphi)$ is also even. Since

$$w'(\varphi) = \alpha[-h(2-\alpha, \varphi) + h(\alpha, \varphi)] < 0$$

on $(\pi/\alpha, \pi)$, we get that $w'(\varphi) > 0$ on $(-\pi, -\pi/\alpha)$, which is what was required to prove.

If $\alpha = 1$ ($u = 0$, $v = \pi$), then the sign of β does not influence the values of the functions $a(\varphi)$ and $b(\varphi)$. We have that

$$\begin{aligned} a'(\varphi)/a(\varphi) &= 1/\varphi - 2\cot\varphi + \varphi/\sin^2\varphi \\ &= (\varphi\sin^2\varphi)^{-1}(\sin\varphi - \varphi\cos\varphi)^2 + \varphi > 0, \end{aligned}$$

$$w'(\varphi) = 2\cot\varphi - 2\varphi/\sin^2\varphi = (\sin 2\varphi - 2\varphi)/\sin^2\varphi < 0.$$

REMARK. By a direct substitution and a subsequent simple computation it is possible to verify that $\varepsilon(\alpha)w(u) > 0$ and $\varepsilon(\alpha)w(v) < 0$ in all cases considered (except for the trivial case $\alpha < 1$, $\beta = -1$ and the case $\alpha > 1$, $\beta = 1$). This means that the function $\varepsilon(\alpha)w(\varphi)$, and with it $\varepsilon(\alpha)b(\varphi)$, changes sign precisely once from plus to minus at some point $u < \sigma < v$.

Let us proceed now to the proof of the theorem itself. If $\alpha < 1$ and $\beta = -1$ or $\alpha > 1$ and $\beta = 1$, then $g(x, \alpha, \beta)$ cannot have a maximum on the semi-axis $x \geq 0$. In the first case this is obvious, and in the second it follows from Lemmas 2.7.3 and 2.7.4. We consider the cases $\alpha < 1$, $\beta = 1$ and $\alpha > 1$, $\beta = -1$. On the semi-axis $x > 0$ the density $g(x, \alpha, \beta)$ can have only an odd number of extremal points, and the number of minima is less by 1 than the number of maxima. This follows from the fact that $g(0, \alpha, 1) = 0$ in the first case, while $g'(0, \alpha, -1) > 0$ in the second (by Lemma 2.7.4).

Let $x_0 > 0$ be one of the extremal points of $g(x, \alpha, \beta)$, i.e., $g'(x_0, \alpha, \beta) = 0$. We show that $g''(x_0, \alpha, \beta)$ is necessarily negative at this point. By (2.2.35), for $x > 0$

$$\begin{aligned} g''(x, \alpha, \beta) &= \frac{2}{\pi}kk' \int_u^v b\exp(-k^\alpha a)\,d\varphi \\ &\quad - \frac{\alpha}{\pi}k^{\alpha+1}k' \int_u^v ab\exp(-k^\alpha a)\,d\varphi \\ &= 2\left(\frac{k'}{k}\right) g'(x, \alpha, \beta) + \frac{\alpha}{\pi}k^{\alpha+1}(-\varepsilon(\alpha)k') \\ &\quad \times \int_u^v a(\varepsilon(\alpha)b)\exp(-k^\alpha a)\,d\varphi. \end{aligned}$$

The first term vanishes at $x = x_0$, while the second term, in which $-\varepsilon(\alpha)k'(x) > 0$, can be estimated as follows for $\alpha \neq 1$ and $x > 0$. By Lemma 2.7.5, in the cases we are considering the function $a(\varphi) > 0$ is monotonically increasing on

(u, v), while $\varepsilon(\alpha)b(\varphi)$ is monotonically decreasing and changes sign at a point $u < \sigma < v$. Consequently,

$$\int_u^\sigma a(\varepsilon b)\exp(-k^\alpha a)\,d\varphi + \int_\sigma^v a(\varepsilon b)\exp(-k^\alpha a)\,d\varphi$$
$$< a(\sigma)\int_u^\sigma \varepsilon b\exp(-k^\alpha a)\,d\varphi + a(\sigma)\int_\sigma^v \varepsilon b\exp(-k^\alpha a)\,d\varphi$$
$$= a(\sigma)\int_u^v \varepsilon b\exp(-k^\alpha a)\,d\varphi.$$

Hence,

$$g''(x_0,\alpha,\beta) < \frac{\alpha}{\pi}k^{\alpha+1}(-\varepsilon(\alpha)k')\int_u^v \varepsilon(\alpha)b\exp(-k^\alpha a)\,d\varphi$$
$$= -\alpha k^{\alpha-1}k'g'(x_0,\alpha,\beta) = 0$$

at the point $x = x_0$, which is what was required.

Thus, $g(x, \alpha, \beta)$ cannot have local minima on the semi-axis $x > 0$, but has a single maximum.

Combining the established facts, we conclude that in the case $\alpha \neq 1$, $\beta = \pm 1$ the density $g(x, \alpha, \beta)$ has exactly one maximum at the point $x = m(\alpha, \beta) = -m(\alpha, -\beta)$; moreover, $m(\alpha, 1) > 0$ if $\alpha < 1$ and $m(\alpha, 1) < 0$ if $\alpha > 1$.

The case $\alpha = 1$ admits similar analysis. However, it is not necessary to do this. The fact of the matter is that the property of unimodality for $G(x, \alpha, \beta)$ is not violated under any linear transformation of the argument of G, hence not in passing from one parametrization form to another; for example, not in passing from G_B to G_M and back. Therefore, $G_M(x, \alpha, \beta_M, \gamma_M, \lambda_M) = G(x, \alpha, \beta_M, \gamma_B, \lambda_B)$ is a unimodal distribution for any sets of parameters. Let us fix β_M, γ_M, and λ_M and let α go to 1. Then, by Property 2.4,

$$G_M(x, \alpha, \beta_M, \gamma_M, \lambda_M) \to G_M(x, 1, \beta_M, \gamma_M, \lambda_M).$$

The distribution $G_M(x, 1, \beta_M, \gamma_M, \lambda_M)$, as a limit of unimodal distributions, is itself unimodal for any β_M, γ_M, and λ_M. An obvious consequence is the unimodality of the distributions $G(x, 1, \beta)$, $\beta = \pm 1$.

This conceptually interesting and purely analytic proof has not yet been extended to the whole family of stable laws. The proof of Yamazato's theorem, to which we now come, is based on completely different considerations which turned out to be so general that in the final analysis they led to the establishment of unimodality for all laws in the class L. The proof reveals that the mainspring of the mechanism for forming the property of unimodality in infinitely divisible laws is a property of distributions discovered as far back as 1956 by Ibragimov [33] and called by him strong unimodality.

A distribution function $F(x)$ is said to be *strongly unimodal* if its composition with any unimodal distribution is unimodal. It was established in [33] that F is strongly unimodal if and only if the function $-\log F'(x)$ is convex, and that the set of all strongly unimodal distributions is closed with respect to complete convergence. A normal distribution is an example of a strongly unimodal distribution.

THEOREM 2.7.6. *Each stable distribution is unimodal.*

We precede the proof of the theorem by several lemmas. Consider an infinitely divisible distribution $F_n(x)$ with characteristic function

$$\mathfrak{f}_n(t) = \exp\left\{\int_0^\infty (e^{itx} - 1)\nu_n(x)\frac{dx}{x}\right\}, \tag{2.7.11}$$

where $\nu_n(x)$ is a step function of the form

$$\nu_n(x) = \begin{cases} \xi_1 + \cdots + \xi_n & \text{if } 0 < x < cn^{-1}, \\ \xi_k + \cdots + \xi_n & \text{if } c(k-1)n^{-1} \le x < ckn^{-1}, \\ 0 & \text{if } c \le x, \end{cases}$$

which is determined by a set of positive numbers $\xi_1, \dots, \xi_n$ with $\xi = \xi_1 + \cdots + \xi_n \ge 4$ and a positive number c. As $t \to \infty$

$$\begin{aligned} |\mathfrak{f}_n(t)| &= \exp\left\{\sum_{k=1}^n \xi_k \int_0^t \left(\cos\frac{ck}{n}x - 1\right)\frac{dx}{x}\right\} \\ &\le \exp\left\{\sum_{k=1}^n \xi_k \int_1^t \cos\left(\frac{ck}{n}x\right)\frac{dx}{x} - \xi\log t\right\} \le \frac{\pi n}{2c}t^{-4}; \end{aligned}$$

hence the distribution F_n is absolutely continuous, and its density p_n has a continuous and uniformly bounded derivative on the whole axis. Each function F_n is easily seen to be the weak limit as $\varepsilon \to 0$ of the generalized Poisson distributions F_n^ε with characteristic functions

$$\mathfrak{f}_n^\varepsilon(t) = \exp\left\{\int_\varepsilon^\infty (e^{itx} - 1)\nu_n(x)\frac{dx}{x}\right\}, \qquad \varepsilon > 0.$$

And since each of the distributions $F_n^\varepsilon(x)$ is concentrated on the semi-axis $x \ge 0$, the same can be said of their limit distribution F_n.

LEMMA 2.7.6. *Any infinitely divisible distribution with finite expectation and characteristic function of the form*

$$\mathfrak{f}(t) = \exp\left\{\int_{x\neq 0} (e^{itx} - 1)\,dH(x)\right\},$$

where the spectral function H is unbounded and absolutely continuous, has a density p that is a solution of the integral equation

$$xp(x) = \int_{y \neq 0} p(x-y)yH'(y)\,dy. \tag{2.7.12}$$

PROOF. The fact that a distribution with the indicated properties of H has a density follows from results in [100]. As is clear from [47], the existence of the expectation of this distribution is equivalent to the convergence of the integral $\int |x|\,dH(x)$.

Let us consider the Fourier transform of the function $xp(x)$. We have that

$$\begin{aligned}\int e^{itx}xp(x)\,dx &= -t\mathfrak{f}'(t) = -i\mathfrak{f}(t) - \frac{d}{dt}\int_{x\neq 0}(e^{itx}-1)H'\,dx \\ &= \mathfrak{f}(t)\int_{x\neq 0} e^{itx}xH'(x)\,dx \\ &= \int e^{itx}\,dx \int_{y\neq 0} p(x-y)yH'(y)\,dy.\end{aligned}$$

Comparison of the Fourier transforms at the beginning and the end of the chain of equalities gives us (2.7.12).

REMARK. The distribution F_n obviously satisfies the conditions of the lemma, and thus the corresponding density p_n is a solution of (2.7.12) with the function $H'(x) = \nu_n(x)/x$. Differentiating this equality, we get after a simple transformation

$$xp_n'(x) = (\xi - 1)p_n(x) - \sum_{k=1}^{n} \xi_k p_n\left(x - \frac{ck}{n}\right). \tag{2.7.13}$$

LEMMA 2.7.7. *The distribution F_n is unimodal with mode at the point $m_n > c/n$.*

PROOF. Since $p_n(x) = 0$ for $x \leq 0$, (2.7.13) has the form $xp_n'(x) = (\xi - 1)p_n(x)$ on the interval $0 < x \leq c/n$, and this implies that

$$p_n(x) = c_1 x^{\xi - 1}, \qquad 0 < x \leq c/n. \tag{2.7.14}$$

It follows from (2.7.13) that $p_n(x) \to 0$ as $x \to \infty$. Together with (2.7.14), this means that $p_n(x)$ has at least one relative maximum on the semi-axis $x > c/n$. Denote by x_0 the infimum of the set A of all relative maxima on the semi-axis $x > c/n$. Two situations are possible.

Case 1. The point x_0 is an isolated point of A, or x_0 is itself a relative maximum. Assume that $p_n(x)$ is not nonincreasing on the semi-axis $x \geq x_0$. Then on this semi-axis there is at least one relative minimum of $p_n(x)$. Let x_1 stand for the infimum of all such relative minima. Obviously, $x_1 > x_0 > c/n$ and $p_n'(x_1) = p_n'(x_0) = 0$.

Consequently, we can make the following assertions:

1°. $p_n(x)$ is strictly increasing on $(0, x_0)$.

2°. $(\xi - 1)p_n(x_0) - \sum_{k=1}^{n} \xi_k p_n(x_0 - ck/n) = 0$.

3°. $p_n(x)$ is strictly decreasing on (x_0, x_1).

4°. $(\xi - 1)p_n(x_1) - \sum_{k=1}^{n} \xi_k p_n(x_1 - ck/n) = 0$.

If $x_1 - c \geq x_0$, then $x_1 - ck/n \geq x_0$ for all values $k \leq n$. From this and from properties 3° and 4° we get

$$0 = (\xi - 1)p_n(x_1) - \sum_{k=1}^{n} \xi_k p_n(x_1 - ck/n)$$
$$= \sum_{k=1}^{n} \xi_k (p_n(x_1) - p_n(x_1 - ck/n)) - p_n(x_1) < 0,$$

which is impossible.

Suppose now that $x_1 - c < x_0$. This means that there is a $j < n$ such that $x_1 - ck/n \geq x_0$ if $k \leq j$ and $x_1 - ck/n < x_0$ if $j < k \leq n$. In this case properties 3° and 4° lead to the inequality

$$\xi_{j+1} p_n(x_1 - c(f+1)/n) + \cdots + \xi_n p_n(x_1 - c)$$
$$< (\xi_{j+1} + \cdots + \xi_n - 1)p_n(x_1).$$

Since $x_0 - ck/n < x_1 - ck/n < x_0$ for $j < k \leq n$, it follows from property 1° that

$$p_n(x_0 - ck/n) < p_n(x_1 - ck/n), \qquad j < k \leq n.$$

Together with the inequality $p_n(x_0 - ck/n) \leq p_n(x_0)$, $1 \leq k \leq j$ (which also follows from 1°), the preceding two inequalities give us that

$$(\xi - 1)p_n(x_0) > \sum_{k=1}^{n} \xi_k p_n(x_0 - ck/n),$$

which contradicts 2°.

Case 2. Suppose that x_0 is a limit point of the set A but is not in A. In this case it is possible to choose a relative maximum point $x_0^* > x_0$ and a larger relative minimum point $x_1 > x_0^*$ such that $x_1 - c/n < x_0$. Since $p_n(x)$ is strictly increasing on $(0, x_0)$, it follows that $p_n(x_0^* - ck/n) < p_n(x_1 - ck/n)$ for $1 \leq k \leq n$. Moreover, $p_n'(x_0^*) = p_n'(x_1) = 0$ and $p_n(x_0^*) > p_n(x_1)$. We have

$$(\xi - 1)p_n(x_1) < (\xi - 1)p_n(x_0^*)$$
$$= \sum_{k=1}^{n} \xi_k p_n(x_0^* - ck/n) < \sum_{k=1}^{n} \xi_k p_n(x_1 - ck/n),$$

which contradicts property 4°. Consequently, the distribution F_n has a unique mode at the point $m_n = x_0 > c/n$.

LEMMA 2.7.8. *The density $p_n(x)$ is logarithmically concave on the interval $(0, m_n)$.*

PROOF. Recall that the condition $\xi > 4$ occurs in the definition of F_n. It follows from the continuity of $p_n'(x)$ on the whole axis and equation (2.7.13) that $p_n''(x)$ is continuous on the semi-axis $x > 0$. Differenitating both sides of (2.7.13), we get that

$$xp_n''(x) = (\xi - 2)p_n'(x) - \sum_{k=1}^{n} \xi_k p_n'(x - ck/n).$$

From this and (2.7.13) we obtain

$$\begin{aligned} xB(x) &= x[(p_n'(x))^2 - p_n(x)p_n''(x)] \\ &= p_n(x)p_n'(x) + \sum_{n=1}^{n} \xi_k [p_n(x)p_n'(x - ck/n) \\ &\qquad - p_n'(x)p_n(x - ck/n)]. \end{aligned} \tag{2.7.15}$$

It is not hard to see that $B(x)$ is continuous on the semi-axis $x > 0$ and positive in the interval $0 < x \le c/n$. We show that $B(x) > 0$ also for $c/n < x \le m_n$. Indeed, suppose not, and let $x_* = \min\{x\colon B(x) = 0, \, x > c/n\}$. Then $B(x) > 0$ for $0 < x < x_*$, $p_n''(x_*) \ge 0$, and $c/n < x_* \le m_n$. Consider the following two possible cases.

1*. For some k with $1 \le k \le n$,

$$\Delta = p_n(x_*)p_n'(x_* - ck/n) - p_n'(x_*)p_n(x_* - ck/n) < 0.$$

2*. $\Delta \ge 0$ for any k with $1 \le k \le n$.

In the first case $ck/n \le x_*$ and $p_n(x_* - ck/n) > 0$. Consequently,

$$\frac{p_n'(x_* - ck/n)}{p_n(x_* - ck/n)} < \frac{p_n'(x_*)}{p_n(x_*)}.$$

This means that there is a point x' between $x_* - ck/n$ and x_* such that $(p_n'/p_n)'(x') > 0$, which is equivalent to $B(x') < 0$. However, this contradicts the choice of x_*.

In the second case, as is clear from (2.7.15),

$$p_n(x_*)p'(x_* - ck/n) - p_n'(x_*)p_n(x_* - ck/n) = 0$$

for all k. Since $c/n < x_*$, it follows that $p_n(x_* - c/n) > 0$, and hence

$$\frac{p_n'(x_* - c/n)}{p_n(x_* - c/n)} = \frac{p_n'(x_*)}{p_n(x_*)}.$$

Hence there exists a point x' in the interval $(x_* - c/n, x_*)$ at which $(p_n'/p_n)(x') = 0$, i.e., $B(x') = 0$, which again contradicts the choice of x_*. It has been proved that $B(x) > 0$ in the interval $0 < x < m_n$, and this is clearly equivalent to the assertion of the lemma.

LEMMA 2.7.9. *Suppose that absolutely continuous distributions with densities $p(x)$ and $q(x)$ have the following properties:*

a) *$p'(x)$ and $q'(x)$ exist and are continuous on the whole axis.*

b) *$p(x) = 0$ for $x > 0$ and $q(x) = 0$ for $x < 0$.*

c) *$p(x)$ and $q(x)$ are unimodel with modes at the respective points $-a < 0$ and $b > 0$.*

d) *$p(x)$ and $q(x)$ are logarithmically concave on the respective intervals $(-a, 0)$ and $(0, b)$.*

Then the convolution of these distributions is unimodal.

PROOF. Two cases are possible: $a < b$ or $a \geq b$. The symmetric nature of the conditions tells us that we can confinc ourselves to an analysis of one of these cases, say the first. Consider the convolution density

$$s(x) = \int_x^{\infty} p(x-y)q(y)\,dy = \int_{-\infty}^{x} q(x-y)p(y)\,dy.$$

The density $s(x)$ also has a continuous and also bounded derivative

$$s'(x) = \int_x^{\infty} p(x-y)q'(y)\,dy = \int_{-\infty}^{x} q(x-y)p'(y)\,dy. \tag{2.7.16}$$

It is clear that the derivative is of constant sign on the semi-axes $(-\infty, a)$ and (b, ∞); therefore, it suffices to prove uniqueness for the maximum of $s(x)$ for $-a < x < b$. We can assert the following.

1°. If $s'(x) \leq 0$ for some $x \in (0, b-a)$, then $s'(y) \leq 0$ for all $y \in (x, a-b)$.

2°. If $s'(x) \geq 0$ for some $x \in (-a, b-a)$, then $s'(y) \geq 0$ for all $y \in (-a, x)$.

Indeed, suppose that $0 < \varepsilon < a$, and let

$$A_\varepsilon(x) = \begin{cases} q(x+\varepsilon)/q(x) & \text{if } q(x) > 0, \\ 0 & \text{if } q(x) = 0. \end{cases}$$

The function $A_\varepsilon(x)$ is continuous, because $q(x)$ is continuous. Since $q(x)$ is logarithmically concave on $(0, b)$, $A_\varepsilon(x)$ is nonincreasing on $(0, b-\varepsilon)$. Further, since $q(x)$ is nondecreasing on $(0, b)$ and nonincreasing on (b, ∞), $A_\varepsilon(x)$ is also nonincreasing on $(b-\varepsilon, a)$. Therefore, $A_\varepsilon(x) \leq 1$ for $x > b$, and $A_\varepsilon(x) \geq 1$ for $x < b - \varepsilon$.

Suppose that the point $x_0 \in (0, b-a)$ is such that $s'(x_0) \leq 0$. By (2.7.16),

$$\begin{aligned} s'(x_0+\varepsilon) = &\int_{-\infty}^{-a} p'(y)A_\varepsilon(x_0-y)q(x_0-y)\,dy \\ &+ \int_{-a}^{0} p'(y)A_\varepsilon(x_0-y)q(x_0-y)\,dy. \end{aligned} \tag{2.7.17}$$

Since A_ε is continuous and nonnegative, the function $p'(x)$ has constant sign on each of the intervals $(-\infty, a)$ and $(-a, 0)$, and the function $p'(y)q(x_0-y)$ is integrable, it follows from (2.7.17) and the integral mean value theorem that

$$q'(x_0+\varepsilon) = A_\varepsilon(x_0+\eta_1)\int_{-\infty}^{a} p'(y)q(x_0-y)\,dy + A_\varepsilon(x_0+\eta_2)\int_{-a}^{0} p'(y)q(x_0-y)\,dy, \tag{2.7.18}$$

where $x_0 \le x_0+\eta_2 \le x_0+a < b$ and $x_0+a \le x_0+\eta_1$. If we choose $\varepsilon > 0$ such that $b-a-x_0 > 0$, then $x_0+\eta_2 < b-\varepsilon$ and, consequently, $A_\varepsilon(x_0+\eta_2) \ge 1$. If $x_0+\eta_1 < a$, then, since $A_\varepsilon(x)$ is nonincreasing on $(0,b)$,

$$A_\varepsilon(x_0+\eta_2) \ge A_\varepsilon(x_0+\eta_1).$$

But if $x_0+\eta_1 \ge a$, then

$$A_\varepsilon(x_0+\eta_2) \ge 1 \ge A_\varepsilon(x_0+\eta_1).$$

It follows from (2.7.16) and (2.7.18) that $s'(x_0+\varepsilon) \le 0$ for $0 < \varepsilon < b-a-x_0$, i.e., assertion 1° is valid.

Further, let $x_0 \in (-a, b-a)$ be such that $s'(x_0) \ge 0$. By (2.7.16), for all $\varepsilon < x_0+a$

$$s'(x_0-\varepsilon) = [A_\varepsilon(x_0-\varepsilon+\eta_1)]^{-1}\int_{-\infty}^{-a} p'(y)q(x_0-y)\,dy + [A_\varepsilon(x_0-\varepsilon+\eta_2)]^{-1}\int_{-a}^{x_0-\varepsilon} p'(y)q(x_0-y)\,dy,$$

where $0 \le x_0-\eta_2-\varepsilon \le x_0+a-\varepsilon \le x_0+\eta_1-\varepsilon$ and $x_0+d-\varepsilon < b-\varepsilon$. By the same arguments as above, this gives us the inequality $s'(x_0-\varepsilon) \ge 0$ for $0 < \varepsilon < x_0+a$, and with it the confirmation of 2°.

Note that the condition $b-a>0$ was not used in the course of the proof. Hence, by the symmetry of the condition in 2°, we can get the following assertion from it as a corollary.

3°. If $s'(x) \le 0$ for some $x \in (b-a, b)$, then $s'(y) \le 0$ for any $y \in (x, b)$.

The condition that $s(x)$ is unimodal is now obtained as a simple consequence of properties 1°–3° (under the assumption that $b-a>0$).

We now proceed to the proof of Theorem 2.7.6 itself. It is clear that we can confine ourselves to standard distributions $G(x,\alpha,\beta)$ without loss of generality. According to (2.2.5), the characteristic functions of these distributions have the form

$$\log \mathfrak{g}(t,\alpha,\beta) = ita + \alpha C_1\int_0^{\infty}(e^{itx}-1-it\sin x)x^{-\alpha}\frac{dx}{x} + \alpha C_2\int_{-\infty}^{0}(e^{itx}-1-it\sin x)|x|^{-\alpha}\frac{dx}{x}$$

when $\alpha \neq 2$. The functions $x^{-\alpha}$ $(x > 0)$ and $|x|^{-\alpha}$ $(x < 0)$ can obviously be approximated by functions of type ν_n in such a way that the corresponding sequences of distributions $F_n^+(x)$ $(x > 0)$ and $F_n^-(x)$ $(x < 0)$ have the following property for any t when the constants a_n are suitably chosen:

$$ita_n + \int_0^\infty (e^{itx} - 1)\nu_n^+(x)\frac{dx}{x} + \int_{-\infty}^0 (e^{itx} - 1)\nu_n^-(x)\frac{dx}{n}$$

$$\to \log \mathfrak{g}(t, \alpha, \beta) \quad \text{as } n \to \infty. \tag{2.7.19}$$

If n is sufficiently large, then the distribution $F_n^+(x)$ and the distribution $1 - F_n^-(-x)$ fall under Lemmas 2.7.7 and 2.7.8, and, consequently, F_n^+ and F_n^- satisfy the conditions of Lemma 2.7.9. Therefore, the distribution $F_n^+ * F_n^-$ is unimodal for any sufficiently large n. This and (2.7.19) tell us that the limit distribution $G(x, \alpha, \beta)$ is also unimodal.

REMARK. The method of proof does not allow us to see where the mode of $G(x, \alpha, \beta)$ is. This deficiency of Theorem 2.7.6 (in comparison with Theorem 2.7.5) can be removed in the case $\alpha \neq 1$ by employing the same argument used in the proof of Theorem 2.7.5. Namely, for any admissible values $\alpha \neq 1$ and β,

$$g'(0, \alpha, \beta) = \tfrac{1}{2\pi}\Gamma(1 + 2/\alpha)\sin[\pi\beta K(\alpha)/\alpha].$$

Since the mode $m(\alpha, \beta)$ of the distribution $G(x, \alpha, \beta)$ is unique, this implies that in the case $\alpha \neq 1$

$$\operatorname{sgn} m(\alpha, \beta) = \operatorname{sgn}(1 - \alpha) \operatorname{sgn} \beta. \tag{2.7.20}$$

It is easy to derive a connection between the mode $m(\alpha, \beta, \gamma, \lambda)$ of $G(x, \alpha, \beta, \gamma, \lambda)$ and the mode $m(\alpha, \beta)$ of the corresponding standard distribution from the relation (2.1.2) by using the formulas (I.18) and (I.19) for passing from the form (A) to the form (B). Thus, in the case $\alpha \neq 1$

$$m(\alpha, \beta, \gamma, \lambda) = \left(\lambda \cos\left(\tfrac{\pi}{2}\beta K(\alpha)\right)\right)^{1/\alpha} m(\alpha, \beta) + \gamma\lambda.$$

This equality, together with (2.7.20), leads to the following conclusion about the location of the mode $m(\alpha, \beta, \gamma, \lambda)$ on the real axis in the case when $\alpha \neq 1$ and $\beta > 0$. Namely,

$$\begin{aligned} m(\alpha, \beta, \gamma, \lambda) &> \gamma\lambda \quad \text{if } \alpha < 1, \\ m(\gamma, \beta, \gamma, \lambda) &< \gamma\lambda \quad \text{if } \alpha > 1. \end{aligned} \tag{2.7.21}$$

In particular, the mode of the unbiased distribution (i.e., the distribution having parameter $\gamma = 0$) in the case $\alpha \neq 1$ is always on the same semi-axis as the mode of the corresponding standard law. This conclusion, as well as the inequality (2.7.21), remains valid also for the form (A).

TABLE OF VALUES OF THE FUNCTION $g'(0,1,\beta)$

β	$g'(0,1,\beta)$	β	$g'(0,1,\beta)$
0.05	− 0.004	0.55	− 0.022
0.10	− 0.007	0.60	− 0.022
0.15	− 0.011	0.65	− 0.021
0.20	− 0.014	0.70	− 0.021
0.25	− 0.016	0.75	− 0.020
0.30	− 0.018	0.80	− 0.019
0.35	− 0.020	0.85	− 0.018
0.40	− 0.021	0.90	− 0.017
0.45	− 0.021	0.95	− 0.016
0.50	− 0.022	1.00	− 0.015

REMARK. The errors in these values of $g'(0,1,\beta)$ do not exceed 0.0005 in absolute value. The computations were carried out on the basis of the following representation of this function:

$$g'(0,1,\beta) = \frac{1}{2\beta^2}\int_{-1}^{1} U(U-1)(\exp(-U)\,d\varphi,$$

where

$$U = \frac{\pi}{2}\frac{(1+\beta\varphi)}{\cos(\pi\varphi/2)}\exp\left\{\frac{1}{2}\pi(\varphi+1/\beta)\tan(\pi\varphi/2)\right\}.$$

Unfortunately, the case $\alpha = 1$, $\beta \neq 0$ cannot be cleared up so easily because of the lack of a sufficiently simple expression for $g'(a,1,\beta)$, which would enable us to establish the sign of the derivative at least for the one point a of the real axis. At the same time, computer computations of the values of $g'(0,1,\beta)$ for $0 < \beta \leq 1$ (see the table above) show that $g'(0,1,\beta) < 0$ and, consequently, $m(1,\beta) < 0$ for $\beta > 0$. From this, the fact that

$$m(1,\beta,\gamma,\lambda) = \tfrac{\pi}{2}\lambda m(1,\beta) + \gamma\lambda + \beta\lambda\log\left(\tfrac{\pi}{2}\lambda\right)$$

(see (2.1.2) and (I.19)), and the fact that, according to (2.1.3),

$$m(\alpha,-\beta,-\gamma,\lambda) = -m(\alpha,\beta,\gamma,\lambda)$$

for all admissible parameter values, we get the estimates

$$\begin{aligned} m(1,\beta,\gamma,\lambda) &< \gamma\lambda + \beta\lambda\log(\pi\lambda/2) \quad \text{if } \beta > 0,\\ m(1,\beta,\gamma,\lambda) &> \gamma\lambda + \beta\lambda\log(\pi\lambda/2) \quad \text{if } \beta < 0. \end{aligned}$$

A number of interesting observations about the modes $m = m(\alpha,\beta,\gamma,\lambda)$ of stable laws are contained in recent papers of Ken-iti Sato* dealing with the study of properties of modes.

---

*The author has at his disposal only preprints of these papers (published by Nagoya University) under the titles:

1) *Bounds of modes and unimodal processes with independent increments*;

2) *Behavior of modes of a class of processes with independent increments*.

(i) The mode m, as a function of λ, is either convex or concave on the semi-axis $\lambda > 0$. More precisely, the sign of the second derivative of m with respect to λ coincides with the sign of β, i.e.,

$$\operatorname{sgn} m''(\alpha, \beta, \gamma, \lambda) = \operatorname{sgn} \beta.$$

The form of the first derivative m' with respect to λ enables us to conclude that, as λ varies from 0 to ∞, the mode $m(\alpha, \beta, \gamma, \lambda)$ changes sign (moreover, only once) always in the case $\alpha = 1$, and if and only if $\gamma\beta(1-\alpha) < 0$ in the case $\alpha \neq 1$.

(ii) It is possible to determine precise bounds for the variation of the mode $m(1, \beta)$, with account taken of the fact that $m(1, -\beta) = -m(1, \beta)$. Namely,

$$|m(1, \beta)| \leq A_p B_p^{1/p} (2p^{-1} + (1-p)^{-1} + 4\pi^{-2}\beta^2(2-p)^{-3})^{1/p},$$

where $0 < p < 1$,

$$B_p = \tfrac{2}{\pi}\Gamma(1+p) \sin \tfrac{\pi}{2} p,$$

and the quantity A_p, which is nonincreasing with p (it does not have a sufficiently simple explicit expression) can be estimated above by the quantity

$$A_p^* = \inf\{A_{q,p} \colon 0 < q < \min(1, p)\},$$

$$A_{q,p} = \left[\frac{q + 1 + (p+1)^{q/p}}{q + 1 - (p+1)^{q/p}}\right]^{1/q}.$$

Some additional information about localization of the mode $m(\alpha, \beta)$ can be extracted from results in [71] (see Theorems 4.1 and 6.1). Let $v_\alpha = |1 - \Gamma(1+\alpha)| \sin \pi\alpha/2$.

If $0 < \alpha < 1$, then

$$m(\alpha, -1) \leq m(\alpha, \beta) \leq m(\alpha, 1),$$

with the inequalities strict for $|\beta| \neq 1$, and

$$\left(\tfrac{1}{\alpha}\Gamma(1-\alpha)\right)^{-1/\alpha} < m(\alpha, 1) + v_\alpha < \left(\tfrac{1}{\alpha}\Gamma(2-\alpha)\right)^{-1/\alpha}.$$

If $1 < \alpha < 2$, then

$$m(\alpha, 1) \leq m(\alpha, \beta) \leq m(\alpha, -1),$$

with the inequalities strict for $|\beta| \neq 1$, and

$$L((\alpha/\Gamma(-\alpha))^{1/\alpha}) < m(\alpha, 1) + v_\alpha < \Gamma(1+\alpha) \sin \tfrac{\pi}{2}\alpha,$$

where

$$L(x) = -x + \frac{1}{\Gamma(-\alpha)} \left[\int_0^x \frac{u^{2-\alpha}\, du}{1+u^2} - \int_x^\infty \frac{u^{-\alpha}\, du}{1+u^2} - x^{1-\alpha}\right].$$

We remark further that the cases $\alpha \neq 2$, $\beta = 0$ and $\alpha = 2$ corresponding to stable distributions symmetric with respect to the line $x = \gamma\lambda$ are the only ones so far in which the exact value of the mode is known: $m(\alpha, 0, \gamma, \lambda) = \gamma\lambda$.

Summarizing the above facts, we can form the following general idea about the behavior of the densities of stable laws.

1. The graphs of the densities

$$g(x, \alpha, \beta, \gamma, \lambda) \quad \text{and} \quad g(x, \alpha, -\beta, \gamma, \lambda)$$

are symmetric with respect to the line $x = \gamma\lambda$, while the graphs of the densities

$$g(x, \alpha, \beta, \gamma, \lambda) \quad \text{and} \quad g(x, \alpha, -\beta, \gamma, \lambda)$$

are symmetric with respect to the ordinate axis. In particular, if $\beta = 0$, then the graph of $g(x, \alpha, 0, \gamma, \lambda)$ is symmetric with respect to the line $x = \gamma\lambda$. Considering that the behavior of the density in the case $\alpha = 2$ is clear without any special investigation, these properties allow us to confine ourselves to the case $0 < \alpha < 2$, $\beta \geq 0$, $\gamma \geq 0$ when analyzing the behavior of the graphs of the densities.

2. If $0 < \alpha < 1$ and $\beta = 1$, then $g(x, \alpha, 1, \gamma, \lambda)$ is nonzero only at points of the semi-axis $x > \gamma\lambda$. In all other cases $g(x, \alpha, \alpha, \gamma, \lambda)$ is nonzero at each point of the real axis.

3. The graphs of the densities $g(x, \alpha, \beta, \gamma, \lambda)$ are unimodal and have a unique mode, which is located to the right of the point $\gamma\lambda$ if $0 < \alpha < 1$ and $\beta > 0$, and to the left of $\gamma\lambda$ if $1 < \alpha < 2$.

4. If $-1 < \beta \leq 1$, then $g(x, \alpha, \beta, \gamma, \lambda)$ decreases like $\text{const}\, x^{-1-\alpha}$ as $x \to \infty$. But if $\beta = -1$, then $g(x, \alpha, \beta, \gamma, \lambda)$ decreases exponentially in the case $\alpha < 1$ as $x \to \gamma\lambda - 0$ and in the case $\alpha \geq 1$ as $x \to \infty$. More precisely,

$$\log g(x, \alpha, \beta, \gamma, \lambda) \sim -\text{const}\, \xi(x, \alpha)$$

(the function ξ was defined in §2.5).

We supplement the facts proved above by some more properties (proved in [40]) of densities of stable laws.

5. If $\alpha < 1$, then for any positive $\lambda_1 \neq \lambda_2$ the graphs of the densities $g(x, \alpha, \beta, \gamma, \lambda_1)$ and $g(x, \alpha, 1, 0, \lambda_2)$ intersect only once on the semi-axis $x > 0$ where they are concentrated.

6. For any $0 < \alpha \leq 2$ and any positive $\lambda_1 \neq \lambda_2$ the graphs of the densities $g(x, \alpha, 0, 0, \lambda_1)$ and $g(x, \alpha, 0, 0, \lambda_2)$ intersect at only two points (symmetrically located with respect to the point $x = 0$, of course).

The graphs of densities of stable laws given in Figures 2–4 illustrate the properties listed above.*

*The graphs in Figures 2–4 are taken from graphs of densities of stable laws contained in the survey article [31] by Holt and Crow. These figures are reproduced here with the permission of the authors of [31].

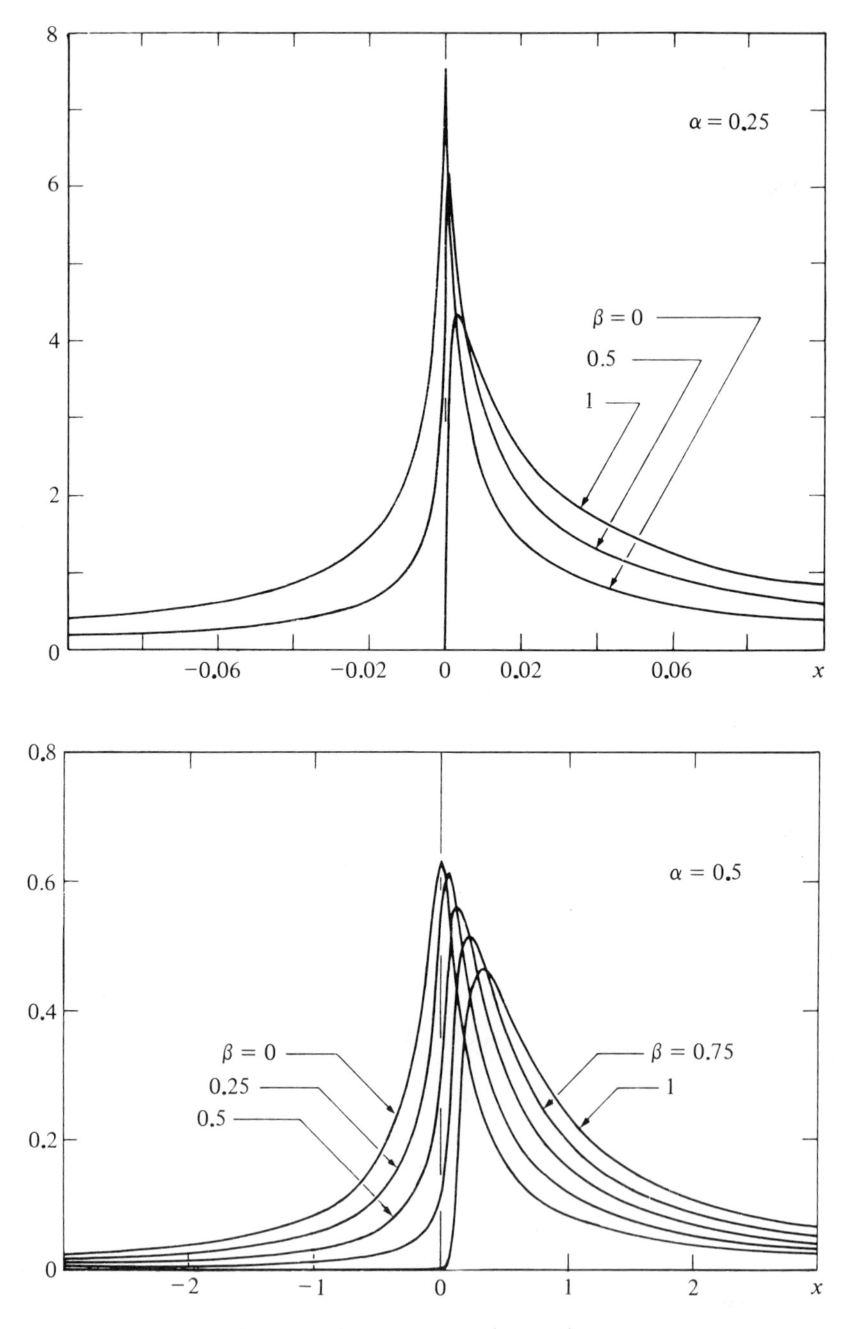

FIGURE 2. Graphs of densities $g_A(x, \alpha, \beta)$ corresponding to the values $\alpha = 0.25$ and $\alpha = 0.5$.

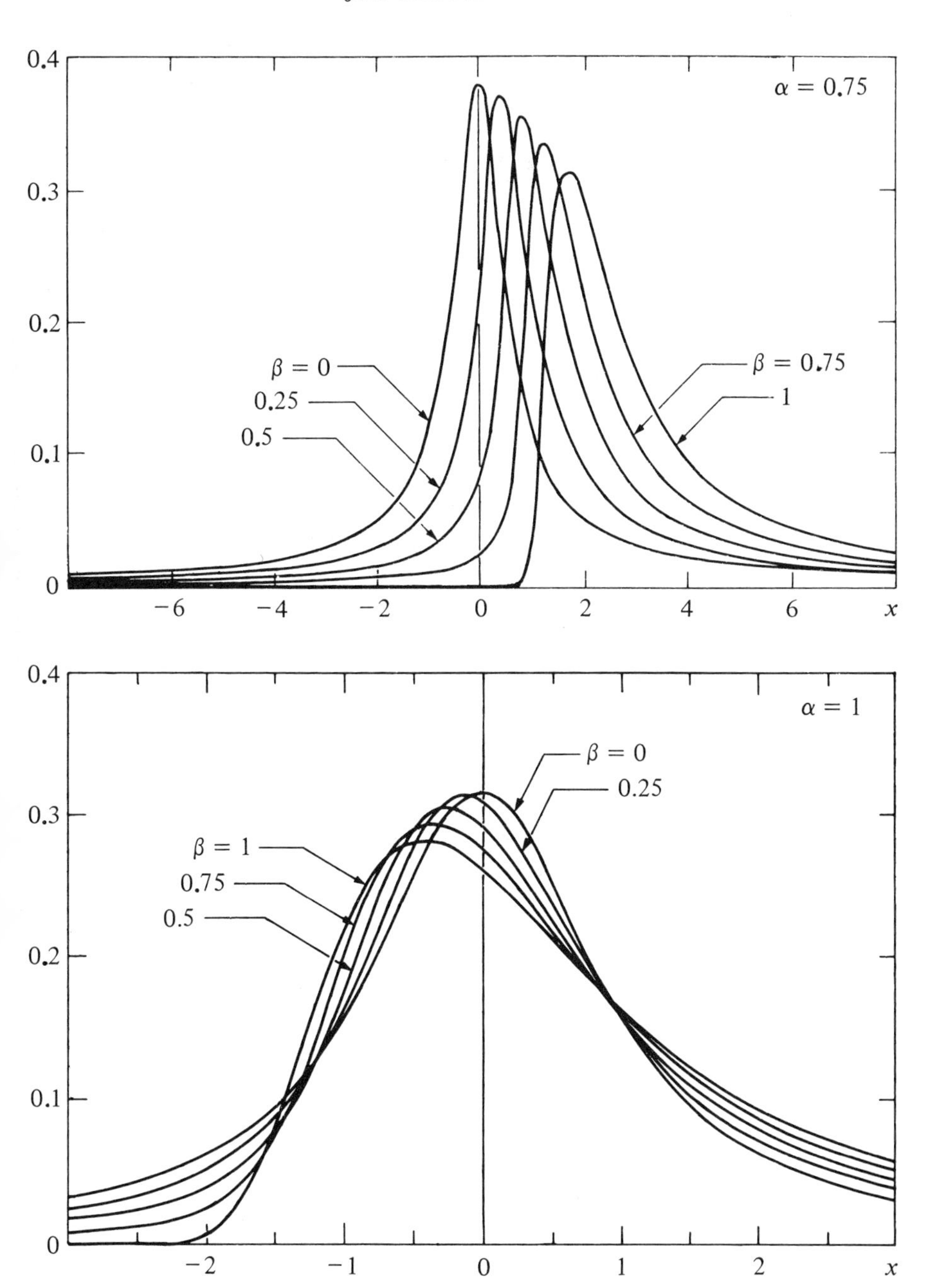

FIGURE 3. Graphs of densities $g_A(x, \alpha, \beta)$ corresponding to the values $\alpha = 0.75$ and $\alpha = 1$.

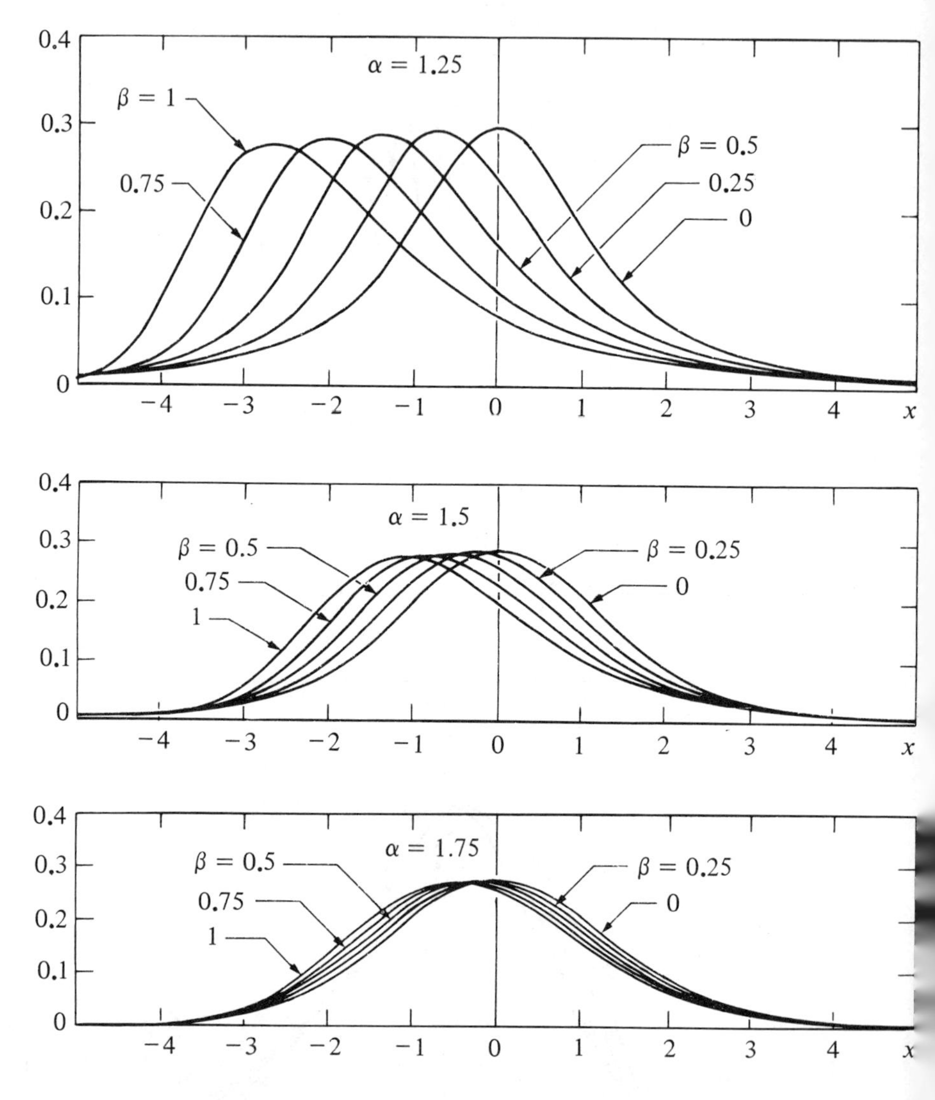

FIGURE 4. Graphs of densities $g_A(x, \alpha, \beta)$ corresponding to the values $\alpha = 1.25$, $\alpha = 1.5$, and $\alpha = 1.75$.

The paper [40] cited contains another observation of interest to us, based on the following fact (which is proved there).

7. *Suppose that X is a positive random variable with absolutely continuous distribution. Then the following assertions are equivalent:*

a) *The distribution of the random variable* $\log X$ *is unimodal.*

b) *There exists at least one version $p(x)$ of the density of X (one version differs from another at most on a set of Lebesgue measure zero) such that for any $c > 0$ the graphs of the densities $cp(cx)$ and $p(x)$ intersect at precisely one point on the semi-axis $x > 0$.*

This criterion and property 5 obviously imply that in the case $0 < \alpha < 1$ the density

$$\exp(x)g(\exp(x), \alpha, 1), \qquad -\infty < x < \infty,$$

of the random variable $\log Y(\alpha, 1, 0, 1)$ is unimodal.

The recent article [247] studies a certain property of the density of a stable law which refines the concept of unimodality, and the following fact is discovered.

8. The density $p(x)$ of a distribution will be said to be *bell-shaped* if $p(x)$ has infinitely many derivatives at all points of the real axis, and for any integer k the derivative $p^{(k)}(x)$ has exactly k zeros on the support of the distribution.

Then the densities of all stable laws are bell-shaped in this sense. What is more, the number of positive zeros and the number of negative zeros of $g^{(k)}(x, \alpha, \beta)$ can be computed in the case $\alpha \neq 1$.

§2.8. Stable distributions as solutions of integral, integrodifferential, and differential equations

In the Introduction we gave a canonical form of expression for the characteristic function $g(t, \alpha, \beta)$ (valid both for the form (A) and for the form (B)), and we shall transform it somewhat:

$$\mathfrak{g}(t) = \mathfrak{g}(t, \alpha, \beta) = \exp\left\{\int_{x\neq 0} (e^{itx} - 1 - it\sin x)\, dH(x)\right\}$$
$$= \exp\left\{-\int_{x\neq 0} (e^{itx} - 1 - it\sin x)R(x)\frac{dx}{x}\right\}, \tag{2.8.1}$$

where $R(x) = -xH'(x)$, $x \neq 0$, is a function that is nondecreasing on the semi-axes $x < 0$ and $x > 0$. It has the simplest form if the parametrization is taken in the form (A):

$$R(x) = \begin{cases} -\frac{1}{\pi}\Gamma(1+\alpha)\sin(\pi\alpha/2)(1+\beta)x^{-\alpha}, & x > 0, \\ \frac{1}{\pi}\Gamma(1+\alpha)\sin(\pi\alpha/2)(1-\beta)|x|^{-\alpha}, & x < 0. \end{cases} \tag{2.8.2}$$

The function $R(x)$ corresponding to the form (B) is obtained by replacing the parameter $\beta = \beta_A$ in (2.8.2) by its expression in terms of α and β_B. Below, let

$$\mathrm{Si}(x) = \int_0^x \frac{\sin t}{t}\, dt$$

be the integral sine function.

THEOREM 2.8.1. *For either of the two forms of parametrization (A) or (B) the distribution function $G(x,\alpha,\beta)$ and the density $g(x,\alpha,\beta)$ satisfy the equations*

$$xG'(x,\alpha,\beta) = -\int_{y\neq 0} (G'(x,\alpha,\beta)\operatorname{Si}(y) - G(x-y,\alpha,\beta) + G(x,\alpha,\beta))\,dR(y), \qquad (2.8.3)$$

$$xg(x,\alpha,\beta) = \int_{y\neq 0} \left(g(x,\alpha,\beta)\frac{\sin y}{y} - g(x-y,\alpha,\beta) \right) R(y)\,dy. \qquad (2.8.4)$$

PROOF. We transform (2.8.1), integrating the integral in the exponential by parts and considering that

$$\left| \int_0^u (e^{itv} - 1 - it\sin v)\frac{dv}{v} \right| = O(u^2) \quad \text{as } u \to 0,$$

$$\left| \int_{1\leq|v|\leq u} (e^{itv} - 1 - it\sin v)\frac{dv}{v} \right| = O(\log u) \quad \text{as } u \to \infty.$$

As a result,

$$\mathfrak{g}(t) = \exp\left\{ \int_{y\neq 0} \left[\int_0^{ty} (e^{iv}-1)\frac{dv}{v} - it\operatorname{Si}(y) \right] dR(y) \right\}. \qquad (2.8.5)$$

Differentiating both sides of (2.8.5) with respect to $t \neq 0$, we find that

$$\mathfrak{g}'(t) = \frac{1}{t}\mathfrak{g}(t)\int (e^{ity} - 1 - it\operatorname{Si}(y))\,dR(y).$$

From this, after the change of variables $t = s/x$, $x \neq 0$, we get

$$\frac{d}{dx}\mathfrak{g}\left(\frac{s}{x}\right) = -\frac{1}{x}\mathfrak{g}\left(\frac{s}{x}\right)\int \left(e^{isy/x} - 1 - \frac{is}{x}\operatorname{Si}(y) \right) dR(y). \qquad (2.8.6)$$

We transform the inversion formula

$$g(x,\alpha,\beta) = \frac{1}{2\pi}\int e^{-itx}\mathfrak{g}(t)\,dt, \qquad x \neq 0,$$

by carrying out the change of variable $tx = s$:

$$xg(x,\alpha,\beta) = \frac{\operatorname{sgn} x}{2\pi}\int e^{-is}\mathfrak{g}\left(\frac{s}{x}\right) ds.$$

Differentiating both sides of this equality with respect to x and replacing $(d/dx)\mathfrak{g}(s/x)$ by its expression (2.8.6), we get that

$$(xg(x,\alpha,\beta))' = -\frac{\operatorname{sgn} x}{2\pi}\int e^{-is}\mathfrak{g}\left(\frac{s}{x}\right)\frac{ds}{x}\int \left(e^{isy/x} - 1 - \frac{is}{x}\operatorname{Si}(y) \right) dR(y).$$

A change in the order of integration and the inverse change of variable $s = tx$ leads to the equality

$$(xg(x,\alpha,\beta))' = -\frac{1}{2\pi}\int dR(y)\int (e^{ity} - 1 - it\,\mathrm{Si}(y))e^{-itx}\mathfrak{g}(t)\,dt.$$

The inside integral is a linear combination of integrals, each of which is none other than the inversion formula for the density or its derivative, i.e.,

$$(xg(x,\alpha,\beta))' = -\int (g(x-y,\alpha,\beta) - g(x,\alpha,\beta) + g'(x,\alpha,\beta)\,\mathrm{Si}(y))\,dR(y). \tag{2.8.7}$$

The equality (2.8.3) is obtained by integration of both sides of the equality with respect to x from $-\infty$ to x and use of the fact that $xg(x,\alpha,\beta) \to 0$ as $x \to -\infty$. Integration by parts in (2.8.3) leads us to (2.8.4).

REMARK 1. Although Theorem 2.8.1 is connected with stable distributions normalized by the conditions $\gamma = 0$ and $\lambda = 1$, the assertion of the theorem extends easily to the general case. Indeed, since

$$G(x,\alpha,\beta,\gamma,\lambda) = G((x-l)\lambda^{-1/\alpha},\alpha,\beta,0,1)$$

by (2.1.2), where the quantity l is connected with β, γ, and λ by simple formulas, we get the equations for $G(x,\alpha,\beta,\gamma,\lambda)$ and $g(x,\alpha,\beta,\gamma,\lambda)$ from (2.8.3) and (2.8.4) by replacing x by $(x-l)\lambda^{-1/\alpha}$.

REMARK 2. Note that in the course of the proof we obtained an integrodifferential equation (2.8.7) for the densities of stable laws.

Integral and integrodifferential equations for stable distributions can differ (sometimes very strongly) in form, though they are all equivalent in the final analysis. The integrodifferential equation (for densities with $\alpha \neq 1$) obtained in [60] is apparently the closest to equation (2.8.7), which has taken from [40] along with (2.8.3) and (2.8.4).

Below we shall meet a certain integrodifferential equation differing strongly from those given in Theorem 2.8.1. A particular merit of it is that differential equations for densities in the case of rational $\alpha \neq 1$ can be derived from it in a fairly simple way. To formulate this equation we must (following [59]) introduce the concept of fractional integration and differentiation.

The fractional integral of order $r > 0$ of a function $h(t)$ on the semi-axis $t > 0$ is defined to be the function

$$I^r h(x) = \frac{e^{-i\pi r}}{\Gamma(r)}\int_x^\infty (t-x)^{r-1}h(t)\,dt$$

(of course, $h(t)$ must be such that the integral on the right-hand side exists). If r is an integer, then the operator I^r coincides to within a sign with the operator of r-fold integration, i.e.,

$$I^r h(x) = -\int_x^\infty I^{r-1} h(t)\,dt.$$

Here I^0 is understood to be the identity operator ($I^0 h(x) = h(x)$).

The operator I^{-r} of fractional differentiation of order $r > 0$ is understood as the inverse of the operator I^r, i.e.,

$$I^r(I^{-r}h(x)) = h(x).$$

If r is an integer, then $I^{-r} = (d/dx)^{-r}$. In the case when $r > 0$ is not an integer the operator I^{-r}, like I^r, can be given an integral form. Namely, if $0 < r < 1$, then

$$I^{-r}h(x) = \frac{e^{i\pi r}}{\Gamma(-r)}\int_x^\infty [h(t) - h(x)](t-x)^{-r-1}\,dt.$$

But if $r \geq 1$, then

$$I^{-r}h(x) = I^{-n}(I^{-(r-n)}h(x)),$$

where n is the integer part of r. The set $\{I^r,\ -\infty < r < \infty\}$ of operators thus defined forms an abelian group ([2.24]).

The operators I^r can be used to write a number of equations for the densities of stable distributions in the class $\mathfrak{W}$. In the functions $\psi(x) = xg(x, \alpha, 0)$, $x > 0$, considered below, the densities $g(x, \alpha, \theta)$ are parametrized in the form (C) (but we omit the index C).

We mention several special properties of the operators I^r that are needed in what follows.

LEMMA 2.8.1. *Let z be a complex number. If $r \geq 0$ and $\operatorname{Re} z > 0$, or if $r < 0$ and $\operatorname{Re} z \geq 0$, then*

$$I^r \exp(-zx) = \exp(-i\pi r - zx)z^{-r}. \tag{2.8.8}$$

In particular, for $r > 0$ and $z = -it$, $t > 0$,

$$I^{-r}\exp(itx) = \exp(i\pi r/2 + itx)t^r. \tag{2.8.9}$$

Let s be a real number, and n the smallest integer such that $n + s > 0$. Then for any $r < s$

$$I^r x^{-s} = \exp(-i\pi r)\frac{s(s+1)\cdots(s+n-1)\Gamma(s-r)}{\Gamma(s+n)}x^{r-s}. \tag{2.8.10}$$

In particular, if $s > 0$, then for any $r < s$

$$I^r x^{-s} = \exp(i\pi r)\frac{\Gamma(s-r)}{\Gamma(s)}x^{r-s}. \tag{2.8.11}$$

PROOF. The case $r = 0$ clearly does not need to be analyzed. If $r > 0$ and $\operatorname{Re} z > 0$, then

$$I^r \exp(-zx) = \frac{\exp(-i\pi r)}{\Gamma(r)} \int_x^\infty e^{-zt}(t-x)^{r-1}\,dt$$
$$= \frac{\exp(-i\pi r)}{\Gamma(r)} e^{-zx} \int_0^\infty e^{-zt} t^{r-1}\,dt.$$

It is not hard to see that (2.8.8) follows from this.

If $-1 < r < 0$ and $\operatorname{Re} z \geq 0$, integration by parts given us that

$$I^r \exp(-zx) = \frac{\exp(-i\pi r)}{\Gamma(r)} \int_x^\infty (e^{-zt} - e^{-zx})(t-x)^{r-1}\,dt$$
$$= \frac{\exp(-i\pi r)}{\Gamma(1+r)} z \int_x^\infty e^{-zt}(t-x)^r\,dt.$$

The integral is then transformed as in the preceding case.

If r is a negative integer, then I^r is the operator of r-fold differentiation, and verification of (2.8.8) does not involve difficulties.

But if r is negative and not an integer, then, choosing a positive integer n such that $-1 < r + n < 0$, we can factor the operator, $I^r = I^{r+n} I^{-n}$, with the result that

$$I^r \exp(-zx) = I^{r+n} \frac{d^n}{dx^n} \exp(-zx)$$
$$= \exp(i\pi n) z^n I^{r+n} \exp(-zx),$$

i.e., we arrive at a case already treated.

The verification of (2.8.10) is by analogous arguments.

THEOREM 2.8.2. *Let $x > 0$, and let $\alpha \neq 1$ and θ be some pair of admissible parameter values. Then for any $r > -1/\alpha$ the function $\psi(x) = xg(x, \alpha, \theta)$ is the real part of a function $\chi(\xi)$, $\xi = x^{-\alpha}$, satisfying the equation* ([2.25])

$$xI^{-\alpha r}(x^{-1}\chi(x^{-\alpha}))$$
$$= \exp(-i\pi r + i\pi\alpha(1-\theta)r/2)\xi^r I^{-r}\chi(\xi). \tag{2.8.12}$$

PROOF. Consider the function

$$\chi(\xi) = \frac{1}{\pi} \int_0^\infty \exp(it - \xi t^\alpha \exp(i\pi\alpha\theta/2))\,dt. \tag{2.8.13}$$

A simple transformation of the integral in (2.2.1a) shows that $\psi(x) = \operatorname{Re}\chi(\xi)$. We show that $\chi(\xi)$ satisfies (2.8.12). Using (2.8.8), we find that

$$I^{-r}\chi(\xi) = \frac{1}{\pi} \int_0^\infty \exp(it) I^{-r} \exp\left(-\xi t^\alpha \exp(i\pi\alpha\theta/2)\right)\,dt$$
$$= \exp(i\pi r + i\pi\alpha\theta r/2)$$
$$\times \frac{1}{\pi} \int_0^\infty t^{\alpha r} \exp(it - \xi t^\alpha \exp(i\pi\alpha\theta/2))\,dt. \tag{2.8.14}$$

Further, since

$$\mu(x) = x^{-1}\chi(x^{-\alpha}) = \frac{1}{\pi}\int_0^\infty \exp(itx - t^\alpha \exp(i\pi\alpha\theta/2))\,dt, \tag{2.8.15}$$

it follows from (2.8.9) that

$$\begin{aligned} xI^{-\alpha r}\mu(x) &= \frac{x}{\pi}\int_0^\infty \exp(-t^\alpha \exp(i\pi\alpha\theta/2))I^{-\alpha r}\exp(itx)\,dt \\ &= \exp(i\pi\alpha r/2)\frac{1}{\pi}\int_0^\infty t^{\alpha r}\exp(itx - t^\alpha\exp(i\pi\alpha\theta/2))\,dtx \\ &= \xi^r \exp(i\pi\alpha r/2)\frac{1}{\pi}\int_0^\infty t^{\alpha r}\exp(it - \xi t^\alpha \exp(i\pi\alpha\theta/2))\,dt. \end{aligned} \tag{2.8.16}$$

Comparison of (2.8.14) and (2.8.16) confirms (2.8.12). The condition $\alpha r > 1$ ensures the existence of the integrals in these equalities.

A slight change in the arguments enables us to get another equation of type (2.8.12).

Let us integrate by parts in the expression (2.8.13) for the function $\chi(\xi)$:

$$\begin{aligned} \chi(\xi) = \frac{i}{\pi} + \frac{\alpha\xi}{\pi}\exp\left(i\frac{\pi}{2}(\alpha\theta - 1)\right) \\ \times \int_0^\infty t^{\alpha-1}\exp\left(it - \xi t^\alpha \exp\left(i\frac{\pi}{\alpha}\alpha\theta\right)\right)\,dt. \end{aligned} \tag{2.8.17}$$

Let $\tau(\xi) = \xi^{-1}\chi(\xi)$. The functions τ, χ, and ψ are connected by the following relations:

$$\begin{aligned} \tau(\xi) = x^\alpha\chi(x^{-\alpha}) = x^{1+\alpha}\mu(x), \\ \mu(x) = \xi^{1/\alpha}\chi(\xi) = \xi^{1+1/\alpha}\tau(\xi), \end{aligned} \tag{2.8.18}$$

and the differential operators with respect to the variables x and ξ are connected by the equalities

$$\frac{d}{d\xi} = -\frac{1}{\alpha}x^{1+\alpha}\frac{d}{dx}, \qquad \frac{d}{dx} = -\alpha\xi^{1+1/\alpha}\frac{d}{d\xi}. \tag{2.8.19}$$

THEOREM 2.8.3. *For any $r > -1$ and any pair of admissible values of the parameters $\alpha \neq 1$ and θ the function $x^{1+\alpha}g(x,\alpha,\theta)$ is the real part of a function $\tau(\xi)$ satisfying the equation*

$$\begin{aligned} &\xi^{r+1}I^{-r}\tau(\xi) + \tfrac{i}{\pi}\Gamma(1_r)\exp(i\pi(1+r)) \\ &\quad = \alpha\exp(i\pi r - i\tfrac{\pi}{2}\alpha(r+1)(1-\theta))I^{1-\alpha(r+1)}(x^{-(1+\alpha)}\tau(x^{-\alpha})). \end{aligned} \tag{2.8.20}$$

PROOF. According to (2.8.17),

$$\tau(\xi) - \xi^{-1}/i\pi = \frac{\alpha}{i\pi}\exp\left(i\frac{\pi}{2}\alpha\theta\right)\int_0^\infty t^{\alpha-1}\exp\left(it - \xi t^\alpha\exp\left(i\frac{\pi}{2}\alpha\theta\right)\right)\,dt.$$

From this, using properties (2.8.8) and (2.8.11) of the operators I^r, we find that

$$\begin{aligned}
&I^{-r}(\tau(\xi) - \xi^{-1}/i\pi) \\
&\quad = I^{-r}\tau(\xi) + \frac{1}{i\pi}\exp(i\pi r)\Gamma(1+r)\xi^{-(1+r)} \\
&\quad = \frac{\alpha}{i\pi}\exp\left(i\pi r + i\frac{\pi}{2}\alpha\theta(1+r)\right) \\
&\qquad \times \int_0^\infty t^{\alpha(r+1)-1}\exp\left(it - \xi t^\alpha \exp\left(i\frac{\pi}{2}\alpha\theta\right)\right)dt. \qquad (2.8.21)
\end{aligned}$$

On the other hand, according to (2.8.15) and (2.8.9),

$$\begin{aligned}
I^{1-\alpha(r+1)}\mu(x) &= I^{1-\alpha(r+1)}(x^{-(1+\alpha)}\tau(x^{-\alpha})) \\
&= \exp\left(i\frac{\pi}{2}(\alpha(r+1)-1)\right) \\
&\quad \times \frac{1}{\pi}\int_0^\infty t^{\alpha(r+1)-1}\exp\left(itx - t^\alpha\exp(i\pi\alpha\theta/2)\right)dt \\
&= \xi^{r+1}\exp\left(i\frac{\pi}{2}(\alpha(r+1)-1)\right) \\
&\quad \times \frac{1}{\pi}\int_0^\infty t^{\alpha(r+1)-1}\exp(it - \xi t^\alpha\exp(i\pi\alpha\theta/2))\,dt. \qquad (2.8.22)
\end{aligned}$$

Comparison of (2.8.21) and (2.8.22) gives us (2.8.20).

COROLLARY. *Let $r = 1/\alpha - 1$. In this case* (2.8.20) *simplifies*:

$$\begin{aligned}
&I^{1-1/\alpha}\tau(\xi) + \tfrac{i}{\pi}\Gamma(1/\alpha)\exp(i\pi/\alpha)\xi^{-1/\alpha} \\
&\quad = \alpha\exp(i\pi(1-\alpha)/\alpha - i\pi(1-\theta)/2)\xi\tau(\xi). \qquad (2.8.23)
\end{aligned}$$

We single out another corollary as an independent theorem, since it is connected with the case when the integrodifferential equation (2.8.20) becomes an ordinary differential equation ([2.26]).

THEOREM 2.8.4. *If $\alpha = p/q$ is a rational number different from 1, then for any pair α, θ of admissible parameter values the density $g(x, \alpha, \theta)$ is the real part of a function $\mu(x)$ satisfying the equation*

$$\begin{aligned}
&q^q\left(x^{1+p/q}\frac{d}{dx}\right)^{q-1}(x^{1+p/q}\mu(x)) \\
&\quad = p^q\exp\left(-i\frac{\pi}{2}p(1-\theta)\right)x^p\left(\frac{d}{dx}\right)^{p-1}\mu(x) + i\frac{\Gamma(1+q)}{\pi}p^{q-1}x^p. \qquad (2.8.24)
\end{aligned}$$

The proof of (2.8.24) reduces to a transformation of (2.8.20) in the case when $\alpha = p/q$ and $r = q-1$. The passage from the variable ξ to x is by means of (2.8.18) and (2.8.19). With their help, equation (2.8.24) can be given another form if we write it for the function $\tau(\xi)$ and pass from the variable x to ξ:

$$q^p \xi^q \left(\frac{d}{d\xi}\right)^{q-1} \tau(\xi) + i\frac{(-1)^q}{\pi}\Gamma(q)q^p$$
$$= (-1)^{p+q}p^p \exp\left(-i\frac{\pi}{2}p(1-\theta)\right)\left(\xi^{1+q/p}\frac{d}{dx}\right)^{p-1}(\xi^{1+q/p}\tau(\xi)). \quad (2.8.25)$$

It is not hard to see that the variant (2.8.24) looks simpler than (2.8.25) if $q < p$, and more complicated if $q > p$. This is very clear, for example, in the case $p = 1$, when (2.8.25) is transformed into the equation (2.27)

$$q\left(\frac{d}{d\xi}\right)^{q-1}\tau(\xi) + (-1)^q \exp\left(-i\frac{\pi}{2}(1-\theta)\right)\xi\tau(\xi)$$
$$+ i(-1)^q\frac{\Gamma(q+1)}{\pi}\xi^{-q} = 0. \quad (2.8.26)$$

The above equations, beginning with (2.8.12), have a common feature. They are connected not with the density $g(x, \alpha, \beta)$ itself but with a certain complex-valued function whose real part is expressed in terms of the density. This means that each of these equations is, generally speaking, a system of two equations for two functions, of which only one is of interest to us. The linear nature of both equations allows us to write an equation for each of these functions, but only at the cost of complicating the equation. Thus, the order of equations (2.8.24) and (2.8.25) is equal to $\max(p-1,\ q-1)$, while the order of the equation for the density $g(x, \alpha, \theta) = \operatorname{Re}\mu(x)$ in the general case is $2\max(p-1,\ q-1)$. Sometimes, however, the complication does not occur, because the coefficients of the operators turn out to be real.

We consider this phenomenon in more detail for the example of equation (2.8.20) and its particular cases (2.8.24) and (2.8.25). The number

$$\exp\left(i\pi r - i\tfrac{\pi}{2}\alpha(r+1)(1-\theta)\right)$$

is the factor in (2.8.20) with which we shall be concerned. It is real if and only if the number $2r - \alpha(r+1)(1-\theta)$ is even, i.e., is equal to $2k$ for some integer k:

$$2r - \alpha(r+1)(1-\theta) = 2k.$$

From this we obtain

$$\theta = 1 - 2(k-r)/\alpha(r+1). \quad (2.8.27)$$

Determination of the numbers θ satisfying (2.8.27) means requiring that θ belongs to the domain of admissible values, i.e., $|\theta| \le \theta_\alpha$. It is not hard to compute that this is equivalent to the conditions

$$\begin{array}{ll} r \le k \le r + \alpha(r+1) & \text{if } 0 < \alpha < 1, \\ (\alpha-1)(r+1) + r \le k \le 2r+1 & \text{if } 1 < \alpha \le 2. \end{array}$$

For equations (2.8.24) and (2.8.25) with $\alpha = p/q \neq 1$ and $r = q - 1$ the representation (2.8.27) is equivalent (since r is an integer) to

$$\theta = \theta^{(k)} = 1 - 2k/p \tag{2.8.28}$$

with the following conditions on the integer k:

$$\begin{aligned} 0 \leq k \leq p \quad &\text{if } p < q, \\ p - q \leq k \leq q \quad &\text{if } p > q. \end{aligned}$$

For each pair $\alpha = p/q$, $\theta^{(k)}$ satisfying (2.8.28) equations (2.8.24)–(2.8.26) split into pairs of mutually unconnected equations for the real and imaginary parts of the corresponding function. Furthermore, the most interesting equation for the real part is always homogeneous.

It is not hard to see that for each $\alpha = p/q \neq 1$ there are at least two values θ satisfying (2.8.28): $\theta = \theta_\alpha$ and $\theta = -\theta_\alpha$. The set of $\alpha = p/q \neq 1$ for which there are no cases of splitting of the differential equations other than cases of extremal laws is made up of numbers of the form $1/(n+1)$ and $(2n+1)/(n+1)$, $n = 1, 2, \ldots$.

If the order of the equations is at most two, then the form of the equations themselves allows us to hope that they can be solved, at least with the help of some special functions. Let us consider these cases.

1. If $\alpha = \frac{1}{2}$ and θ is any admissible value, then from (2.8.26) we get

$$\tau'(\xi) - \tfrac{1}{2}\exp(i\pi\rho)\xi\tau(\xi) + i\xi^{-2}/\pi = 0.$$

This equation can be solved without difficulty when we recall that $\tau(\infty) = 0$:

$$\tau(\xi) = \frac{i}{\pi}\exp\left(\frac{\xi^2}{4}e^{i\pi\rho}\right)\int_\xi^\infty \exp\left(-\frac{t^2}{4}e^{i\pi\rho}\right)\frac{dt}{t^2}.$$

Consequently,

$$\begin{aligned} x^{3/2}g(x, 1/2, \theta) &= \operatorname{Re}\tau(\xi) \\ &= -\frac{1}{\pi}\operatorname{Im}\left(z\exp(z^2/4)\int_z^\infty \exp(-t^2/4)\frac{dt}{t^2}\right), \end{aligned} \tag{2.8.29}$$

where $z = x^{-1/2}\exp(i\pi\rho/2)$.

In particular, if $\theta = 0$ (i.e., $\rho = \frac{1}{2}$, which corresponds to a symmetric distribution), then after suuitable transformations (2.8.29) takes the form

$$\begin{aligned} g\left(x, \frac{1}{2}, 0\right) = \frac{x^{-3/2}}{2\sqrt{2\pi}}\Bigg[&\cos\frac{1}{4x}\left(\frac{1}{2} - C\left(\frac{1}{x}\right)\right) \\ &+ \sin\frac{1}{4x}\left(\frac{1}{2} - S\left(\frac{1}{x}\right)\right)\Bigg], \end{aligned} \tag{2.8.30}$$

where

$$C(u) = \frac{1}{\sqrt{2\pi}} \int_0^u \frac{\cos t}{\sqrt{t}}\, dt, \qquad S(u) = \frac{1}{\sqrt{2\pi}} \int_0^u \frac{\sin t}{\sqrt{t}}\, dt$$

are special function called Fresnel integrals ([2.28]).

2. If $\alpha = \frac{1}{3}$ and $\theta = 1$ (a case in which equation (2.8.26) splits), then for $y(\xi) = \operatorname{Re} \tau(\xi)$ we get the equation

$$y''(\xi) - \tfrac{1}{3}\xi y(\xi) = 0.$$

This equation is solved (to within a constant multiple) by the so-called Airy integral, which can be expressed, in turn, in terms of the Macdonald function of order $\frac{1}{3}$:

$$y(\xi) = 3c \int_0^\infty \cos(t^3 + t\xi)\, dt = c\sqrt{\xi} K_{1/3}\left(2\xi^{3/2}/3\sqrt{3}\right).$$

To establish the value c we are aided by the expansion (2.4.8) of the function $x^{4/3} g(x, \frac{1}{3}, 1)$ in powers of $\xi = x^{-1/3}$. On the one hand (see 8.4.32(5) in [27])

$$\sqrt{\xi} K_{1/3}\left(2\xi^{3/2}/3\sqrt{3}\right) \to \sqrt{3}\Gamma(1/3)/2 \quad \text{as } \xi \to 0,$$

and on the other hand

$$y(\xi) = x^{4/3} g\left(x, \frac{1}{3}, 1\right) \to \frac{1}{3\pi} \frac{\sqrt{3}}{2} \Gamma\left(\frac{1}{3}\right)$$

as $\xi \to 0$; therefore, $c = 1/3\pi$. Consequently,

$$g\left(x, \frac{1}{3}, 1\right) = \frac{x^{-3/2}}{3\pi} K_{1/3}\left(\frac{2}{3\sqrt{3}} x^{-1/2}\right). \tag{2.8.31}$$

There are four more cases when the density $g(x, \alpha, \theta)$ can be expressed in terms of special functions.

3. If $\alpha = \frac{2}{3}$ and $\theta = 1$, it follows from (2.8.25) that $y(\xi) = \operatorname{Re} \tau(\xi)$ satisfies the equation

$$y''(\xi) + \tfrac{4}{9}\xi^2 y''(\xi) + \tfrac{10}{9}\xi y(\xi) = 0.$$

4. If $\alpha = \frac{2}{3}$ and $\theta = 0$, then for $y(\xi) = \operatorname{Re} \tau(\xi)$ we have that

$$y''(\xi) - \tfrac{4}{9}\xi^2 y'(\xi) - \tfrac{10}{9}\xi y(\xi) = 0.$$

5. If $\alpha = \frac{3}{2}$ and $\theta = \theta_{3/2} = \frac{1}{3}$, then, by (2.8.26), the function $y(x) = \operatorname{Re} \mu(x)$ is a solution of the equation

$$y''(x) + \tfrac{4}{9}x^2 y'(x) + \tfrac{10}{9}xy(x) = 0.$$

6. If $\alpha = \frac{3}{2}$ and $\theta = -\frac{1}{3}$, then for $y(x) = \operatorname{Re} \mu(x)$ we have

$$y''(x) - \tfrac{4}{9}x^2 y'(x) - \tfrac{10}{9}xy(x) = 0.$$

Recall that in the last two cases the value $\theta = \frac{1}{3}$ corresponds to the value $\beta = -1$ in the forms (A) and (B), while $\theta = -\frac{1}{3}$ corresponds to the value $\beta = 1$ in the same forms.

All these equations are connected with a single type of special functions, the so-called Whittaker functions $W_{k,m}(x)$. We confine ourselves to a treatment of the particular case $\alpha = \frac{2}{3}$, $\theta = 0$. The other cases are handled similarly.

It is known (see, for example, 2.273(5) in [39]) that in the chosen case a function of the form

$$y(\xi) = C\xi^{-1} \exp\left(\tfrac{2}{27}\xi^3\right) W_{-1/2,1/6}\left(\tfrac{4}{27}\xi^3\right)$$

is a solution of the equation. As in the analysis of the preceding case, the value of the constant C can be determined by comparing the asymptotic behavior of $y(\xi)$ as $\xi = x^{-2/3} \to \infty$, obtained in two different ways. On the one hand (see 9.227 in [27])

$$y(\xi) \sim C\frac{3\sqrt{3}}{2}\xi^{-(1+3/2)} = C\frac{3\sqrt{3}}{2}x^{1+2/3}.$$

On the other hand, on the basis of (2.2.11)

$$y(\xi) = x^{1+2/3} g\left(x, \frac{2}{3}, 0\right) \sim x^{1+2/3} g\left(0, \frac{2}{3}, 0\right) = \frac{3}{4\sqrt{\pi}} x^{1+2/3}.$$

This implies that $C = 1/(2\sqrt{3\pi})$, i.e., for any $x > 0$

$$\begin{aligned} g\left(x, \frac{2}{3}, 0\right) &= g\left(-x, \frac{2}{3}, 0\right) \\ &= \frac{x^1}{2\sqrt{3\pi}} \exp\left(\frac{2}{27}x^{-2}\right) W_{-1/2,1/6}\left(\frac{4}{27}x^{-2}\right). \end{aligned} \tag{2.8.32}$$

In the remaining cases we obtain in the same way the following expressions for the densities on the semi-axis $x > 0$ (2.29):

$$g\left(x, \frac{2}{3}, 1\right) = \frac{x^{-1}}{\sqrt{3\pi}} \exp\left(-\frac{2}{27}x^{-2}\right) W_{1/2,1/6}\left(\frac{4}{27}x^{-2}\right), \tag{2.8.33}$$

$$g\left(x, \frac{3}{2}, \frac{1}{3}\right) = \frac{x^{-1}}{\sqrt{3\pi}} \exp\left(-\frac{2}{27}x^{3}\right) W_{1/2,1/6}\left(\frac{4}{27}x^{3}\right), \tag{2.8.34}$$

$$g\left(x, \frac{3}{2}, -\frac{1}{3}\right) = \frac{x^{-1}}{2\sqrt{3\pi}} \exp\left(\frac{2}{27}x^{3}\right) W_{-1/2,1/6}\left(\frac{4}{27}x^{3}\right). \tag{2.8.35}$$

According to the duality law (2.3.3), the densities under consideration on the positive semi-axis must be connected by the equalities

$$xg\left(x, \tfrac{3}{2}, \tfrac{1}{3}\right) = x^{-3/2} g\left(x^{-3/2}, \tfrac{2}{3}, 1\right),$$

$$xg\left(x, \tfrac{3}{2}, -\tfrac{1}{3}\right) = x^{-3/2} g\left(x^{-3/2}, \tfrac{2}{3}, 0\right),$$

and this can indeed be observed in the above formulas (therefore, in particular, we can obtain (2.8.35) as a consequence of (2.8.32), and (2.8.33) as a consequence of (2.8.34)).

The integral representation

$$W_{-1/2,1/6}(z) = \frac{z^{-1/2}}{\Gamma(7/6)} e^{-z/2} \int_0^\infty e^{-t} t^{1/6} \left(1 + \frac{t}{z}\right)^{-5/6} dt$$

of the function $W_{-1/2,1/6}(z)$ in the z-plane with a cut along the negative semi-axis (see 9.221 in [27]) shows that this is a multi-valued function with third-order algebraic branch points $z = 0$ and $z = \infty$. Consequently, the function $W_{-1/2,1/6}(z^3)$ is analytic in the whole complex plane, and we can speak of its values $W_{-1/2,1/6}(z^3)$ on the negative semi-axis, although the formula given is unsuitable for expressing them. This consideration, together with the consequence

$$g\left(x, \tfrac{3}{2}, \tfrac{1}{3}\right) = g\left(-x, \tfrac{3}{2}, -\tfrac{1}{3}\right)$$

of (2.1.3), leads to new expressions for the densities $g(x, \frac{3}{2}, \frac{1}{3})$ and $g(x, \frac{2}{3}, 1)$:

$$g\left(x, \frac{3}{2}, \frac{1}{3}\right) = -\frac{x^{-1}}{2\sqrt{3\pi}} \exp\left(\frac{2}{27}x^3\right) W_{-1/2,1/6}\left(-\frac{4}{27}x^3\right),$$

$$g\left(x, \frac{2}{3}, 1\right) = \frac{x^{-1}}{2\sqrt{3\pi}} \exp\left(\frac{2}{27}x^{-2}\right) W_{-1/2,1/6}\left(-\frac{4}{27}x^{-2}\right).$$

A comparison of these expressions with (2.8.34) and (2.8.33) shows that in the complex plane

$$W_{-1/2,1/6}(-z^3) = -2W_{-1/2,1/6}(z^3).$$

We conclude this section with another theorem.

Consider the function

$$S_\alpha(\xi, \eta) = \begin{cases} xg(x, \alpha, \beta) & \text{if } \alpha \neq 1,\ x > 0, \\ \beta g(x, \alpha, \beta) & \text{if } \alpha = 1,\ \beta > 0, \end{cases}$$

where the variables ξ and η are connected with x and β by the relations

$$\xi = \begin{cases} -\log x & \text{if } \alpha \neq 1, \\ x/\beta - \log \beta & \text{if } \alpha = 1, \end{cases}$$

$$\eta = \begin{cases} (\pi/2)\beta K(\alpha)/\alpha & \text{if } \alpha \neq 1, \\ 2/2\beta & \text{if } \alpha = 1. \end{cases}$$

Let $z = \xi + i\eta$ and define the function $\Psi_\alpha(z)$ as follows: if $0 < \alpha < 1$, then

$$\Psi_\alpha(z) = \frac{i}{\pi} \int_0^\infty \exp\left(-t - t^\alpha \exp\left(-i\frac{\pi}{2}\alpha + \alpha z\right)\right) dt;$$

if $1 < \alpha \le 2$, then

$$\Psi_\alpha(z) = \frac{1}{\pi}\int_0^\infty \exp(ite^{-z} - z - t^\alpha)\,dt;$$

and if $\alpha = 1$, then

$$\Psi_1(z) = \frac{i}{\pi}\int_0^\infty \exp\left(-\left(\frac{\pi}{2} + z\right)t - t\log t\right)\,dt.$$

Returning to the representations (2.2.8)–(2.2.10) of the densities $g(x, \alpha, \beta)$, we see that

$$S_\alpha(\xi, \eta) = \operatorname{Re}\Psi_\alpha(z). \tag{2.8.36}$$

It is obvious that $\Psi_\alpha(x)$ is an entire analytic function for all α.

We thus arrive at the following assertion.

THEOREM 2.8.5. *For each admissible value of α the function $S_\alpha(\xi, \eta)$ is a solution of the first boundary value problem of the Laplace equation $\Delta S_\alpha = 0$ in the strip*

$$-\infty < \xi < \infty, \quad -\pi\theta_\alpha/2 \le \eta \le \pi\theta_\alpha/2$$

for the case $\alpha \ne 1$, and in the half-plane $-\infty < \xi < \infty$, $\eta \ge \pi/2$ for the case $\alpha = 1$. Furthermore, the boundary conditions are

$$S_\alpha(\pm\infty, \eta) = 0,$$

$$S_\alpha(\xi, \pm\pi\theta_\alpha/2) = e^{-\xi}g(e^{-\xi}, \alpha, \pm\operatorname{sgn}(1 - \alpha))$$

for $\alpha \ne 1$, and

$$S_1(\pm\infty, \eta) = S_1(\xi, \infty) = 0, \qquad S_1(\xi, \pi/2) = g(\xi, 1, 1)$$

for $\alpha = 1$.

REMARK. Here the appearance of the domain of variation of the variables ξ and η is connected, of course, with the restrictions to which the parameter β is subjected. But if $S_\alpha(\xi, \eta)$ is defined directly as the real part of $\Psi_\alpha(z)$, then it is harmonic at each point of the (ξ, η)-plane.

§2.9. Stable laws as functions of parameters

Of the four parameters α, β, γ, and λ determining the distributions of class $\mathfrak{S}$, the last two play the role of shift and scale change parameters. It is clear from property (2.1.1) that

$$G = G(x, \alpha, \beta, \gamma, \lambda) = G((x - l)\lambda^{-1/\alpha}, \alpha, \beta, 0, 1),$$

where l is uniquely determined by the parameter values. Therefore, for fixed α and β the distribution G admits a representation as a superposition of a standard distribution $G(x, \alpha, \beta)$ and a linear function $(x - l)\lambda^{-1/\alpha}$, with a relatively simple dependence of l on the parameters. Thus, the study of the properties of the distribution G as a function of γ and λ reduces in fact to the study of the properties of $G(x, \alpha, \beta)$ as a function of x for fixed α and β and as a function of x, α, and β in the general case.

It follows from Theorem 2.8.5 that the densities $g(x, \alpha, \beta)$ are entire functions of the modified variables ξ and η, which are variables on an equal footing in the sense that they are the real and imaginary parts of the single variable z in (2.8.36).

An analysis of the function $\Psi_\alpha(z)$ shows that its order as an entire function is infinite for any α. This indicates that $S_\alpha(\xi, \eta)$ also has infinite order as an entire function of ξ or η.

The integral expressions for $\Psi_\alpha(z)$ enable us easily to obtain a series expansion of it in powers of z, and from that it is the very easy to find an expansion in a double series of powers of ξ and η, after which separation of the real part of the double series gives us an analogous expansion of the function $S_\alpha(\xi, \eta)$ in a double series. However, we do not give these fairly unwieldy formulas here; they are not hard to get by following the remarks above.

Regardless of the choice of the parametrization system, the distributions, in the family $\mathfrak{S}$ are continuous with respect to the parameter α in the intervals $(0, 1)$ and $(1, 2)$. The distributions $G(x, \alpha, \beta)$ corresponding to the parametrization systems (M) (for the whole family S), (C), and (E) (for the subset W of it) turn out to be jointly continuous in all variables in the domain where they vary. The question of the behavior of stable distributions when α is near zero and of the distributions $G_M(x, \alpha, \beta)$ near $\alpha = 1$ merit extra attention here, the latter in connection with the desire for good approximations of the distributions $G_M(x, \alpha, \beta)$ in terms of $G_M(x, 1, \beta)$.

Let $0 < \alpha < 1$, and consider random variables $Y(\alpha, \theta) = Y(\alpha, \beta, 0, 1)$, $\theta = \beta K(\alpha)/\alpha = \beta$, and $Z(\alpha, \rho)$, $\rho = (1+\theta)/2$, with the respective distributions $G(x, \alpha, \beta)$ and $(G(x, \alpha, \beta) - G(0, \alpha, \beta))/\rho$. Denote by U_θ a random variable taking the values ± 1 with respective probabilities $(1+\theta)/2$, and by E a random variable independent of U_θ and with the standard exponential distribution (i.e., with mean value equal to 1).

THEOREM 2.9.1. *The following relations hold for any sequence of pairs of admissible values $\alpha < 1$ and $\rho' = (1 + \theta')/2$ such that $\alpha \to 0$ and $\rho' \to \rho = (1 + \theta)/2$:*

$$Y^\alpha(\alpha, \theta') = |Y(\alpha, \theta')|^\alpha \operatorname{sgn} Y(\alpha, \theta') \xrightarrow{d} U_\theta/E, \tag{2.9.1}$$

$$Z^\alpha(\alpha, \rho') \xrightarrow{d} 1/E. \tag{2.9.2}$$

NOTE. A modified concept of power is used on the left-hand side of (2.9.1), and it is *not* understood (as we stipulated in §2.1) that the separately written random variables $|Y(\alpha,\theta')|^\alpha$ and $\operatorname{sgn} Y(\alpha,\theta')$ are independent. There are more details on the modified power and on the random variables $Y(\alpha,\theta)$ and $Z(\alpha,\rho)$ at the beginning of Chapter 3 and in §3.1.

PROOF. The elements of the characteristic transform of the distribution of $Y^\alpha(\alpha,\theta')$ have the form

$$w_k(\alpha s,\alpha,\theta') = \frac{\cos\frac{\pi}{2}(k-\alpha s\theta')\Gamma(1-s)}{\cos\frac{\pi}{2}(k-\alpha s)\Gamma(1-\alpha s)}, \qquad k=0,1.$$

For fixed $-1<s<1$ it is obvious that as $\alpha\to 0$ and $\theta'\to 0$

$$\begin{aligned} w_0(\alpha s,\alpha,\theta') &\to \Gamma(1-s) = w_0(s),\\ w_1(\alpha s,\alpha,\theta') &\to \theta\Gamma(1-s) = w_1(s). \end{aligned} \tag{2.9.3}$$

We show that the pair w_0, w_1 corresponds to the characteristic transform of the random variable U_θ/E. Indeed,

$$\mathsf{E}|U_\theta/E|^s = \mathsf{E}(E^{-1})^s = \int_0^\infty x^{-s}e^{-x}\,dx = w_0(s),$$

$$\begin{aligned} \mathsf{E}|U_\theta/E|^s\operatorname{sgn}(U_\theta/E) &= \mathsf{E}(E^{-1})^s\operatorname{sgn} U_\theta = (\mathsf{E}\operatorname{sgn} U_\theta)(\mathsf{E}(E^{-1})^s)\\ &= \theta w_0(s) = w_1(s). \end{aligned}$$

Consequently, (2.9.3) is equivalent to (2.9.1). The relation (2.9.2) can be proved similarly ([2.30]).

The distribution functions of the random variable $Y^\alpha(\alpha,\theta)$, the cutoff $Z^\alpha(\alpha,\rho)$ of it, and their limiting random variables U_θ/E and $1/E$ as $\alpha\to 0$ are as follows:

$$\begin{aligned} F_\alpha(x) &= G(|x|^{1/\alpha}\operatorname{sgn} x,\alpha,\theta), && x\in R^1,\\ \hat F_\alpha(x) &= \tfrac{1}{\rho}[G(x^{1/\alpha},\alpha,\rho)-(1-\rho)], && x>0,\\ F_0(x) &= \tfrac12(1-\theta+(\operatorname{sgn} x+\theta)\exp(-1/|x|)), && x\in R^1,\\ \hat F_0(x) &= \exp(-1/x), && x>0. \end{aligned}$$

The relations (2.9.1) and (2.9.2) can be refined by constructing asymptotic expansions of $F_\alpha(x)$ and $\hat F_\alpha(x)$ in series of powers of α as $\alpha\to 0$. To do this we need to introduce some new notation. Let $a_1^{(0)} = a_1^{(1)} = b_1 = -\mathbb{C}$, where $\mathbb{C}$ is the Euler constant, and let

$$a_n^{(k)} = \frac{2^n(2^n-1)^{1-k}|B_n|}{n\Gamma(n+1)}\left(\frac{\pi}{2}\right)^n(1-\theta^n) - \frac{\varsigma(n)}{n},$$

$$b_n = \frac{2^n|B_n|}{n\Gamma(n+1)}\pi^n(1-\rho^n) - \frac{\varsigma(n)}{n}$$

for $n \geq 2$ and $k = 0, 1$, where the B_n are the Bernoulli numbers, and $\varsigma(n)$ is the value at n of the Riemann ς-function. With the help of $a_n^{(k)}$ and b_n we determine the quantities $C_n^{(k)}$ and D_n by

$$\exp\left(\sum_{n=1}^{\infty} a_n^{(k)} s^n\right) = \sum_{n=0}^{\infty} C_m^{(k)} \frac{s^m}{m!},$$
$$\exp\left(\sum_{n=1}^{\infty} b_n s^n\right) = \sum_{m=0}^{\infty} D_m \frac{s^m}{m!}.$$

Clearly, the quantities D_m are formed from the quantities b_n by the same rule as the $C_m^{(k)}$ are formed from the $a_n^{(k)}$. In both cases the so-called Bell polynomials $C_n(u_1, \dots, u_n)$ realize this rule (an explicit expression for them is given in §3.6). Namely,

$$C_n^{(k)} = C_n(1!a_1^{(k)}, \dots, n!a_n^{(k)}),$$
$$D_n = C_n(1!b_1, \dots, n!b_n).$$

We define the polynomials $P_n(x)$, $n = 0, 1, 2, \dots$, by the recursion relation

$$P_0(x) \equiv 1, \quad P_{n+1}(x) = xP_n(x) + xP_n'(x), \quad n \geq 0.$$

It is not hard to compute that the generating function of these polynomials is equal to

$$\sum_{n=0}^{\infty} P_n(x) \frac{z^n}{n!} = \exp(x(z^z - 1)).$$

Consequently, $P_n(x)$ can if desired be interpreted as the nth-order moment of the Poisson distribution with step 1 and mean value $x > 0$.

THEOREM 2.9.2. *Let $-1 \leq \theta \leq 1$ and $x \in R^1$ be fixed numbers. Then the following asymptotic expansions are valid as $\alpha \to 0$* ([2.31]): *for any real $x \neq 0$*

$$F_\alpha(x) \sim F_0(x) + \frac{1}{2} \sum_{n=1}^{\infty} (C_n^{(0)} + \theta C_n^{(1)} \operatorname{sgn} x) P_n\left(-\frac{1}{|x|}\right) \exp\left(-\frac{1}{|x|}\right) \frac{\alpha^n}{n}!; \tag{2.9.4}$$

and for any $x > 0$

$$\hat{F}_\alpha(x) \sim \hat{F}_0(x) + \sum_{n=1}^{\infty} D_n P_n(-1/x) \exp(-1/x) \alpha^n / n!. \tag{2.9.5}$$

PROOF. We transform the function $w_k(\alpha s, \alpha, \theta)$ by using the quantities $a_n^{(k)}$ introduced above:

$$
\begin{aligned}
w_k(\alpha s, \alpha, \theta) &= \frac{\cos\frac{\pi}{2}(k - \alpha\theta s)}{\cos\frac{\pi}{2}(k - \alpha s)} \frac{\Gamma(1-s)}{\Gamma(1-\alpha s)} \\
&= \exp\left(\sum_{n=1}^{\infty} a_n^{(k)} (s\alpha)^n\right) \Gamma(1-s) \\
&= \theta^k \sum_{n=0}^{\infty} C_n^{(k)} \frac{(\alpha s)^n}{n!} \Gamma(1-s), \qquad k = 0, 1.
\end{aligned} \tag{2.9.6}
$$

Here, $\theta^0 = 1$ if $\theta = 0$; see §3.6 about the transform (2.9.6). Adding w_0 and w_1 and subtracting w_1 from w_0, we get that for $-1 < s < 1$

$$
\begin{aligned}
2\int_0^{\infty} x^s \, dF_\alpha(x) &= \sum_{n=0}^{\infty} (C_n^{(0)} + \theta C_n^{(1)}) \frac{\alpha^n}{n!} s^n \Gamma(1-s), \\
2\int_{-\infty}^{0} (-x)^s \, dF_\alpha(x) &= \sum_{n=0}^{\infty} (C_n^{(0)} - \theta C_n^{(1)}) \frac{\alpha^n}{n!} s^n \Gamma(1-s).
\end{aligned} \tag{2.9.7}
$$

The convergence of the series is a consequence of the analyticity in the disk $|s| < 1$ of the functions multiplying $\Gamma(1-s)$ in the expression for $w_k(\alpha s, \alpha, \theta)$. Further,

$$
\begin{aligned}
s^n \Gamma(1-s) &= s^n \int_0^{\infty} x^s \, d\exp\left(-\frac{1}{x}\right) \\
&= (-1)^n \int_0^{\infty} x^s \, d\left(x\frac{d}{dx}\right)^n \exp\left(-\frac{1}{x}\right).
\end{aligned}
$$

Since

$$
\begin{aligned}
&(-1)^n \left(x\frac{d}{dx}\right)^n \exp\left(-\frac{1}{x}\right) \\
&\quad = \left(u\frac{d}{du}\right)^n e^u \bigg|_{u=-1/x} = P_n\left(-\frac{1}{x}\right) \exp\left(-\frac{1}{x}\right),
\end{aligned}
$$

it follows that

$$
s^n \Gamma(1-s) = \int_0^{\infty} x^s \, d[P_n(-1/x)\exp(-1/x)].
$$

In a completely analogous way,

$$
s^n \Gamma(1-s) = \int_{-\infty}^{n} (-x)^s \, d[P_n(1/x)\exp(1/x)].
$$

After inversion of the Mellin transforms, these equalities together with (2.9.7) to the formal equality (2.9.4). The formal relation thus obtained has the rigorous meaning of an asymptotic expansion. This fact is proved by the standard method used in such cases. The proof of the next result, Theorem 2.9.3, in which an analogous problem is solved, can serve as an example. It should just be taken into account that by setting $s = it$ ($t \in R^1$) we can interpret the Mellin transform as a characteristic function. The inverse Fourier transforms of the individual terms exist because of well-known properties of the function Γ:

$$|\Gamma(1-it)| \sim \sqrt{2\pi}|t|^{1/2}\exp(-\pi|t|/2) \quad \text{as } |t| \to \infty.$$

The relation (2.9.5) can be proved with analogous arguments by starting from the equality (see (2.6.20) and (3.0.2))

$$\begin{aligned} M(\alpha s, \alpha, \rho) &= \frac{\sin \pi\rho\alpha s}{\rho \sin \pi\alpha s} \frac{\Gamma(1-s)}{\Gamma(1-\alpha s)} \\ &= \exp\left(\sum_{n=1}^{\infty} b_n(\alpha s)^n\right)\Gamma(1-s) \\ &= \sum_{n=0}^{\infty} D_n \frac{\alpha^n}{n!} s^n \Gamma(1-s). \end{aligned}$$

The asymptotic behavior of the distribution $G_M(x, \alpha, \beta)$ near the point $\alpha = 1$ (where it is continuous, as mentioned in §I.3) will be studied on the basis of the characteristic function $\mathfrak{g}_M(t, \alpha, \beta)$ corresponding to it.

Let us consider the difference

$$\begin{aligned} \Delta(t) &= \log \mathfrak{g}_M(t, \alpha, \beta) - \log \mathfrak{g}_M(t, 1, \beta) \\ &= -|t|(|t|^{\alpha-1} - 1) + it(|t|^{\alpha-1} - 1)\beta \tan \tfrac{\pi}{2}\alpha + it\beta\tfrac{2}{\pi}\log|t|. \end{aligned}$$

For brevity we let $\varepsilon = \alpha - 1$ and expand the last expression in a series of powers of ε:

$$\begin{aligned} \Delta(t) &= -|t|(\exp(\varepsilon \log|t|) - 1) + it\log|t|\beta\left(\tfrac{2}{\pi} - \varepsilon\cot\tfrac{\pi}{2}\varepsilon\right) \\ &\quad - it(\exp(\varepsilon\log|t|) - 1 - \varepsilon\log|t|)\beta\cot\tfrac{\pi}{2}\varepsilon \\ &= -(|t|\log|t| + it\beta\tfrac{1}{\pi}\log^2|t|)\varepsilon + \cdots \\ &= -\sum_{n=1}^{\infty} h_n\varepsilon^n. \end{aligned}$$

Consequently,

$$\begin{aligned} \mathfrak{g}_M(t, \alpha, \beta) &= \mathfrak{g}_M(t, 1, \beta)\exp\left(-\sum_{n=1}^{\infty} h_n\varepsilon^n\right) \\ &= \mathfrak{g}_M(t, 1, \beta)(1 - h_1\varepsilon - (2h_2 - h_1^2)\varepsilon^2/2 - \cdots). \end{aligned}$$

Formal inversion of this equality yields an expansion of the density $g_M(x, \alpha, \beta)$ in an asymptotic series of powers of ε as $\varepsilon \to 0$:

$$g_M(x, \alpha, \beta) \sim g_M(x, 1, \beta) - H_1(x, \beta)\varepsilon - \cdots. \tag{2.9.8}$$

Although the functions $H_1, H_2, \ldots$ can be expressed in terms of the distribution $G_M(x, 1, \beta)$ and its multiple derivatives (including mixed derivatives) with respect to the variables x and β, these expressions are so unwieldy that we confine ourselves to computing only the first term of the asymptotic series (2.9.8) [2.32], i.e., the function $H_1(x, \beta)$ corresponding to

$$h_1 = |t| \log |t| + it\frac{\beta}{\pi} \log^2 |t|.$$

THEOREM 2.9.3. *Let $x \in R^1$ and $-1 \leq \beta \leq 1$. Then as $\alpha \to 1$*

$$\begin{aligned}(g_M(x, \alpha, \beta) - g_M(x, 1, \beta))(1 - \alpha)^{-1} &\to H_1(x, \beta) \\ &= \frac{\pi}{4}\beta\frac{\partial^2}{\partial\beta^2}G_M(x, 1, \beta) + \frac{\pi}{2}\frac{\partial}{\partial\beta}G_M(x, 1, \beta) \\ &\quad - g_M(x, 1, \beta) - \left(\frac{\pi}{2}x + \beta\right)\frac{\partial}{\partial\beta}g_M(x, 1, \beta).\end{aligned} \tag{2.9.9}$$

PROOF. By the inversion formula for the densities,

$$\begin{aligned}\Xi_\alpha &= g_M(x, \alpha, \beta) - g_M(x, 1, \beta) \\ &= \frac{1}{2\pi}\int e^{-itx}(\mathfrak{g}_M(t, \alpha, \beta) - \mathfrak{g}_M(t, 1, \beta))\, dt \\ &= \frac{1}{2\pi}\int e^{-itx}\mathfrak{g}_M(t, 1, \beta)\Delta(t)\, dt + \delta,\end{aligned} \tag{2.9.10}$$

where

$$\delta = \frac{1}{2\pi}\int e^{-itx}\mathfrak{g}_M(t, 1, \beta)(e^\Delta - 1 - \Delta)\, dt.$$

Let us estimate $|\delta|$, considering that $|\varepsilon| < \frac{1}{2}$. To do this we first estimate $|\Delta(t)|$:

$$\begin{aligned}|\Delta| &\leq \max(|t|, |t|^{2/3})|\log |t|\,|\varepsilon + |t|\,|\log |t|\,|\varepsilon^2 + \max(|t|, |t|^{3/2})\log^2 |t|\varepsilon \\ &< 3|t|(1 + |t| + \log^2 |t|)\varepsilon.\end{aligned}$$

Since $|\mathfrak{g}_M(t, 1, \beta)| = \exp(-|t|)$ and $\operatorname{Re}\Delta = -|t|^\alpha + |t|$, it follows that

$$\begin{aligned}|\mathfrak{g}_M(t, 1, \beta)(e^\Delta - 1 - \Delta)| &\leq \exp(-|t|)\max(1, \exp\operatorname{Re}\Delta)|\Delta|^2/2 \\ &< \max(\exp(-|t|), \exp(-|t|^\alpha))|\Delta|^2/2.\end{aligned}$$

Consequently, $|\delta| = O(\varepsilon^2)$. Further, it is not hard to compute that

$$\Delta(t) = -h_1(t)\varepsilon + \eta[(1+t^2)(1+\log^4|t|)]\varepsilon^2,$$

where $|\eta| < \text{const}$.

From this and (2.9.10), together with the estimate for δ, we get

$$\begin{aligned}\Xi_\alpha &= -\frac{\varepsilon}{2\pi}\int e^{-itx}\mathfrak{g}_M(t,1,\beta)h_1(t)\,dt + O(\varepsilon^2)\\ &= -H_1(x,\beta)\varepsilon + O(\varepsilon^2).\end{aligned}$$

It remains for us to compute the integral $H_1(x,\beta)$. According to (2.1.2), for any $\lambda > 0$

$$\mathfrak{g}_M^\lambda = \mathfrak{g}_M^\lambda(t,1,\beta) = \int e^{itx}\,dG_M\left(\frac{x}{\lambda} - \frac{2}{\pi}\beta\log\lambda, 1, \beta\right).$$

This implies that

$$\frac{1-\mathfrak{g}_M^\lambda}{it} = \int e^{itx}\left[G_M\left(\frac{x}{\lambda} - \frac{2}{\pi}\beta\log\lambda, 1, \beta\right) - D(x)\right]dx, \qquad (2.9.11)$$

where $D(x) = (1+\operatorname{sgn} x)/2$.

For brevity let

$$\psi = -\log\mathfrak{g}_M(t,1,\beta) = |t| + it\tfrac{2}{\pi}\beta\log|t|.$$

Differentiating both sides of (2.9.11) with respect to λ and then letting $\lambda = 1$, we arrive at the equality

$$\frac{\psi}{it}\exp(-\psi) = \int e^{itx}\left(x + \frac{2}{\pi}\beta\right)\mathfrak{g}_M(x,1,\beta)\,dx. \qquad (2.9.12)$$

We now differentiate both sides of (2.9.12) with respect to β:

$$\begin{aligned}&\frac{2}{\pi}\log|t|e^{-\psi} - \frac{2}{\pi}\log|t|\psi e^{-\psi}\\ &\qquad = \int e^{itx}\left(\frac{2}{\pi}\mathfrak{g}_M + \left(x + \frac{2}{\pi}\beta\right)\frac{\partial}{\partial\beta}\mathfrak{g}_M\right)dx. \qquad (2.9.13)\end{aligned}$$

On the other hand, setting $\lambda = 1$ in (2.9.11) and differentiating both sides with respect to β, we get

$$\frac{2}{\pi}\log|t|e^{-\psi} = \int e^{itx}\frac{\partial}{\partial\beta}G_M(x,1,\beta)\,dx. \qquad (2.9.14)$$

Repeated differentiation of (2.9.14) with respect to β gives us that

$$-\frac{4}{\pi^2}it\log^2|t|e^{-\psi} = \int e^{itx}\frac{\partial^2}{\partial\beta^2}G_M(x,1,\beta)\,dx. \qquad (2.9.15)$$

Equalities (2.9.13)–(2.9.15) enable us to represent as a Fourier transform the function

$$\mathfrak{g}_M(t,1,\beta)h_1(t) = \log|t|\psi e^{-\psi} - (\beta/\pi)it\log^2|t|e^{-\psi}.$$

This yields the expression (2.9.9) for the function $H_1(x,\beta)$.

§2.10. Densities of stable distributions as a class of special functions

Many facts indicate that, by virtue of their richness of analytic properties, the functions $g(x,\alpha,\beta)$ merit being distinguished as an independent class and accorded "civil rights" in the theory of special functions. Some of these facts are presented below.

The functions $g(x,\alpha,1)$ with $0<\alpha<1$ turn out to be useful in the theory of Laplace transforms and the operational calculus connected with it. Let

$$V(s) = \int_0^\infty \exp(-sx)v(x)\,dx, \qquad \operatorname{Re} s \ge 0,$$

denote the Laplace transform of a function $v(x)$ (briefly: $V(s) \rightharpoonup v(x)$). The following connection formulas are well known (for example, see [19] or [15]). If $V(x) \rightharpoonup v(x)$, then, for instance,

$$V\left(\sqrt{s}\right) \rightharpoonup \frac{1}{2\sqrt{\pi}}x^{-3/2}\int_0^\infty \exp\left(-\frac{u^2}{4x}\right)v(u)\,du; \tag{2.10.1}$$

for any number $c>0$

$$V(cs+\sqrt{s}) \rightharpoonup \frac{1}{2\sqrt{\pi}}\int_0^{x/c} u(x-cu)^{-3/2}\exp\left(-\frac{u^2}{4(x-cu)}\right)v(u)\,du. \tag{2.10.2}$$

The functions $g(x,\alpha,1)$ with $0<\alpha<1$ provide an opportunity for generalizing these relations.

THEOREM 2.10.1. *If $V(s) \rightharpoonup v(x)$, then the following relations hold for any $0<\alpha<1$ and any $c>0$* (2.33) :

$$V(s^\alpha) \rightharpoonup u^{-1/\alpha}g(xu^{-1/\alpha},\alpha,1)v(u)\,du, \tag{2.10.3}$$

$$V(cs+s^\alpha) \rightharpoonup \int_0^{x/c} u^{-1/\alpha}g(n^{-1/\alpha}(x-cu),\alpha,1)v(u)\,du. \tag{2.10.4}$$

In particular, for $\alpha=\frac{1}{3}$

$$V(s^{1/3}) \rightharpoonup \frac{x^{-3/2}}{3\pi}\int_0^\infty u^{3/2}K_{1/3}\left(\frac{2}{3\sqrt{3}}u^{3/2}x^{-1/2}\right)v(u)\,du, \tag{2.10.5}$$

$$V(cs+s^{1/3}) \rightarrow \frac{1}{3\pi}\int_0^x \left(\frac{u}{x-cu}\right)^{3/2} \times K_{1/3}\left(\frac{2}{3\sqrt{3}}\left(\frac{u^2}{x-cu}\right)^{1/2}\right) v(u)\,du, \qquad (2.10.6)$$

where $K_{1/3}$ is the Macdonald function of order $1/3$.

PROOF. By (2.6.10) and (2.1.2), for any $x>0$

$$\exp(-xs^\alpha) = \int_0^\infty e^{-su} x^{-1/\alpha} g(x^{-1/\alpha}u, \alpha, 1)\,du. \qquad (2.10.7)$$

Using this equality and the integral expression for $V(s^\alpha)$, we obtain (2.10.3) after changing the order of integration.

Further, by the same equality (2.10.7),

$$\begin{aligned} V(cs+s^\alpha) &= \int_0^\infty \exp(-csy - s^\alpha y)v(y)\,dy \\ &= \int_0^\infty\int_0^\infty \exp(-s(cy+u))y^{-1/\alpha}g(y^{-1/\alpha}u,\alpha,1)\,dy\,du. \end{aligned}$$

The change of variable $u = x - cy$ gives us an integral of the form

$$\int_0^\infty \exp(-sx)U(x)\,dx,$$

where the function $U(x)$ coincides with the right-hand side of (2.10.4).

The relations (2.10.6) and (2.10.7) are obtained from (2.10.3) and (2.10.4) after we replace $g(u, \frac{1}{3}, 1)$ by its expression in (2.8.31).

We say in §2.8 that the functions $g(x,\alpha,\theta)$ are solutions of various types of integral and integrodifferential equations, and even of special types of ordinary differential equations in the case of rational $\alpha \neq 1$. And though the function $g(x,\alpha,\theta)$ represents only one of the solutions, a detailed analysis of its analytic extensions may possibly reveal (as can be seen by the example of an analysis of the Bessel and Whittaker equations) other linearly independent solutions.

Here it is apparently the equations with the densities $g(x,p/q,1)$ $(p<q)$ as solutions which should be considered first and foremost.

A good illustration is the case $p=1$ and $q=n+1\geq 2$ connected with the equation (see (2.8.26))

$$y^{(n)}(\xi) = (-1)^n \xi y(\xi)/(n+1), \qquad \xi > 0,$$

which as a solution the function

$$y(\xi) = \xi^{-(n+2)} g(\xi^{-(n+1)} 1/(n+1), 1).$$

Of interest is the connection between the densities of extremal stable laws and the Mittag-Leffler function ([1.5])

$$E_\sigma(x) = \sum_{n=0}^{\infty} \frac{x^n}{\Gamma(n\sigma+1)}, \qquad \sigma > 0.$$

THEOREM 2.10.2. *For any* $0 < \alpha < 1$ *and any complex* s

$$\alpha E_\alpha(-s) = \int_0^\infty \exp(-sx) x^{-1-1/\alpha} g(x^{-1/\alpha}, \alpha, 1)\, dx. \tag{2.10.8}$$

If $\frac{1}{2} \le \alpha < 1$, *then* ([2.34])

$$\alpha E_\alpha(-s) = \int_0^\infty \exp(-sx) g(x, 1/\alpha, -1)\, dx. \tag{2.10.9}$$

PROOF. Let $s > 0$. It is known [4] that in this case the function $E_\alpha(-s)$ has the representation

$$E_\alpha(-s) = \frac{1}{2\pi i\alpha} \int_L \exp(z^{1/\alpha}) \frac{dz}{z+s}, \tag{2.10.10}$$

where the contour of integration L consists of the following three parts ($z = x + iy$): the line L_1 given by $y = -(\tan\varphi)x$, where x varies from $x = +\infty$ to $x = h$, with $h > 0$ and $\pi\alpha/2 < \varphi < \pi\alpha$; the circular arc L_2 given by $|z| = h/\cos\varphi$, $-\varphi \le \arg z \le \varphi$; and the reflection L_3 of L_1 with respect to the x-axis. We replace $(z+s)^{-1}$ in (2.10.10) by the equivalent integral

$$\int_0^\infty \exp(-(z+s)u)\, du.$$

The double integral thus obtained converges absolutely; hence we can change the order of integration. Then

$$E_\alpha(-s) = \int_0^\infty \exp(-su) f_\alpha(u)\, du,$$

where the function $f_\alpha(u)$ can be transformed by integration by parts:

$$\begin{aligned} f_\alpha(u) &= \frac{1}{2\pi i\alpha} \int_L \exp(z^{1/\alpha} - zu)\, dz \\ &= \frac{1}{2\pi i\alpha u} \int_L \exp(z^{1/\alpha} - zu)\, dz^{1/\alpha}. \end{aligned}$$

We make a change of variable, setting $z = \varsigma^\alpha/u$:

$$f_\alpha(u) = \frac{1}{\alpha} u^{-1-1/\alpha} \left(\frac{1}{2\pi i} \int_{L^*} \exp(-\varsigma^\alpha + \varsigma u^{-1/\alpha})\, d\varsigma \right).$$

The contour L^* is the image of L under this change of variable. The integral in parentheses is the inverse Laplace transform of the function $\exp(-s^\alpha)$ $(0 < \alpha < 1)$ and, consequently, represents the function $g(u^{-1/\alpha}, \alpha, 1)$ according to (2.6.10).

The equality (2.10.9) is obtained from (2.10.8) by using the duality law for the densities (2.3.3) and the fact that the pair $\alpha < 1$, $\theta = 1$ in the form (C) corresponds to the pair $\alpha' = 1/\alpha$, $\beta = -1$ in the form (B).

The left-hand sides of (2.10.8) and (2.10.9) can be extended analytically from the half-line $s > 0$ to the whole complex plane, as is clear from the definition of the function $E_\alpha(x)$. On the other hand, $x^{-1-1/\alpha} g(x^{-1/\alpha}, \alpha, 1)$ decreases with increasing x more rapidly than any function $\exp(-cx)$, $c > 0$. Therefore, the right-hand side can also be extended to the whole complex plane, i.e., the equalities (2.10.8) and (2.10.9) are valid for any complex s.

Another interesting connection between the functions $E_\sigma(x)$ and $g(x, \alpha, 1)$ reveals itself when the following equality (obtained in [32] for any $\sigma > 0$) is generalized:

$$E_{\sigma/2}(x) = \frac{1}{\sqrt{\pi}} \int_0^\infty E_\sigma(xu^\sigma) \exp\left(-\frac{u^2}{4}\right) du. \tag{2.10.11}$$

THEOREM 2.10.3. *Suppose that* $0 < \alpha < 1$ *and* $\sigma > 0$. *Then for any complex* s

$$E_{\alpha\sigma}(s) = \int_0^\infty E_\sigma(su^{-\alpha\sigma}) g(u, \alpha, 1)\, du. \tag{2.10.12}$$

PROOF. We assume first that $s = -x < 0$ and compute the Mellin transform of the left and right sides of (2.10.12).

Let $0 < \alpha < 1$ and $0 < p < 1$. Using the representation (2.10.8) and the explicit form (3.0.2) of the Mellin transform $M(s, \alpha, 1)$ of $g(x, \alpha, 1)$, we find that

$$\begin{aligned}
\int_0^\infty x^{p-1} E_\alpha(-x)\, dx &= \frac{1}{\alpha} \int_0^\infty u^{-1-1/\alpha} g(u^{-1/\alpha}, \alpha, 1)\, du \int_0^\infty e^{-xu} x^{p-1}\, dx \\
&= \Gamma(p) \int_0^\infty u^{-p}\, d(1 - G(u^{-1/\alpha}, \alpha, 1)) \\
&= \frac{\Gamma(p)\Gamma(1-p)}{\Gamma(1-\alpha p)}.
\end{aligned}$$

Consequently, the Mellin transform of the left-hand side of (2.10.12) is equal to $\Gamma(p)\Gamma(1-p)/\Gamma(1-\alpha\sigma p)$. On the right-hand side we have the M-convolution of two functions (see Chapter 3 about this). Therefore, if $0 < \sigma < 1$, then the Mellin transform of the right-hand side is equal to

$$\frac{\Gamma(1-\sigma p)}{\Gamma(1-\alpha\sigma p)} \frac{\Gamma(p)\Gamma(1-p)}{\Gamma(1-\sigma p)} = \frac{\Gamma(p)\Gamma(1-p)}{\Gamma(1-\alpha p)}.$$

Thus, the Mellin transforms of the right- and left-hand sides of (2.10.12) coincide. Consequently, (2.10.12) itself is true in this particular case when $0 < \sigma < 1$. It now remains to note that $E_\sigma(s)$ represents an analytic function of σ in the half-plane $\operatorname{Re}\sigma > 0$, and we can thus extend (2.10.12) not only to the whole complex s-plane but to the whole semi-axis $\sigma > 0$.

There is one more consideration which speaks for regarding the densities $g(x, \alpha, \beta)$ as a class of special functions. The so-called Airy function (or Airy integral) is well known in the theory of special functions. We have already encountered this function at the end of §2.8. It has the form

$$A(x) = \frac{1}{\pi} \int_0^\infty \cos(t^3 + tx)\, dt = \frac{1}{2\pi} \int \exp(-it^3 - itx)\, dt \tag{2.10.13}$$

and can be expressed in terms of Bessel functions as follows: for $x > 0$

$$\begin{aligned} A(x) &= \frac{\sqrt{x}}{3\sqrt{3}} \left[I_{-1/3}\left(\frac{2x\sqrt{x}}{3\sqrt{3}}\right) - I_{1/3}\left(\frac{2x\sqrt{x}}{3\sqrt{3}}\right) \right] \\ &= \frac{\sqrt{x}}{3\pi} K_{1/3}\left(\frac{2x\sqrt{x}}{3\sqrt{3}}\right) > 0, \\ A(-x) &= \frac{\sqrt{x}}{3\sqrt{3}} \left[J_{-1/3}\left(\frac{2x\sqrt{x}}{3\sqrt{3}}\right) + J_{1/3}\left(\frac{2x\sqrt{x}}{3\sqrt{3}}\right) \right]. \end{aligned}$$

The asymptotic behavior of the functions $J_\nu(x)$ and $K_\nu(x)$ is well known (see 8.351(1) and 8.452(6) in [27]), and it makes it possible for us to trace the behavior of the function $A(x)$ as $x \to \infty$ and $x \to -\infty$. Namely, as $x \to \infty$

$$\begin{aligned} A(x) &\sim \frac{(3x)^{-1/4}}{2\sqrt{\pi}} \exp\left(-\frac{2x\sqrt{x}}{3\sqrt{3}}\right), \\ A(-x) &\sim \frac{(3x)^{-1/4}}{\sqrt{3\pi}} \cos\left(\frac{2x\sqrt{x}}{3\sqrt{3}} - \frac{\pi}{4}\right). \end{aligned}$$

Thus, $A(x)$ is positive on the semi-axis $x > 0$ and decreases very rapidly as $x \to \infty$. At the same time, it is of variable sign on the negative semi-axis and decreases no more rapidly than a power of x.

A formula for inverting the Fourier transform of the function $\exp(-it^3)$ can be seen in (2.10.13). But then

$$\int \exp(itx) A(x)\, dx = \exp\{(it)^3\}. \tag{2.10.14}$$

It is clear from the asymptotic expressions for $A(x)$ that the Fourier transform (2.10.14) admits an analytic extension with respect to t into the half-plane $\operatorname{Im} t \leq 0$. But this is equivalent to the existence of the two-sided Laplace transform

$$\int \exp(sx) A(x)\, dx = \exp(s^3)$$

in the half-plane $\operatorname{Re} s \geq 0$. This formula is very reminiscent of (2.6.10) in the case $1 < \alpha \leq 2$ and suggests regarding the function $A(x)$ as a "trans-stable distribution" corresponding to the value $\alpha = 3$. This interpretation is further corroborated by the fact that on the semi-axis $x > 0$ the function $A(x)$ turns out to be connected, not unlike the duality law, with the stable distribution $g(x, \frac{1}{3}, 1)$:

$$xA(x) = x^{-3} g(x^{-3}, \tfrac{1}{3}, 1). \tag{2.10.15}$$

The relation (2.3.3) embraces only the part of the stable laws with $\alpha < 1$. The equality (2.10.15) suggests using the other part of them to define "trans-stable" measures (more precisely, charges) under the assumption that the densities A corresponding to them form a three-parameter set of functions with the property that

$$A(x\lambda^{1/\alpha}, \alpha, \theta, \lambda) = x^{-1-\alpha} g(x^{-\alpha}, \alpha', \theta') \tag{2.10.16}$$

for any $x > 0$, $\lambda > 0$, $\alpha' = 1/\alpha < 1$, and $|\theta'| \leq 1$, where $1 + \theta = \alpha'(1 + \theta')$.

The fact that the right-hand side of (2.10.16) is an entire analytic function enables us to extend A analytically from the semi-axis $x > 0$ to the whole x-axis.

The next section is devoted to the realization of these ideas.

§2.11. Trans-stable functions and trans-stable distributions

The class $\mathfrak{W}$ of strictly stable laws is the starting object for the following generalizations. As we know (Theorem C.4), each law in this class is characterized in the form (C) by three parameters $(\alpha, \theta, \lambda)$, or by the three parameters (α, ρ, λ), where $\rho = (1 + \theta)/2$. In this case the parameter λ is purely a scale parameter, because

$$g_C(x, \alpha, \rho, \lambda) = \lambda^{-1/\alpha} g_C(x\lambda^{-1/\alpha}, \alpha, \rho, 1).$$

Therefore, without any detriment to the purposes of the projected generalization we can set $\lambda = 1$ and operate with only the two parameters α and ρ. According to (2.4.6) and (2.4.8), the densities $g_C(x, \alpha, \rho)$ can be represented by convergent power series. We use these representations here in a somewhat extended variant, including also the case $\alpha = 1$. Namely:

If $0 < \alpha < 1$, $0 \leq \rho \leq 1$, and $x > 0$, or if $\alpha = 1$, $0 < \rho < 1$, and $x > 1$, then

$$g_C(x, \alpha, \rho) = \frac{1}{\pi} \sum_{n=1}^{\infty} (-1)^{n-1} \frac{\Gamma(n\alpha + 1)}{\Gamma(n + 1)} \sin(\pi\alpha\rho n) x^{-\alpha n - 1}. \tag{2.11.1}$$

If $1 < \alpha \leq 2$, $1 - 1/\alpha \leq \rho \leq 1/\alpha$, and $x \in R^1$, or if $\alpha = 1$, $0 < \rho < 1$, and $|x| < 1$, then

$$g_C(x, \alpha, \rho) = \frac{1}{\pi} \sum_{n=1}^{\infty} (-1)^{n-1} \frac{\Gamma(n/\alpha + 1)}{\Gamma(n+1)} \sin(\pi \rho n) x^{n-1}. \tag{2.11.2}$$

It is not hard to see that the series (2.11.1) and (2.11.2) remain convergent if, keeping the restrictions on the variation of x, we extend the domain of variation of (α, ρ) in the first case to the strip $0 < \alpha \leq 1$, $\rho \in R^1$, and in the second case to the half-plane $\alpha \geq 1$, $\rho \in R^1$.

It is thus possible to define the functions $g(x, \alpha, \rho)$, $\alpha > 0$, $\rho \in R^1$, for the indicated values of x in the representations of the densities (2.11.1) and (2.11.2).

This enables us, in turn, to define in the complex plane $\mathbf{Z}$ (possibly with cuts) the family of analytic functions

$$\mathfrak{T} = \{\sigma(z, \alpha, \rho) \colon \alpha > 0,\ \rho \in R^1\},$$

by setting them equal to the corresponding functions $g(x, \alpha, \rho)$ on the parts of the real axis $x = \operatorname{Re} z$ where the latter were defined. It is easy to verify that in the case $\alpha = 1$ the analytic functions defined by means of (2.11.1) and (2.11.2) coincide if the parameter ρ is given the same value, i.e., $\sigma(z, 1, \rho)$ represents a single analytic function, independent of the way in which it was defined.

We mention several simple properties of the functions in $\mathfrak{T}$.

1*. If $\alpha > 1$, then, for any ρ, $\sigma(z, \alpha, \rho)$ is an entire analytic function.

2*. If $\alpha = 1$ and ρ is not an integer, then $\sigma(z, 1, \rho)$ is a meromorphic analytic function on $\mathbf{Z}$ with two simple conjugate poles that are solutions of the equation (see (2.3.5a))

$$z^2 + 2z \cos \pi\rho + 1 = 0.$$

The function $\sigma(z, 1, \rho)$ is equal to zero for ρ and integer.

3*. If $0 < \alpha < 1$, $x > 0$, and $\alpha\rho$ is an integer, then $\sigma(x, \alpha, \rho) = 0$.

4*. If $\alpha > 1$ and $x > 0$ or if $\alpha = 1$ and $0 < x < 1$, then

$$x^{-\alpha} \sigma(x^{-\alpha}, 1/\alpha, \alpha\rho) = x\sigma(x, \alpha, \rho). \tag{2.11.3}$$

This equality is established by comparing the power series representing its left and right sides.

Properties 3* and 1* enable us to understand what the functions $\sigma(z, \alpha, \rho)$ are in the case when $\alpha < 1$ and $\alpha\rho$ is not an integer. Indeed, by (2.11.3), if $0 < \alpha < 1$ and $\alpha\rho$ is not an integer, then

$$\sigma(x, \alpha, \rho) = x^{-\alpha} \sigma(x^{-\alpha}, 1/\alpha, \alpha\rho). \tag{2.11.4}$$

According to the property 1, the function $\sigma(z, 1/\alpha, \alpha\rho)$ is entire, and z^α is the elementary multivalued function on $\mathbf{Z}$ with branch points at $z = 0$ and $z = \infty$. Consequently, the superposition of these functions on the right-hand side of (2.11.4), and with it the analytic extension of the left-hand side, i.e., the function $\sigma(z, \alpha, \rho)$, is a multivalued analytic function on $\mathbf{Z}$ with branch points $z = 0$ and $z = \infty$.

In the case when $0 < \alpha < 1$ and $\alpha\rho$ is not an integer, $\sigma(z, \alpha, \rho)$ wll be understood to be the principal branch of this function defined in the whole complex plane with a cut along the ray $\arg z = \pi$ (the lower edge of the cut is included in the domain of definition).

DEFINITION. Let $\mathfrak{T}_0$ be the set of analytic functions formed by all the functions $\sigma(z, \alpha, \rho)$ in $\mathfrak{T}$ with $\alpha \geq 1$ or with $0 < \alpha < 1$ and $\alpha\rho$ an integer, together with the principal branches of all the functions $\sigma(z, \alpha, \rho)$ in $\mathfrak{T}$ with $0 < \alpha < 1$ and $\alpha\rho$ not an integer.

The functions in $\mathfrak{T}_0$ are called *trans-stable functions.*

Let us proceed to the properties of trans-stable functions ([2.35]).

PROPERTY 1. *Each function $\sigma(z, \alpha, \rho) \in \mathfrak{T}_0$ is periodic in the variable ρ, with period $T_\alpha = 2\min(1, 1/\alpha)$, i.e.,*

$$\sigma(z, \alpha, \rho + T_\alpha) = \sigma(z, \alpha, \rho).$$

PROPERTY 2. *For any function $\sigma \in \mathfrak{T}_0$*

$$\sigma(z, \alpha, -\rho) = -\sigma(z, \alpha, \rho). \tag{2.11.5}$$

PROPERTY 3. *If $\alpha \geq 1$, then for any complex number z*

$$\sigma(-z, \alpha, \rho) = \sigma(z, \alpha, 1 - \rho), \tag{2.11.6}$$

$$z\sigma(z, \alpha, \rho) = z^{-\alpha}\sigma(z^{-\alpha}, 1/\alpha, \alpha\rho). \tag{2.11.7}$$

The left-hand side of (2.11.7) is an entire function, and the right-hand side is a superposition of principal branches of multivalued functions. If these functions are defined on complex plans with identical cuts, then their superposition has removable singular points on the whole curve of the cut; these singularities can be removed by appropriately extending the definition of the superposition. This gives an entire analytic functionon the right-hand side of (2.11.7).

Properties 1–3 can be verified by elementary transformations of the series representing the functions in $\mathfrak{T}_0$.

PROPERTY 4. $\sigma(x, \alpha, \rho) = 0$ *on the semi-axis $x = \operatorname{Re} z > 0$ if and only if $\rho\min(1, \alpha)$ is an integer.*

The sufficiency of this condition is obvious. The necessity becomes clear if we write out the first term of the asymptotic expression for the function $\sigma(x, \alpha, \rho)$ as $x \to \infty$ in the case $\alpha \leq 1$ and as $x \to 0$ in the case $\alpha \geq 1$.

PROPERTY 5. *Suppose that $\alpha > 0$, $0 \leq \rho \leq T_\alpha$, and $x > 0$. Then $\sigma(x, \alpha, \rho) \geq 0$ if and only if $0 \leq \rho \leq \min(1, 1/\alpha)$.*

This property can be established as a consequence of the general duality relation (2.11.7), the positivity of the function $\sigma(x, \alpha, \rho) = g_C(x, \alpha, \rho)$ in the case when $0 < \alpha < 1$ and $0 \leq \rho \leq 1$, and the periodicity of $\sigma(x, \alpha, \rho)$ with respect to the parameter ρ.

Figure 4a should help familiarize the reader with these properties; the domains corresponding to various properties are indicated there in the half-plane of values of the parameters (α, ρ). The numbers with asterisks correspond to the domains obtained in passing from $\alpha \geq 1$ to $\alpha^* = 1/\alpha$ in the duality relation (2.11.7). Here one must bear in mind that the mapping of the domains takes place in the form of a curved reflection with respect to the line $\alpha = 1$, i.e., the point $(2, \frac{1}{2})$ passes into $(\frac{1}{2}, 1)$, the curve $\alpha \geq 1$, $\rho = 1/\alpha$ passes into the segment $0 < \alpha \leq 1$, $\rho = 1$, and so on.

The domains marked by the numbers $\mathbb{1}$ and $\mathbb{1}^*$ are doubled because of the periodicity property of the function $\sigma(z, \alpha, \rho)$. However, while in the case $\alpha \geq 1$ we have the period $T_\alpha = 2$ and the domain $\mathbb{1}$ is simply shifted, in the case $0 < \alpha < 1$ the period T_α is $2/\alpha$ and the duplicates of the domain $\mathbb{1}^*$ are distorted.

We should note especially the equality (2.11.6). It shows that the values of ρ symmetric with respect to the half-line $\alpha \geq 1$, $\rho = \frac{1}{2}$ correspond, as it were, to mirror reflections of the functions σ. At the same time, the functions σ with values $0 < \alpha < 1$ do not have this property. This may seem strange, since the densities $g_C(x, \alpha, \rho)$ with values $0 < \alpha < 1$ have the following property (see (2.1.3)):

$$\begin{aligned} g_C(x, \alpha, \rho) &= g(x, \alpha, \theta) \\ &= g(-x, \alpha, -\theta) = g_C(-x, \alpha, 1-\rho). \end{aligned}$$

The disagreement between the properties of the functions g_C and σ is explained by the fact that σ is a branch of a single analytic function, while $g_C(x, \alpha, \rho)$ is a patching together of the halves $g_C(x, \alpha, \rho)$, $x > 0$, and $g_C(x, \alpha, \rho) = g_C(-x, \alpha, 1-\rho)$, $x < 0$, of different analytic functions. The duality relation (2.11.7) connects the halves (on the semi-axes $x > 0$ and $x < 0$) of one and the same analytic function $\sigma(x, \alpha, \rho)$, $\alpha > 1$, with different analytic functions σ having one and the same value of the parameter $\alpha' = 1/\alpha$, but different values of the second parameter ρ'.

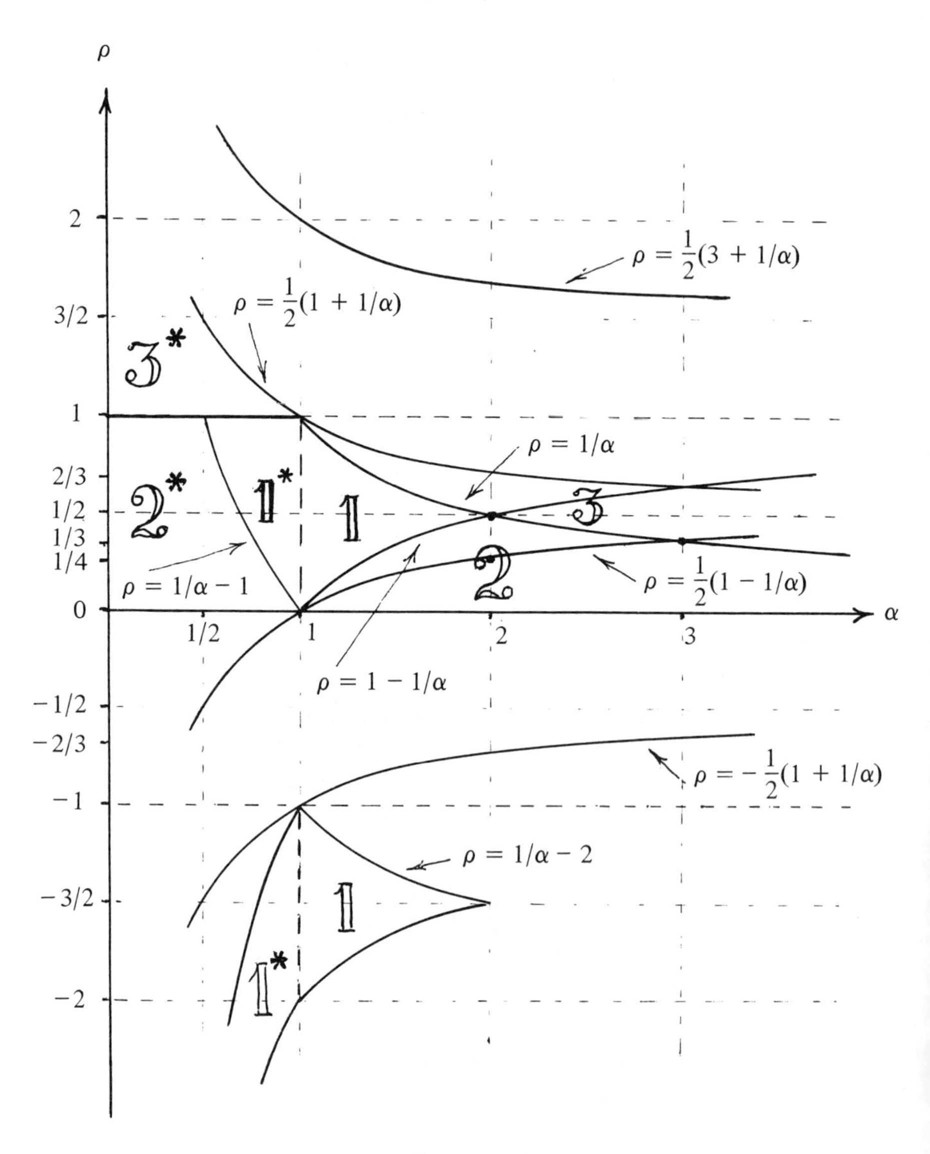

FIGURE 4a

PROPERTY 6. *Suppose that $\alpha \geq 1$ and $|\rho| \leq 1$, i.e., we consider points (α, ρ) in a strip where the periodicity with respect to ρ does not yet manifest itself. The Fourier transform $\hat{\sigma}(t, \alpha, \rho)$ of the function $\sigma(x, \alpha, \rho)$ exists if and only if $|\rho \pm \frac{1}{2}| \leq (2\alpha)^{-1}$, which corresponds to the domain bounded by the curves $\rho = (1 + 1/\alpha)/2$ and $\rho = (1 - 1/\alpha)/2$ and the domain symmetric to it* *with respect to the α-axis. In the first domain*

$$\log \hat{\sigma}(t, \alpha, \rho) = -|t|^\alpha \exp\{i\pi\alpha(\tfrac{1}{2} - \rho) \operatorname{sgn} t\}. \tag{2.11.8}$$

In particular, $\hat{\sigma}(t, \alpha, \frac{1}{2}) = \exp(-|t|^\alpha)$, $\alpha \geq 1$. The function σ corresponding to the parameters $(\alpha, \frac{1}{2})$ are symmetric.

For the points (α, ρ) on the curves $\rho = (1 \pm 1/\alpha)/2$ the Fourier transforms $\hat{\sigma}$ of the functions σ have, respectively, the forms

$$\hat{\sigma}(t, \alpha, \rho) = \exp(\pm i|t|^\alpha \operatorname{sgn} t). \tag{2.11.9}$$

The existence of the Fourier integral $\hat{\sigma}$ depends directly on the behavior of σ as $x \to \infty$ and as $x \to -\infty$. The bounds $\rho = \frac{1}{2}(1+1/\alpha)$ and $\rho = -\frac{1}{2}(1+1/\alpha)$ distinguish in the half-plane $\alpha \geq 1$ a domain in whose interior $\sigma(x, \alpha, \rho) \to 0$ as $x \to \infty$. The proof of this fact is based on the use of integral representations of the type (2.2.8) and (2.2.10) for the functions σ. According to (2.11.6), $\sigma(x, \alpha, \rho) \to 0$ as $x \to -\infty$ interior to the domain bounded by $\alpha \geq 1$, $\rho = 1$ and $\alpha \geq 1$, $\rho = (1 - 1/\alpha)/2$ and interior to the domain symmetric to it with respect to the α-axis.

In only four cases do the densities of stable laws have explicit expressions in terms of elementary functions (we have in mind, of course, the main parameters α and β). They were listed in §2.2. In some cases the densities of stable laws can be expressed with the help of special functions (see §2.8).

So far, among the trans-stable functions that are not densities of stable laws no examples analogous to the four mentioned above are known, with the exception of the obvious negative variants which appear due to Property 2. However, it is sometimes possible to express the functions σ in terms of special functions. We already encountered one such example in the preceding section: the Airy integral $A(x) = \sigma(x, 3, \frac{1}{3})$.

We indicate another case when the function σ can be written with the help of special functions: Fresnel integrals. These integrals were used for writing the densities $g(x, \frac{1}{2}, 0)$ in §2.8 (formula (2.8.30)).

By using the duality relation and (2.8.30) it is possible to write the function $\sigma(x, 2, \frac{1}{4})$ as follows:

$$\begin{aligned}
\sigma\left(x, 2, \frac{1}{4}\right) &= x^{-3}\sigma\left(x^{-2}, \frac{1}{2}, \frac{1}{2}\right) \\
&= x^{-3} g_C\left(x^{-2}, \frac{1}{2}, \frac{1}{2}\right) = x^{-3} g\left(x^{-2}, \frac{1}{2}, 0\right) \\
&= \frac{1}{2\sqrt{2\pi}}\left[\left(\frac{1}{2} - C(x^2)\right)\cos\frac{x^2}{4} + \left(\frac{1}{2} - S(x^2)\right)\sin\frac{x^2}{4}\right].
\end{aligned}$$

We note that to the function $\sigma(x, 2, \frac{1}{4})$, like $\sigma(x, 3, \frac{1}{3})$, there corresponds a Fourier transform of the form (2.11.9):

$$\hat{\sigma}(t, 2, \tfrac{1}{4}) = \exp(-it^2 \operatorname{sgn} t). \tag{2.11.10}$$

PROPERTY 7. *In the case when $\alpha > 1$ and $|\rho| \leq 1$ the function $|\sigma(x, \alpha, \rho)|$ decreases as $x \to \infty$ more rapidly than $\exp(-cx)$ for any $c > 0$ if and only if $\alpha|\rho| = 1$ (i.e., the points (α, ρ) must lie on the curves $\rho = \pm 1/\alpha$). The symmetry property* (2.11.6) *indicates that the function $\sigma(x, \alpha, \rho)$ decreases this rapidly as $x \to -\infty$ only if the point (α, ρ) lies on the curves $|\rho| = 1 - 1/\alpha$.*

This implies the following facts (in the strip $\alpha \geq 1$, $|\rho| \leq 1$):

a) The function $|\sigma(x, \alpha, \rho)|$ is decreasing more rapidly than an exponential both as $x \to \infty$ and as $x \to -\infty$ only in the cases when the points (α, ρ) lie on the intersection of the curves $|\rho| = 1/\alpha$ and $|\rho| = 1 - 1/\alpha$, i.e., when $\alpha = 1$, $\rho = 0$ or when $\alpha = 2$, $\rho = \pm\frac{1}{2}$. The first point corresponds to the case $\sigma(x, 1, 0) = 0$, and the two other cases correspond to the density of the normal law $\sigma(x, 2, \frac{1}{2})$ and its negative variant

$$\sigma(x, 2, -\tfrac{1}{2}) = -\sigma(x, 2, \tfrac{1}{2}).$$

b) Let s be a complex number in the half-plane $\operatorname{Re} s \geq 0$. The two-sided Laplace transform

$$\varsigma(s, \alpha, \rho) = \int \exp(sx)\sigma(x, \alpha, \rho)\, dx$$

exists if and only if $1 \leq \alpha \leq 3$ and $|\rho| = 1/\alpha$. In these cases

$$\varsigma(s, \alpha, \pm 1/\alpha) = \pm \exp(s^\alpha).$$

c) For any complex number s the one-sided Laplace transform of $\sigma(x, \alpha, \rho)$ exists if and only if $|\rho| = 1/\alpha$. In these cases

$$\int_0^\infty \exp(sx)\sigma(x, \alpha, \pm 1/\alpha)\, dx = \pm E_{1/\alpha}(s)/\alpha,$$

where $E_a(s)$ is the Mittag-Leffler function (see Theorem 2.10.2).

PROPERTY 8. *Suppose that $\alpha > 0$ and $|\rho| \leq T_\alpha/2$. Consider the Mellin transform of the function $x\sigma(x, \alpha, \rho)$:*

$$R(s, \alpha, \rho) = \int_0^\infty x^s \sigma(x, \alpha, \rho)\, dx.$$

If $0 < \alpha \leq 1$, then $R(s, \alpha, \rho)$ exists for values s in some neighborhood of the point $s = 0$ for any ρ. But if $\alpha > 1$, then $R(s, \alpha, \rho)$ exists for s in some *neighborhood of zero of and only if $|\rho| \le (1+1/\alpha)/2$. If $R(s,\alpha,\rho)$ exists, then it exists for all $-1 < s < \alpha$, and*

$$R(s,\alpha,\rho) = \frac{\sin(\pi\rho s)}{\sin(\pi s)}\frac{\Gamma(1-s/\alpha)}{\Gamma(1-s)}. \tag{2.11.11}$$

The first part of this assertion follows directly from the representation of σ by the series (2.11.1). The second part can be proved like Property 7 with use of the integral expressions for σ. The positivity of $\sigma(x,\alpha,\rho)$ in the case $\alpha > 2$, $0 < \rho \le 1/\alpha$ enables us to introduce a distribution connected with σ. Since, according to (2.11.11),

$$\int_0^\infty \sigma(x,\alpha,\rho)\,dx = \rho > 0,$$

the distribution can be given by the density

$$\rho(x,\alpha,\rho) = \sigma(x,\alpha,\rho)/\rho.$$

This distribution is said to be *trans-stable*, and the random variables corresponding to it are denoted by $Z(\alpha,\rho)$.

The fact that the notation for trans-stable random variables is identical with that for the cutoffs $Z(\alpha,\rho)$ introduced in §3.1 is not accidental.

It turns out that trans-stable random variables satisfy the same relations as the cutoffs (of course, with a corresponding broadening of the domain of variation of the parameters α and ρ). Naturally, this is connected with the fact that trans-stable random variables $Z(\alpha,\rho)$ have transforms $R(s,\alpha,\rho) = \mathsf{E}Z^s(\alpha,\rho)$ of precisely the same form as the transforms $M(s,\alpha,\rho) = \mathsf{E}Z^s(\alpha,\rho)$ for the cutoffs (see (3.0.2)).

As in the case of cutoffs, the definition of trans-stable random variables corresponding to the value $\rho = 0$ can be extended in a natural way by condition (3.1.4) in the case when $\alpha > 2$.

The following can serve as examples of relations between trans-stable random variables analogous to those to be considered in detail in §§3.2–3.4.

If $\alpha^* > 1$, $\alpha > 1$, and $0 \le \rho \le 1/\alpha$, then

$$z(\alpha^*, 1/\alpha^*) z^{1/\alpha^*}(\alpha,\rho) \stackrel{d}{=} z(\alpha\alpha^*, \rho/\alpha^*),$$

$$z(\alpha, 1/\alpha)/z(\alpha,\rho) \stackrel{d}{=} z(1,\rho)$$

(recall that, according to the convention we have adopted, the random variables written especially on the left-hand side of the equalities are assumed to be independent).

CHAPTER 3

Special Properties of Laws in the Class $\mathfrak{W}$

§3.0. Introduction

The analytic properties considered in the preceding chapter for the laws in the class $\mathfrak{S}$ are based on explicit expressions for the characteristic functions of these laws. The fact that the characteristic functions are an analytic tool especially suited for working with sums of independent random variables was reflected in the material of Chapter 2. For example, this is very clear from §2.1, where the properties of stable laws are written in terms of sums of independent random variables.

Mellin transforms and characteristic transforms of probability distributions serve as the analytic basis for the investigations carried out below. These transforms fulfill the same purpose in the multiplication scheme for independent random variables (the M-scheme) that characteristic functions have in the addition scheme. Therefore, a large part of the material in the present chapter involving various relations for distributions in $\mathfrak{W}$ is presented in terms of products of independent random variables.

As we saw in I.5, the form (C) is a natural parametrization system for strictly stable laws. Recall that each distribution $G \in \mathfrak{W}$ is uniquely identified by a triple $(\alpha, \theta, \lambda)$ of parameters varying within bounds given by

$$0 < \alpha \leq 2, \quad |\theta| \leq \theta_\alpha = \min(1, 2/\alpha - 1), \quad \lambda > 0$$

(the case $\alpha = 1$, $|\theta| = 1$ corresponds to a degenerate distribution at the point $x = \theta$), or by a triple (α, ρ, λ) of parameters, where $\rho = (1+\theta)/2$ varies within the bounds

$$1 - \min(1, 1/\alpha) \leq \rho \leq \min(1, 1/\alpha). \tag{3.0.1}$$

In the distributions $G_C(x, \alpha, \theta, \lambda)$ the parameter λ has to do with the choice of the scale on the x-axis. This is shown by the equality

$$G_C(x, \alpha, \theta, \lambda) = G_C(x\lambda^{-1/\alpha}, \alpha, \theta, 1),$$

which is valid for any x and any admissible values of the parameters α, θ, and λ.

If $\alpha \neq 1$, then $\lambda_C = \lambda_B$, and thus the condition $\lambda_C = 1$ distinguished the subclass of standard distributions in $\mathfrak{W}$. In the case $\alpha = 1$ the choice $\lambda_C = 1$ does not correspond to a standard distribution. If $|\theta| \neq 1$, then

$$G_C(x, 1, \theta, 1) = G_B(x, 1, 0, \gamma, (\pi^2/4 + \gamma^2)^{-1/2}),$$

so the set of distributions $G_C(x, 1, \theta, 1)$ standardized by the condition $\lambda_C = 1$ is a set of specially normalized Cauchy distributions whose mode biases do not go outside the interval $(-1, 1)$. The case $|\theta| = 1$ is a limit case for this set and corresponds to the distribution $G_C(x, 1, \theta, 1) = D(x - \theta)$ degenerate at the point θ.

The set of distributions singled out in $\mathfrak{W}$ by the condition $\lambda_C = 1$ is denoted by $\mathfrak{W}_1$. The nature of the material making up the main content of this chapter permits us to consider only the distributions in $\mathfrak{W}_1$ without loss of generality. For brevity let $G_C(x, \alpha, \theta) = G_C(x, \alpha, \theta, 1)$.

Further, the symbol $Y(\alpha, \theta)$, without a subscript C required, will stand for random variables with the distribution $G_C(x, \alpha, \theta)$. Along with the parameter system (α, θ) we use a slight variant of it—the system (α, ρ), where $\rho = (1 + \theta)/2$. To avoid confusion it is well to keep in mind that the letter θ always denotes the parameter in the first system, and ρ the parameter in the second system.

The results in §2.6 give the following expressions for the Mellin transform (normalized by its value at the point $s = 0$ under the assumption that this value is positive) and for the characteristic transform of distributions in $\mathfrak{W}_1$:

$$\begin{aligned} M(s, \alpha, \rho) &= \frac{m_C(s, \alpha, \rho)}{\rho} = \frac{\sin \pi \rho s}{\rho \sin \pi s} \frac{\Gamma(1 - s/\alpha)}{\Gamma(1 - s)}, \\ W(s, \alpha, \theta) &= \begin{pmatrix} w_0(s, \alpha, \theta) & 0 \\ 0 & w_1(s, \alpha, \theta) \end{pmatrix}, \end{aligned} \tag{3.0.2}$$

where

$$\begin{aligned} w_k(s, \alpha, \theta) &= \mathsf{E}|Y(\alpha, \theta)|^s \, (\operatorname{sgn} Y(\alpha, \theta))^k \\ &= \frac{\cos \frac{\pi}{2}(k - \theta s)}{\cos \frac{\pi}{2}(k - s)} \frac{\Gamma(1 - s/\alpha)}{\Gamma(1 - s)}, \qquad k = 0, 1. \end{aligned} \tag{3.0.3}$$

In both transforms s can take any values in the strip $-1 < \operatorname{Re} s < \alpha$.

To avoid awkward formulas it makes sense to agree on some more notation. If X is a random variable, then for any complex number s we stipulate that*

$$X^s = |X|^s \operatorname{sgn} X, \tag{3.0.4}$$

with $0^s = 0$ for any s. If X is nonnegative with probability 1, then this generalized understanding of a power obviously coincides with the traditional one. Powers of random variables in the sense of (3.0.4) clearly have the basic properties of ordinary powers:

$$(X^s)^r = X^{sr}, \qquad X_1^s X_2^s = (X_1 X_2)^s,$$

but there is also an essential difference, since

$$X^0 = \operatorname{sgn} X, \qquad (-X)^s = -X^s$$

for any s.

Here it is appropriate to point out a peculiarity of equalities between random variables in the sense of "$\stackrel{d}{=}$". We have already encountered them in §2.1 and will make repeated use of them in this chapter. If $X' + X \stackrel{d}{=} X'' + X$ or

$$X'X \stackrel{d}{=} X''X, \tag{3.0.5}$$

then it does not follow in general that

$$X' \stackrel{d}{=} X''. \tag{3.0.6}$$

However, if the characteristic function $f_X(t)$ is nonzero for almost all t or, correspondingly, $w_k(t)_X$ (the elements of the characteristic transform $W_X(t)$ are so denoted; see §2.6) are nonzero almost everywhere, then this equality is valid. Of course, the converse is valid without any restrictions with regard to the distribution of X.

The use of relations between random variables (as done in §2.1) in place of writing out relations between the corresponding distributions makes even more sense in the present chapter than in §2.1. The fact is that we shall have to deal repeatedly with products $X = X_1 X_2$ of independent random variables, and the distribution function F of the product X is expressed in terms of the distribution functions F_1 and F_2 of the factors in a form considerably more cumbersome than, say, the convolution of F_1 and F_2. Even in the simplest case when F_1 and F_2 are continuous,

$$F(x) = \int_0^\infty F_1(x/y)\, dF_2(y) + \int_{-\infty}^0 (1 - F_1(x/y))\, dF_2(y). \tag{3.0.7}$$

---

*See the second part of Comment 3.1 at the end of the book.

In what follows we encounter only this case, so those wishing to carry over the relations between random variables to the language of the corresponding distributions should use this equality ([3.1]).*

§3.1. The concept of a cutoff of a random variable

DEFINITION. A *cutoff* of a random variable X with continuous distribution function $F(x)$ not entirely concentrated on the negative semi-axis is defined to be any positive random variable $\hat{X}$ with distribution function ([3.2])

$$\hat{F}(x) = \frac{F(x) - F(0)}{1 - F(0)} = \mathsf{P}(X < x | X \geq 0), \qquad x \geq 0.$$

There are constructions enabling us to construct cutoffs $\hat{X}$ as functions of the random variable X. One such construction is well known:

$$\hat{X} = \hat{F}_{-1}(F(X)) = F_{-1}\left(\frac{F(X) + c}{1 + c}\right), \quad \text{where } c = \frac{F(0)}{1 - F(0)} < \infty.$$

But this construction is trivially nonunique. For example, setting

$$S(x) = \begin{cases} x & \text{if } x \geq 0, \\ F_{-1}(1 - F(x)/c) & \text{if } x < 0, \end{cases}$$

we can assert that the random variable $\hat{X} = S(X)$ is a cutoff of X if $0 < c < \infty$.

Indeed, let $x > 0$. Then

$$\begin{aligned} \mathsf{P}(S(X) < x) &= \mathsf{P}(S(X) < x, X \geq 0) + \mathsf{P}(S(X) < x, X < 0) \\ &= \mathsf{P}(X < x, X \geq 0) \\ &\quad + \mathsf{P}(F_{-1}(1 - F(X)/c) < x, F(X) < F(0)) \\ &= \mathsf{P}(X < x | X \geq 0)(1 - F(0)) \\ &\quad + \mathsf{P}(c(1 - F(x)) < F(X) < F(0)). \end{aligned}$$

Since $\mathsf{P}(X < x | X \geq 0) = \hat{F}(x)$ and $F(X)$ has a uniform distribution on $(0, 1)$, it follows that

$$\begin{aligned} \mathsf{P}(S(X) < x) &= (1 - F(0))\hat{F}(x) = F(0) - c(1 - F(x)) \\ &= (1 - F(0))\hat{F}(x) + F(0)\hat{F}(x) = \hat{F}(x). \end{aligned}$$

Let us single out some useful properties of cutoffs.

1. If a is a positive constant, then

$$\widehat{(aX)} = a\hat{X}.$$

---

*Raised numbers in this form refer to the corresponding notes in the comments at the end of the book.

2. If $X_1 \overset{d}{=} X_2$, then

$$\hat{X}_1 \overset{d}{=} \hat{X}_2, \qquad (\widehat{-X_1}) \overset{d}{=} (\widehat{-X_2}),$$

and conversely (the last equality is not considered in the case when $\mathsf{P}(X_1 > 0) = 1$).

3. Let F_1 and F_2 be continuous distribution functions not entirely concentrated on the negative semi-axis, let $c_k = F_k(0)/(1 - F_k(0))$, and let $\mu(F_1, F_2)$ be one of the three well-known distances: the Lévy metric, the uniform metric, or the total variation distance. Then

$$\mu(\hat{F}_1, \hat{F}_2) \leq (1 + c_1)\mu(F_1, F_2) + |c_2 - c_1|. \tag{3.1.1}$$

From this it follows, in particular, that if $F_0, F_1, \dots$ is a sequence of such distribution functions, then the convergence $\mu(F_n, F_0) \to 0$ as $n \to \infty$ implies $\mu(\hat{F}_n, \hat{F}_0) \to 0$.

4. If the random variable X has a continuous quasi-symmetric distribution, i.e., $F(-x) = c(1 - F(x))$ for $x \geq 0$, where $0 \leq c < \infty$, then $\hat{X} \overset{d}{=} |X|$.

5. In the sets of cutoffs of independent random variables X_1 and X_2 there are cutoffs $\hat{X}_1$ and $\hat{X}_2$ that are independent random variables (for example, $\hat{X}_1 = S(X_1)$ and $\hat{X}_2 = S(X_2)$).

6. Let $\hat{X}$ be a cutoff of a random variable X having a finite Mellin transform $\mathfrak{M}(s) = \mathsf{E}X^s I(X > 0)$ in some strip $-c_1 < \operatorname{Re} s < c_2$, where I is the indicator function of an event. Then

$$\mathfrak{M}(s)/\mathfrak{M}(0) = \mathsf{E}\hat{X}^s. \tag{3.1.2}$$

All the properties except perhaps property 3 are of an elementary nature and do not require special proofs. In property 3 the cases of the uniform metric and the total variation distance are also uncomplicated. Therefore, we give a proof of the inequality (3.1.1) only for the case when μ is the Lévy metric. According to the definition of this metric, if $\varepsilon = \mu(F_1, F_2)$, then for any x

$$F_1(x - \varepsilon) - \varepsilon \leq F_2(x) \leq F_1(x + \varepsilon) + \varepsilon,$$

and it is impossible to decrease ε while preserving the inequalities for all x. Let us multiply the sides of the inequalities by $(1 + c_1)$ and subtract c_1 from them. Considering that $\hat{F}_k(x) = (1 + c_k)F_k(x) - c_k,\ x > 0$, we have that

$$\begin{aligned}\hat{F}_1(x - \varepsilon) - (1 + c_1)\varepsilon &\leq (1 + c_1)F_2(x) - c_1 \\ &= \hat{F}_2(x) + (c_2 - c_1)(1 - F_2(x)) \\ &\leq \hat{F}_1(x + \varepsilon) + (1 + c_1)\varepsilon.\end{aligned}$$

Setting $\delta = (1 + c_1)\varepsilon + |c_2 - c_1|$, we see that for all x

$$\hat{F}_1(x - \delta) - \delta \leq \hat{F}_2(x) \leq \hat{F}_1(x + \delta) + \delta.$$

Consequently, $\mu(F_1, F_2) \leq \delta$.

We shall often have to deal with cutoffs of stably distributed random variables in $\mathfrak{W}_1$; therefore, we adopt a separate notation for them, setting

$$Z(\alpha, \rho) = \hat{Y}(\alpha, \theta).$$

If $\alpha \leq 1, \rho = 0$ or if $\alpha = 1, \rho = 1$, then the cutoff $\hat{Y}(\alpha, \theta)$ is not defined. In the first case the distribution of $Y(\alpha, -1)$ is concentrated on the semi-axis $x < 0$, and in the second case $Y(1,1) \stackrel{d}{=} 1$, i.e., its distribution is discrete.

In these cases the definition of the concept of a cutoff should, of course, be extended in such a way as to preserve the property of continuity of the distributions of the cutoffs with respect to the parameters α and ρ in their whole domain of variation—a property enjoyed by the distributions of the random variables $Y(\alpha, \theta)$.

The case $\alpha = 1$, $\rho = 1$ turns out to be simplest, since we get what we want by setting $Z(1,1) = 1$. Namely,

$$Z(\alpha, \rho) \stackrel{d}{\to} 1 \quad \text{as } \alpha \to 1, \rho \to 1. \tag{3.1.3}$$

In the case $\alpha \leq 1, \rho = 0$ the situation is more complicated, and it turns out to be correct to set $Z(\alpha, 0)$ equal not to zero (as suggests itself) but to the random variable

$$Z(\alpha, 0) = Z(\alpha, 1)E/E, \tag{3.1.4}$$

where E is a random variable with the exponential distribution

$$\mathsf{P}(E > x) = \exp(-x), \qquad x \geq 0.$$

Recall that, according to our permanent condition, all the random variables written separately on the right-hand side of (3.1.4) are assumed to be independent.

In explaining why this extension of the definition of a cutoff is the natural one we are aided by the characteristic transform $W_Z(s, \alpha, \rho)$ of the cutoff $Z(\alpha, \rho)$, which, since $Z(\alpha, \rho) \geq 0$ with probability 1, has the matrix form

$$W_Z(s, \alpha, \rho) = \begin{pmatrix} M(s, \alpha, \rho) & 0 \\ 0 & M(s, \alpha, \rho) \end{pmatrix},$$

where $M(s, \alpha, \rho)$ is the Mellin transform of the distribution of $Z(\alpha, \rho)$.

If $\alpha \leq 1$, then ρ varies from 0 to 1, and for $\rho > 0$

$$M(s, \alpha, \rho) = \mathsf{E}Z^s(\alpha, \rho) = \frac{\sin \pi\rho s}{\rho \sin \pi s} \frac{\Gamma(1 - s/\alpha)}{\Gamma(1 - s)}.$$

If α is fixed and $\rho \to 0$, then clearly

$$M(s, \alpha, \rho) \to \frac{\pi s}{\sin \pi s} M(s, \alpha, 1) = \Gamma(1-s)\Gamma(1+s)M(s, \alpha, 1). \tag{3.1.5}$$

Since $\Gamma(1+s)$ (or $\Gamma(1-s)$) is the Mellin transform of an exponential distribution, i.e., the random variable E (respectively, the Mellin transform of the distribution of the variable $1/E$), (3.14) follows from (3.15).

We now proceed to systematically study the properties of the distributions of the random variables $Y(\alpha,\theta)$ and $Z(\alpha,\rho)$ defined at the beginning of this chapter.

§3.2. The random variables $Y(\alpha,\theta)$ and $Z(\alpha,\rho)$. Equivalence theorems

We begin from the simplest relations. For any pairs (α,θ) of admissible values

$$-Y(\alpha,\theta)\stackrel{d}{=}Y(\alpha,-\theta). \tag{3.2.1}$$

If $|\theta|=1$, then

$$Y(1,\theta)\stackrel{d}{=}\theta. \tag{3.2.2}$$

For any pairs (α,θ_1) and (α,θ_2) of admissible values

$$\frac{Y(\alpha,\theta_1)}{Y(\alpha,\theta_2)}\stackrel{d}{=}\frac{Y(\alpha,\theta_2)}{Y(\alpha,\theta_1)}. \tag{3.2.3}$$

In particular [3.3], for any $-1\le\theta\le 1$

$$Y^{-1}(1,\theta)\stackrel{d}{=}Y(1,\theta).$$

Recall (this was stipulated in §I.3) that, in all the equalities "$\stackrel{d}{=}$" connecting functions of random variables in the sense that their distributions are equal, the random variables written separately on one side of the equality (such as, for example, $Y(\alpha,\theta_1)$ and $Y(\alpha,\theta_2)$ in (3.2.3)) are assumed to be independent.

All these relations can be proved by a single universal method. The characteristic transforms of the distributions of the left and right sides of the equalities are computed and compared. Since the characteristic transforms, like the characteristic functions, uniquely determine the distributions connected with them, coincidence of the characteristic transforms means coincidence of the corresponding distributions.

For instance, using (3.0.3), we find that

$$(-1)^k w_k(s,\alpha,\theta)=w_k(s,\alpha,-\theta),\qquad k=0,1,$$

which implies (3.2.1) (note that (3.2.1) is a particular case of Property 2.2).

Further, if $|\theta|=1$, then

$$w_k(s,1,\theta)=\frac{\cos\frac{\pi}{2}(k-s\theta)}{\cos\frac{\pi}{2}(k-s)}=\theta^k,\qquad k=0,1,$$

which is equivalent to (3.2.2). In the third case

$$w_k(s,\alpha,\theta_1)w_k(-s,\alpha,\theta_2) = \frac{\cos\frac{\pi}{2}(k-s\theta_1)}{\cos\frac{\pi}{2}(k-s)}\frac{\Gamma(1-s/\alpha)}{\Gamma(1-s)} \times \frac{\cos\frac{\pi}{2}(k+s\theta_2)}{\cos\frac{\pi}{2}(k+s)}\frac{\Gamma(1+s/\alpha)}{\Gamma(1+s)}, \qquad k=0,1.$$

Since

$$\frac{\cos\frac{\pi}{2}(k-s\theta)}{\cos\frac{\pi}{2}(k-s)} = \frac{\cos\frac{\pi}{2}(k+s\theta)}{\cos\frac{\pi}{2}(k+s)}$$

for the values $k = 0,1$, we get after the appropriate change of signs and a permutation of the factors that

$$w_k(s,\alpha,\theta_1)w_k(-s,\alpha,\theta_2) = w_k(-s,\alpha,\theta_1)w_k(s,\alpha,\theta_2), \qquad k=0,1,$$

i.e., the distributions of the left and right sides of (3.2.3) coincide. A special case is obtained if we set $\alpha = 1$ and $\theta_1 = 1$ in (3.2.3) and use (3.2.2).

The following relations are among the simplest relations:

If $0 < \alpha < 1$, then $Z(\alpha,1) = Y(\alpha,1)$.

For any admissible values of α

$$Z(\alpha,\tfrac{1}{2}) \stackrel{d}{=} |Y(\alpha,0)|. \tag{3.2.4}$$

For any pairs of admissible values (α,ρ_1) and (α,ρ_2)

$$\frac{Z(\alpha,\rho_1)}{Z(\alpha,\rho_2)} = \frac{Z(\alpha,\rho_2)}{Z(\alpha,\rho_1)}, \tag{3.2.5}$$

except for the case when one of the numbers ρ_1 or ρ_2 is zero. In particular, if $0 < \rho < 1$, then

$$Z^{-1}(1,\rho) \stackrel{d}{=} Z(1,\rho).$$

The first and second equalities are simple consequences of property 4 of cutoffs. The validity of (3.2.4) and (3.2.5) is established by computing and comparing the Mellin transforms of their left and right sides. The random variables $X_1 = |Y(\alpha,\theta)|$ and $X_2 = \operatorname{sgn} Y(\alpha,\theta)$ are independent if and only if either $\theta = 0$ or $0 < \alpha < 1$ and $|\theta| = 1$.

In term of characteristic transforms this property is equivalent to the equality $W_{X_1X_2}(s) = W_{X_1}(s)W_{X_2}(s)$, which in our case is easily seen to reduce to the requirement that

$$w_1(s,\alpha,\theta) = \theta w_0(s,\alpha,\theta),$$

which implies the indicated condition.

THEOREM 3.2.1. *Let (α_1,θ_1) and (α_2,θ_2) be pairs of admissible values of parameters, and let ν be a number such that*

$$\alpha_2(1+|\theta_1|)/2 \le \nu \le (\alpha_1(1+|\theta_2|)/2)^{-1}. \tag{3.2.6}$$

Then

$$Y(\alpha_1,\theta_1)Y^\nu(\alpha_2,\theta_2) \stackrel{d}{=} Y(\alpha_2/\nu,\theta_1)Y^\nu(\alpha_1\nu,\theta_2). \tag{3.2.7}$$

PROOF. Let us compute the characteristic transform of the left-hand side of (3.2.7). Using (2.6.23) and (3.0.3), we have that $(k=0,1)$

$$\begin{aligned} w_k(s,\alpha_1,\theta_1)w_k(s\nu,\alpha_2,\theta_2) = {} & \frac{\cos\frac{\pi}{2}(k-s\theta_1)}{\cos\frac{\pi}{2}(k-s)}\frac{\Gamma(1-s/\alpha_1)}{\Gamma(1-s)} \\ & \times\frac{\cos\frac{\pi}{2}(k-s\nu\theta_2)}{\cos\frac{\pi}{2}(k-s\nu)}\frac{\Gamma(1-s\nu/\alpha_2)}{\Gamma(1-s\nu)}. \end{aligned}$$

Similarly, for the right-hand side

$$\begin{aligned} w_k\left(s,\frac{\alpha_2}{\nu},\theta_1\right)w_k(s\nu,\alpha_1\nu,\theta_2) = {} & \frac{\cos\frac{\pi}{2}(k-\theta_1 s)}{\cos\frac{\pi}{2}(k-s)}\frac{\Gamma(1-s\nu/\alpha_2)}{\Gamma(1-s)} \\ & \times\frac{\cos\frac{\pi}{2}(k-s\nu\theta_2)}{\cos\frac{\pi}{2}(k-s\nu)}\frac{\Gamma(1-s/\alpha_1)}{\Gamma(1-s\nu)}. \end{aligned}$$

It is obvious that the expressions coincide. It remains to verify that the parameters in the right-hand side of (3.2.7) are in the domain of admissible values. We know that $|\theta_j| \le \min(1, 2/\alpha_j - 1)$, and therefore the conditions

$$|\theta_1| \le \min(1, 2\nu/\alpha_2 - 1), \qquad |\theta_2| \le \min(1, 2/(\alpha,\nu) - 1)$$

needed for the right-hand side hold if and only if

$$|\theta_1| \le 2\nu/\alpha_2 - 1, \qquad |\theta_2| \le 2/(\alpha_1\nu) - 1.$$

But these inequalities are equivalent to (3.2.6).

THEOREM 3.2.2. *Let (α_1,θ_1) and (α_2,θ_2) be pairs of admissible values of the parameters, and let μ be any real number such that*

$$|\theta_1|/\min(1,2/\alpha_2-1) \le |\mu| \le \min(1,2/\alpha_1-1)/|\theta_2|. \tag{3.2.8}$$

Then

$$\begin{aligned} Y(\alpha_1,\theta_1)Y^\mu(\alpha_2,\theta_2) & \stackrel{d}{=} Y(\alpha_1,\theta_2\mu)Y^\mu(\alpha_2,\theta_1/\mu) \\ & \stackrel{d}{=} Y(\alpha_1,-\theta_2\mu)Y^\mu(\alpha_2,-\theta_1/\mu). \end{aligned} \tag{3.2.9}$$

The proof of the first part of (3.2.9) is just like that of (3.2.7). The second part is a consequence of (3.2.1).

We mention another relation following from (3.2.9) and (3.2.7).

COROLLARY. *For any pairs* (α_1, θ_1) *and* (α_2, θ_2) *of admissible values of the parameters and any number* $\mu > 0$ *such that*

$$|\theta_1|/\mu \le \min(1, 2/(\alpha_1, \mu) - 1),$$
$$|\theta_2|\mu \le \min(1, 2\mu/\alpha_2 - 1), \tag{3.2.10}$$

the relation

$$Y(\alpha_1, \theta_1)Y^\mu(\alpha_2, \theta_2) \stackrel{d}{=} Y(\alpha_2/\mu, \theta_2\mu)Y^\mu(\alpha_1\mu, \theta_1/\mu)$$
$$\stackrel{d}{=} Y(\alpha_2/\mu, -\theta_2\mu)Y^\mu(\alpha_1\mu, -\theta_1/\mu) \tag{3.2.11}$$

is valid.

The cutoffs $Z(\alpha, \rho)$ satisfy analogous relations.

THEOREM 3.2.3. *For any pairs* (α_1, ρ_1) *and* (α_2, ρ_2) *of admissible values of the parameters and any number* ν *such that*

$$\alpha_2 \left(|\rho_1 - \tfrac{1}{2}| + \tfrac{1}{2}\right) \le \nu \le \left(\alpha_1|\rho_2 - \tfrac{1}{2}| + \alpha_1/2\right)^{-1}, \tag{3.2.12}$$

the relation

$$Z(\alpha_1, \rho_1)Z^\nu(\alpha_2, \rho_2) \stackrel{d}{=} Z(\alpha_2/\nu, \rho_1)Z^\nu(\alpha_1\nu, \rho_2) \tag{3.2.13}$$

is valid.

PROOF. Let us compute and compare the Mellin transforms of the distributions of the left and right sides of (3.2.13). They coincide if and only if the theorem holds, provided, of course, that the parameters of the random variables on the right-hand side of (3.2.13) lie in the domain of admissible values.

According to (3.0.1),

$$M(s, \alpha_1, \rho_1)M(s\nu, \alpha_2, \rho_2) = \frac{\sin \pi\rho_1 s}{\rho_1 \sin \pi s}\frac{\Gamma(1 - s/\alpha_1)}{\Gamma(1 - s)} \times \frac{\sin \pi\rho_2\nu s}{\rho_2 \sin \pi\nu s}\frac{\Gamma(1 - s\nu/\alpha_2)}{\Gamma(1 - s\nu)},$$

$$M\left(s, \frac{\alpha_2}{\nu}, \rho_1\right) M(s\nu, \alpha_1\nu, \rho_2) = \frac{\sin \pi\rho_1 s}{\rho_1 \sin \pi s}\frac{\Gamma(1 - s\nu/\alpha_2)}{\Gamma(1 - s)} \times \frac{\sin \pi\rho_2\nu s}{\rho_2 \sin \pi\nu s}\frac{\Gamma(1 - s/\alpha_1)}{\Gamma(1 - s\nu)}.$$

It is obvious that these transforms are equal. By (3.0.1), the pairs $(\alpha_1/\nu, \rho_1)$ and $(\alpha_1\nu, \rho_2)$ are in the domain of admissible values if and only if

$$|\rho_1 - \tfrac{1}{2}| \le \min(1, \nu/\alpha_2) - \tfrac{1}{2},$$
$$|\rho_2 - \tfrac{1}{2}| \le \min(1, 1/(\alpha_1\nu)) - \tfrac{1}{2}.$$

If we consider that, according to a condition of the theorem,

$$|\rho_2-\tfrac{1}{2}|\le\min(1,1/\alpha_1)-\tfrac{1}{2},$$
$$|\rho_2-\tfrac{1}{2}|\le\min(1,1/\alpha_2)-\tfrac{1}{2},$$

then it becomes apparent that the condition written for ν is equivalent to (3.2.12).

THEOREM 3.2.4. *Let (α_1,ρ_1) and (α_2,ρ_2) be pairs of admissible values of the parameters, and let μ be a real number such that*

$$\begin{aligned}&1-\min(1,1/\alpha_1)\le|\mu|\rho_2\le\min(1,1/\alpha_1),\\&1-\min(1,1/\alpha_2)\le\rho_1/|\mu|\le\min(1,1/\alpha_2).\end{aligned}\tag{3.2.14}$$

Then

$$Z(\alpha_1,\rho_1)Z^{\mu}(\alpha_2,\rho_2)\stackrel{d}{=}Z(\alpha_1,\rho_2|\mu|)Z^{\mu}(\alpha_2,\rho_1/|\mu|).\tag{3.2.15}$$

The proof is by the same method used to prove the preceding theorem. We note a corollary of (3.2.14) and (3.2.15).

COROLLARY. *For any pairs (α_1,ρ_1) and (α_2,ρ_2) of admissible values of the parameters and any $\mu>0$ such that*

$$\begin{aligned}&|\rho_1/\mu-\tfrac{1}{2}|\le\min(1,1/(\alpha_1\mu))-\tfrac{1}{2},\\&|\rho_2\mu-\tfrac{1}{2}|\le\min(1,\mu/\alpha_2)-\tfrac{1}{2},\end{aligned}\tag{3.2.16}$$

the relation

$$Z(\alpha_1,\rho_1)Z^{\mu}(\alpha_2,\rho_2)\stackrel{d}{=}Z(\alpha_2/\mu,\rho_2\mu)Z^{\mu}(\alpha_1\mu,\rho_1/\mu)\tag{3.2.17}$$

is valid.

REMARK. Condition (3.2.16) can be replaced by explicit inequalities determining the intervals of variation. They are not given because of their unwieldiness.

THEOREM 3.2.5. *Let $\alpha\ge1,\rho$ be a pair of admissible values of the parameters. Then*

$$Z(\alpha,\rho)\stackrel{d}{=}Z^{-1/\alpha}(1/\alpha,\alpha\rho).\tag{3.2.18}$$

PROOF. Let us compute the Mellin transforms of the distributions of both sides of (3.2.18). We have that

$$M(s,\alpha,\rho)=\frac{\sin\pi\rho s}{\rho\sin\pi s}\frac{\Gamma(1-s/\alpha)}{\Gamma(1-s)},$$
$$M\left(-\frac{s}{\alpha},\frac{1}{\alpha},\alpha\rho\right)=\frac{\sin\pi\rho s}{\alpha\rho\sin(\pi s/\alpha)}\frac{\Gamma(1+s)}{\Gamma(1+s/\alpha)}.$$

Consider the ratio of these transforms:

$$\Delta = \frac{\alpha \sin(\pi s/\alpha)}{\sin \pi s} \frac{\Gamma(1-s/\alpha)\Gamma(1+s/\alpha)}{\Gamma(1-s)\Gamma(1+s)}.$$

Since $\Gamma(1-s)\Gamma(s) = \pi/\sin \pi s$, it follows that

$$\Delta = \frac{\sin(\pi s/\alpha)}{\sin \pi s} \frac{\sin \pi s}{\sin(\pi s/\alpha)} = 1.$$

The Mellin transforms coincide, so the distributions corresponding to them coincide, i.e., (3.2.18) is valid. According to (3.0.1), the condition that the pair $(1/\alpha, \alpha\rho)$ belong to the domain of admissible values amounts to the inequalities

$$1 - \min(1, \alpha) \leq \alpha\rho \leq \min(1, \alpha),$$

which are obviously equivalent to the condition

$$1 - \min(1, 1/\alpha) \leq \rho \leq \min(1, 1/\alpha).$$

But this is the condition for (α, ρ) to belong to the domain of admissible values if it is taken into account that $\alpha \geq 1$.

REMARK 1. (3.2.18) is none other than the duality relation constituting Theorem 2.3.1. Thus, here we have a second proof of that theorem.

REMARK 2. Choose $\rho = \frac{1}{2}$ in (3.2.18), which corresponds to the value $\theta = 0$. Considering (3.2.4), we get in the case $\alpha \geq 1$ that

$$|Y(\alpha, 0)| \stackrel{d}{=} Z^{-1/\alpha}(1/\alpha, \alpha/2). \tag{3.2.19}$$

With $\alpha = 2$, this gives us for a normally distributed random variable $Y(2,0)$ with expectation 0 and variance 2 that

$$Y\left(\tfrac{1}{2}, 1\right) \stackrel{d}{=} |Y(2,0)|^{-2} \stackrel{d}{=} N^{-2}/2, \tag{3.2.20}$$

where N is a random variable with standard normal distribution.

There is apparently no analogue of (3.2.18) for the random variables $Y(\alpha, \theta)$. However, we have the following relation.

THEOREM 3.2.6. *Let $\alpha \geq 1, \theta$ be a pair of admissible values of the parameters, and $0 \leq \rho \leq 1$ any number.** *Let $\nu = (2\rho)^{-1}$. Then*

$$Z^\nu(1, \rho/\alpha)|Y(\alpha, \theta)| \stackrel{d}{=} Z^\nu(1, \rho)|Y(1/\alpha, \alpha\theta)|^{-1/\alpha}. \tag{3.2.21}$$

PROOF. Let us compute the Mellin transforms of both sides of (2.2.21), recalling that the Mellin transform of the distribution of $|Y(\alpha, \theta)|$ coincides with $w_0(s, \alpha, \theta)$. For the left-hand side,

$$M\left(s\nu, 1, \frac{\rho}{\alpha}\right) w_0(s, \alpha, \theta) = \frac{\alpha \sin(\pi\nu\rho s/\alpha)}{\rho \sin \pi\nu s} \frac{\cos \frac{\pi}{2}\theta s}{\cos \frac{\pi}{2} s} \frac{\Gamma(1-s/\alpha)}{\Gamma(1-s)}.$$

*In this theorem ρ appears as an independent parameter *not* connected with θ by the previously adopted relation $\rho = (1+\theta)/2$.

On the other hand, for the right-hand side,

$$M(s\nu,1,\rho)w_0\left(-\frac{s}{\alpha},\frac{1}{\alpha},\alpha\theta\right)=\frac{\sin\pi\nu\rho s}{\rho\sin\pi s\nu}\frac{\cos\frac{\pi}{2}\theta s}{\cos\frac{\pi}{2}(s/\alpha)}\frac{\Gamma(1+s)}{\Gamma(1+s/\alpha)}.$$

Using the known equalities

$$\Gamma(s+1)=s\Gamma(s)\quad\text{and}\quad\Gamma(s)\Gamma(1-s)=\pi/\sin\pi s$$

and the fact that $\nu=(2\rho)^{-1}$, we find that the ratio of the Mellin transforms computed is equal to

$$\frac{\sin\pi\nu\rho s/\alpha}{\sin\pi\nu s}\frac{\cos\frac{\pi}{2}\theta s}{\cos\frac{\pi}{2}s}\frac{\cos\frac{\pi}{2}s/\alpha}{\cos\frac{\pi}{2}\theta s}\frac{\sin\pi\nu s}{\sin\pi\nu\rho s}\frac{\sin\pi s}{\sin\pi s/\alpha},$$

which, after substitution of the value ν and the obvious cancellations, takes the form

$$\frac{\sin\frac{\pi}{2}s/\alpha\cos\frac{\pi}{2}s\alpha\sin\pi s}{\cos\frac{\pi}{2}s\sin\frac{\pi}{2}s\sin\pi s/\alpha}=1.$$

Since $\rho/\alpha<1$ and $\alpha|\theta|\le\alpha(1/\alpha)=1$, the parameters determining all the random variables in (3.2.21) are in the domain of admissible values.

We mention two particular cases of this theorem.

1. If $\alpha\ge1,\theta=0$, and $\rho=1$, then

$$\begin{aligned}|Y(1/\alpha,0)|^{-1/\alpha}&\stackrel{d}{=}Z^{-1/\alpha}(1/\alpha,1/2)\\&\stackrel{d}{=}Z^{1/2}(1,1/\alpha)|Y(\alpha,0)|\\&\stackrel{d}{=}Z^{1/2}(1,1/\alpha)Z(\alpha,1/2).\end{aligned}\tag{3.2.22}$$

2. If $\alpha\ge1,\theta=1/\alpha$, and $\rho=1$, then

$$\begin{aligned}Y(1/\alpha,1)^{-1/\alpha}&\stackrel{d}{=}Z^{-1/\alpha}(1/\alpha,1)\\&\stackrel{d}{=}Z^{1/2}(1,1/\alpha)|Y(\alpha,1/\alpha)|.\end{aligned}\tag{3.2.23}$$

If now (3.2.18) holds, then, transforming the left-hand side of (3.2.23), we get that

$$Z(\alpha,1/\alpha)\stackrel{d}{=}Z^{1/2}(1,1/\alpha)|Y(\alpha,1/\alpha)|.\tag{3.2.24}$$

The "equivalence-type" relations considered above are not, of course, exhausted by the list given here. It can be extended both by combining the relations of this and the next section, and by finding relations of a new type. For example, among the given relations there were none that would include the following easily verified equalities.

Let α,θ be a pair of admissible values of the parameters. Then

$$Z^{1/2}(1,|\theta|)|Y(\alpha,\theta)|\stackrel{d}{=}Z(\alpha,|\theta|).\tag{3.2.25}$$

If in addition $\alpha < 1$, then

$$|Y(1,\theta)|^2 Z(\alpha, |\theta|) \stackrel{d}{=} Z^2(1, |\theta|) Y(\alpha, 1). \tag{3.2.26}$$

§3.3. The random variables $Y(\alpha, \theta)$ and $Z(\alpha, \rho)$. Multiplication and division theorems

In this section the terminology "multiplication and division theorems" is used more or less conditionally, and its main purpose is to perform a classification in our collection of relations between random variables.

THEOREM 3.3.1. *Let (α, θ) be a pair of admissible values of the parameters, and let $0 < \alpha' \le 1$. Then $(\rho = (1+\theta)/2)$*

$$Y(\alpha,\theta) Y^{1/\alpha}(\alpha', 1) \stackrel{d}{=} Y(\alpha\alpha', \theta), \tag{3.3.1}$$

$$Z(\alpha,\rho) Z^{1/\alpha}(\alpha', 1) \stackrel{d}{=} Z(\alpha\alpha', \rho). \tag{3.3.2}$$

If $1 \le \alpha' \le 2$, then

$$Z(\alpha', 1/\alpha') Z^{1/\alpha'}(\alpha, \rho) \stackrel{d}{=} Z(\alpha\alpha', \rho/\alpha') \tag{3.3.3}$$

for any pair (α, ρ) such that $\alpha\alpha' \le 2$ and $\rho \ge \alpha' - 1/\alpha$.

COROLLARY 1. 1. *If $\alpha \le 1$ and $\alpha' \le 1$, then*

$$Y(\alpha, 1) Y^{1/\alpha}(\alpha', 1) \stackrel{d}{=} Y(\alpha\alpha', 1). \tag{3.3.4}$$

2. *If $\alpha \ge 1, \alpha' \ge 1$, and $\alpha\alpha' \le 2$, then*

$$Z(\alpha', 1/\alpha') Z^{1/\alpha'}(\alpha, 1/\alpha) \stackrel{d}{=} Z(\alpha\alpha', 1/\alpha\alpha'). \tag{3.3.5}$$

COROLLARY 2. *Let $\alpha = \frac{1}{2}, \theta = 1$, and $\alpha' = 1/2^k$ in (3.3.1). Then*

$$Y(1/2, 1) Y^2(1/2^k, 1) \stackrel{d}{=} Y(1/2^{k+1}, 1).$$

By (3.2.20), which implies that $Y(1/2, 1) \stackrel{d}{=} (2N^2)^{-1}$, where N is a random variable with standard normal distribution, this relation gives us (3.4.13).

If $\alpha = 1, \theta = 0$, and $\alpha' = 1/2^k$ in (3.3.1), then the relation

$$N \cdot N^{-1} Y(1/2^k, 1) \stackrel{d}{=} Y(1/2^k, 0)$$

follows from the fact (see (3.3.21) and (3.3.17)) that $N/N \stackrel{d}{=} Y(1, 0)$. It then follows from (3.4.13) that

$$2^{1-2^k} N \cdot N^{-1} N^{-2} N^{-4} \cdots N^{-2^k} \stackrel{d}{=} Y(1/2^k, 0).$$

The last relation, like (3.4.13), was apparently the first relation of that kind known in probability theory. These relations were obtained in 1946 by Brown and Tukey [245].

PROOF OF THE THEOREM. Let us consider (3.3.1). The components of the characteristic transform of its left-hand side have the form

$$\begin{aligned} w_k(s,\alpha,\theta)w_k\left(\frac{s}{\alpha},\alpha',1\right) &= \frac{\cos\frac{\pi}{2}(k-s\theta)}{\cos\frac{\pi}{2}(k-s)}\,\frac{\Gamma(1-s/\alpha)\Gamma(1-s/\alpha\alpha')}{\Gamma(1-s\alpha)\Gamma(1-s/\alpha)} \\ &= \frac{\cos\frac{\pi}{2}(k-s\theta)}{\cos\frac{\pi}{2}(k-s)}\,\frac{\Gamma(1-s/\alpha\alpha')}{\Gamma(1-s)} \\ &= w_k(s,\alpha\alpha',\theta), \qquad k=0,1. \end{aligned}$$

Moreover, considering that $\alpha' \leq 1$, we have that

$$|\theta| \leq \min(1, 2\alpha - 1) \leq \min(1, 2/\alpha\alpha' - 1),$$

i.e., the pair $(\alpha\alpha', \theta)$ belongs to the domain of admissible values. The relations (3.3.2) and (3.3.3) are proved similarly ([3.4]).

THEOREM 3.3.2. *For any pairs (α,θ) and $(1,\theta')$ of admissible values,*

$$Y(\alpha,\theta)Y^{\theta}(1,\theta') \stackrel{d}{=} Y(\alpha,\theta\theta'). \tag{3.3.6}$$

If (α,ρ) and $(1,\rho')$ are pairs of admissible values such that $\rho\rho' \geq 1-1/\alpha$, then

$$Z(\alpha,\rho)Z^{\rho}(1,\rho') \stackrel{d}{=} Z(\alpha,\rho\rho'). \tag{3.3.7}$$

COROLLARY. *For any pair (α,θ) of admissible values,*

$$Y(\alpha,\theta) \stackrel{d}{=} Y(\alpha,\theta_\alpha)Y^{\theta_\alpha}(1,\theta/\theta_\alpha), \tag{3.3.8}$$

$$Z(\alpha,\rho) \stackrel{d}{=} Z(\alpha,\rho_\alpha)Z^{\rho_\alpha}(1,\rho/\rho_\alpha), \tag{3.3.9}$$

where $\theta_\alpha = \min(1, 2/\alpha - 1)$ and $\rho_\alpha = \min(1, 1/\alpha)$.

In particular, if $\alpha \leq 1$, then

$$Y(\alpha,\theta) \stackrel{d}{=} Y(\alpha,1)Y(1,\theta), \qquad Z(\alpha,\rho) \stackrel{d}{=} Z(\alpha,1)Z(1,\rho), \tag{3.3.10}$$

i.e., there is a separation of the dependence of random variables $Y(\alpha,\theta)$ and $Z(\alpha,\rho)$ on the parameters determining them.

The proofs of (3.3.6) and (3.3.7) are carried out in the same way as the proof of the preceding theorem.

THEOREM 3.3.3. *Consider pairs (α_1,θ_1) and (α_2,θ_2) of admissible values of the parameters such that $\alpha_1 \leq 2\alpha_2$.*

Let $\alpha_2 \geq 1$, and define $\theta = \alpha_2\theta_1$ and $\theta' = \theta_2$ if $|\theta_1| \leq 1/\alpha_2$ and $\theta = \alpha_2\theta_2$ and $\theta' = \theta_1$ if $|\theta_2| \leq 2/\alpha_1 - 1/\alpha_2$. Then

$$Z\left(1, \frac{1}{\alpha_2}\right) \frac{Y(\alpha_1, \theta_1)}{Y(\alpha_2, \theta_2)} \stackrel{d}{=} Y^{1/\alpha_2}\left(\frac{\alpha_1}{\alpha_2}, \theta\right) Y(1, \theta') Y\left(1, \frac{1}{\alpha_2}\right), \qquad (3.3.11)$$

$$Z^{1/2}\left(1, \frac{1}{\alpha_2}\right) \left|\frac{Y(\alpha_1, \theta_1)}{Y(\alpha_2, \theta_2)}\right| \stackrel{d}{=} \left|Y\left(\frac{\alpha_1}{\alpha_2}, \theta\right)\right|^{1/\alpha_2} |Y(1, \theta')|. \qquad (3.3.12)$$

Let $\alpha_2 \leq 1$, and define $\theta = \alpha_2\theta_1$ and $\theta' = \theta_2$ in the case $|\theta_1| \leq 2/\alpha_1 - 1/\alpha_2$, and $\theta = \alpha_2\theta_2$ and $\theta' = \theta_1$ in the case $|\theta_2| \leq 2/\alpha_1 - 1/\alpha_2$. Then

$$Y^{1/\alpha_2}(1, \alpha_2) \frac{Y(\alpha_1, \theta_1)}{Y(\alpha_2, \theta_2)} \stackrel{d}{=} Y^{1/\alpha_2}\left(\frac{\alpha_1}{\alpha_2}, \theta\right) Y(1, \theta') Z^{1/\alpha_2}(1, \alpha_2), \qquad (3.3.13)$$

$$\left|\frac{Y(\alpha_1, \theta_1)}{Y(\alpha_2, \theta_2)}\right| \stackrel{d}{=} \left|Y\left(\frac{\alpha_1}{\alpha_2}, \theta\right)\right|^{1/\alpha_2} |Y(1, \theta')| Z^{1/2\alpha_2}(1, \alpha_2). \qquad (3.3.14)$$

PROOF. The conditions on $|\theta_1|$ and $|\theta_2|$ arise as necessary conditions for the values of the pair $(\alpha_1/\alpha_2, \theta)$ of parameters to be admissible, which is equivalent to the requirement that $|\theta| \leq \min(1, 2\alpha_2/\alpha_1 - 1)$. The validity of (3.3.11)–(3.3.14) can be verified by the method already worked out. We confine ourselves to checking two of them: (3.3.11) and (3.3.14). In the first case we compute the characteristic transform of the left-hand side, considering that for the random variable $Z(1, 1/\alpha_2)$ the functions w_0 and w_1 of the characteristic transform coincide with $M(s, \alpha, \rho)$ $(k = 0, 1)$:

$$M\left(s, 1, \frac{1}{\alpha_2}\right) w_k(s, \alpha_1, \theta_1) w_k(-s, \alpha_2, \theta_2)$$
$$= \frac{\alpha_2 \sin \pi s/\alpha_2}{\sin \pi s} \frac{\cos \frac{\pi}{2}(k - s\theta_1)}{\cos \frac{\pi}{2}(k - s)} \frac{\Gamma(1 - s/\alpha_1)}{\Gamma(1 - s)} \frac{\cos \frac{\pi}{2}(k - s\theta_2)}{\cos \frac{\pi}{2}(k - s)} \frac{\Gamma(1 + s/\alpha_2)}{\Gamma(1 + s)}.$$

The right-hand side of (3.3.11) has characteristic transform of the form

$$w_k\left(\frac{s}{\alpha_2}, \frac{\alpha_1}{\alpha_2}, \theta\right) w_k(s, 1, \theta') w_k\left(s, 1, \frac{1}{\alpha_2}\right)$$
$$= \frac{\cos \frac{\pi}{2}(k - s\theta/\alpha_2)}{\cos \frac{\pi}{2}(k - s/\alpha_2)} \frac{\Gamma(1 - s/\alpha_1)}{\Gamma(1 - s/\alpha_2)} \frac{\cos \frac{\pi}{2}(k - \theta' s)}{\cos \frac{\pi}{2}(k - s)} \frac{\cos \frac{\pi}{2}(k - s/\alpha_2)}{\cos \frac{\pi}{2}(k - s)}.$$

Division of the first expression by the second gives us

$$\frac{\sin \pi s/\alpha_2 \cos \frac{\pi}{2}(k - s\theta_1) \cos \frac{\pi}{2}(k - s\theta_2) \Gamma(s/\alpha_2) \Gamma(1 - s/\alpha_2)}{\sin \pi s \cos \frac{\pi}{2}(k - s\theta/\alpha_2) \cos \frac{\pi}{2}(k - \theta' s) \Gamma(s) \Gamma(1 - s)}$$

after some simplifying transformations and cancellations. Since $\Gamma(s)\Gamma(1-s) = \pi/\sin \pi s$, the choice of θ and θ' implies that the last expression is equal to 1.

In proving (3.3.14) it should be recalled that it suffices to compute and compare only the functions w_0 when dealing with positive random variables. For the left-hand side,

$$w_0(s,\alpha_1,\theta_1)w_0(-s,\alpha_2,\theta_2) = \frac{\cos\frac{\pi}{2}\theta_1 s\,\Gamma(1-s/\alpha_1)}{\cos\frac{\pi}{2}s\quad\Gamma(1-s)}\frac{\cos\frac{\pi}{2}\theta_2 s\,\Gamma(1+s/\alpha_2)}{\cos\frac{\pi}{2}s\quad\Gamma(1+s)}.$$

For the right-hand side,

$$\begin{aligned} w_0&\left(\frac{s}{\alpha_2},\frac{\alpha_1}{\alpha_2},\theta\right)w_0(s,1,\theta')w_0\left(\frac{s}{2\alpha_2},1,\alpha_2\right) \\ &= \frac{\cos\frac{\pi}{2}\theta s/\alpha_2\,\Gamma(1-s/\alpha_1)}{\cos\frac{\pi}{2}s/\alpha_2\,\Gamma(1-s/\alpha_2)}\frac{\cos\frac{\pi}{2}\theta' s}{\cos\frac{\pi}{2}s}\frac{\sin\frac{\pi}{2}s}{\alpha_2\sin\frac{\pi}{2}s/\alpha_2}. \end{aligned}$$

The ratio of these expressions has the form

$$\frac{\cos\frac{\pi}{2}\theta_1 s\cos\frac{\pi}{2}\theta_2 s\sin\pi s/\alpha_2\,\Gamma(s/\alpha_2)\Gamma(1-s/\alpha_2)}{\cos\frac{\pi}{2}\theta s/\alpha_2\cos\frac{\pi}{2}\theta' s\sin\pi s\qquad\Gamma(s)\Gamma(1-s)},$$

after transformations and cancellations. By the choice of parameters, a further transformation leads us to the conclusion that this ratio is 1.

THEOREM 3.3.4. *Let (α_1,ρ_1) and (α_2,ρ_2) be pairs of admissible parameters such that $\alpha_1\le 2\alpha_2$ and $1/\alpha_2-1/\alpha_1\le\rho_1\le 1/\alpha_2$, or such that $\alpha_1\le 2\alpha_2$ and $1/\alpha_2-1/\alpha_1\le\rho_2\le 1/\alpha_1$. With $\rho=\alpha_2\rho_1$ and $\rho'=\rho_2$ in the first case and $\rho=\alpha_2\rho_2$ and $\rho'=\rho_1$ in the second case, it follows that*

$$\frac{Z(\alpha_1,\rho_1)}{Z(\alpha_2,\rho_2)} \stackrel{d}{=} Z^{1/\alpha_2}\left(\frac{\alpha_1}{\alpha_2},\rho\right)Z(1,\rho'). \tag{3.3.15}$$

COROLLARIES. 1. *For any admissible values of the parameters α and α'*

$$\frac{Y(\alpha\alpha',1)}{Y(\alpha,1)} \stackrel{d}{=} Z^{1/\alpha}(\alpha',\alpha). \tag{3.3.16}$$

2. *If $\alpha\ge 1$, then for any $\rho\le 1/\alpha$*

$$\frac{Z(\alpha,1/\alpha)}{Z(\alpha,\rho)} \stackrel{d}{=} Z(1,\rho). \tag{3.3.17}$$

3. *If $\alpha\le 1$, then for any $\rho\le 1$*

$$\frac{Z(\alpha,1)}{Z(\alpha,\rho)} \stackrel{d}{=} Z^{1/\alpha}(1,\rho\alpha). \tag{3.3.18}$$

4. *Suppose that $1\le 2\alpha\le 2$. Then for any admissible pairs (α,ρ)*

$$\frac{Z(2\alpha,1/2\alpha)}{Z(\alpha,\rho)} \stackrel{d}{=} |Y(2,0)|^{1/\alpha}Z(1,\rho) \tag{3.3.19}$$

and for any admissible pairs $(2\alpha,\rho)$

$$\frac{Z(2\alpha,\rho)}{Z(\alpha,1/2\alpha)} \stackrel{d}{=} |Y(2,0)|^{1/\alpha}Z(1,\rho). \tag{3.3.20}$$

The proof reduces to a check that the Mellin transforms of the distributions on the left and right sides of (3.3.15) coincide.

In many cases the relations given above between $Y(\alpha,\theta)$ and $Z(\alpha,\rho)$ can facilitate and simplify the computation of the distributions of statistics formed by random variables with stable laws. We illustrate this by two examples.

EXAMPLE 1. Consider independent random variables $Y_1 \stackrel{d}{=} Y_2 \stackrel{d}{=} Y_3 \stackrel{d}{=} Y_4 \stackrel{d}{=} Y_B(\alpha,\beta,\gamma,\lambda)$, and form from them the statistic

$$L = \frac{Y_1 - Y_2}{Y_3 - Y_4}$$

(statistics of this kind are used, for example, in the problem of recovering the type of a distribution). We compute the distribution of L. According to (2.1.6) and (2.1.1),

$$\begin{aligned} Y_1 - Y_2 \stackrel{d}{=} Y_3 - Y_4 &\stackrel{d}{=} Y_B(\alpha,0,0,2\lambda) \\ &\stackrel{d}{=} (2\lambda)^{1/\alpha} Y_B(\alpha,0,0,1) = (2\lambda)^{1/\alpha} Y_C(\alpha,0,1). \end{aligned}$$

Consequently,

$$L \stackrel{d}{=} \frac{Y(\alpha,0)}{Y(\alpha,0)} \stackrel{d}{=} \frac{Z(\alpha,1/2)}{Z(\alpha,1/2)} U, \tag{3.3.21}$$

where U is a random variable taking the values 1 and -1 with probability $\frac{1}{2}$. Since the distribution function $F_L(x)$ of L is symmetric, it is connected with the distribution function $\hat{F}_L(x)$ of a cutoff $\hat{L}$ by the equality

$$\hat{F}_L(x) = 2F_L(x) - 1, \qquad x \geq 0. \tag{3.3.22}$$

Therefore, to compute $F_L(x)$ it suffices to compute $\hat{F}_L(x)$. Since

$$\hat{L} \stackrel{d}{=} \frac{Z(\alpha,1/2)}{Z(\alpha,1/2)},$$

it follows from (3.3.15) that

$$\hat{L} \stackrel{d}{=} Z^{1/\alpha}(1,\alpha/2) Z(1,\tfrac{1}{2}).$$

As is clear from the comment after Theorem 2.3.1, the distribution functions of the random variables $Z^{1/\alpha}(1,\alpha/2)$ and $Z(1,\frac{1}{2})$ are, respectively,

$$F_\alpha(x) = \frac{2}{\alpha}\left[\frac{1}{\pi}\arctan\left(\frac{x^\alpha + \cos\frac{\pi}{2}\alpha}{\sin\frac{\pi}{2}\alpha}\right) - \frac{1}{2}(1-\alpha)\right] \tag{3.3.23}$$

and $F_1(x)$. From this we find that

$$\hat{F}_L(x) = \int_0^\infty F_1(x/y)\, dF_\alpha(y)$$

and the density $p_L(x)$ of L has the form ([3.5])

$$p_L(x) = \frac{1}{2}\int_0^\infty F_1'(x/y)F_\alpha'(y)\frac{dy}{y}$$
$$= \frac{2}{\pi^2}\int_0^\infty \left[\left(\frac{x}{y}\right)^2 + 1\right]^{-1} \frac{y^{\alpha-2}\sin\frac{\pi}{2}\alpha}{y^{2\alpha}+2y^\alpha\cos\frac{\pi}{2}\alpha+1}\,dy.$$

EXAMPLE 2. Suppose that $1 < \alpha \leq 2$, and let $Y_0 \stackrel{d}{=} Y_1 \stackrel{d}{=} \cdots \stackrel{d}{=} Y_n$ be independent random variables distributed like $Y(\alpha, -1, \gamma, \lambda)$. Denote by Z_j a cutoff of the random variable $Y_j - \gamma\lambda$, and consider the statistic

$$Q = \left(\frac{n^\alpha Z_0^{-\alpha}}{Z_1^{-\alpha} + \cdots + Z_n^{-\alpha}}\right)^{1/\alpha}. \tag{3.3.24}$$

According to (2.1.2), the distribution of Z_j does not depend on γ. On the other hand, by the property 1 of cutoffs and the very structure of the statistic Q, its distribution does not depend on λ, i.e.

$$Q \stackrel{d}{=} \left(\frac{n^\alpha Z^{-\alpha}(\alpha, 1/\alpha)}{Z^{-\alpha}(\alpha,1/\alpha) + \cdots + Z^{-\alpha}(\alpha,1/\alpha)}\right)^{1/\alpha}.$$

We compute the distribution of Q. It follows from the duality relation (3.2.18) that $Z^{-\alpha}(\alpha, 1/\alpha) = Y(1/\alpha, 1)$. Moreover, it follows from (2.1.4) that

$$n^{-\alpha}(Y(1/\alpha,1) + \cdots + Y(1/\alpha,1)) \stackrel{d}{=} Y(1/\alpha, 1).$$

Thus,

$$Q \stackrel{d}{=} \left(\frac{Y(1/\alpha,1)}{Y(1/\alpha,1)}\right)^{1/\alpha},$$

which, when the rule (3.3.16) is applied to the ratio in parentheses, gives us that

$$Q \stackrel{d}{=} Z(1, 1/\alpha). \tag{3.3.25}$$

The distribution of the right-hand side of (3.3.25) can be obtained without difficulty from that of the random variable $Z^{1/\alpha}(1, \alpha/2)$ (see (3.2.23)):

$$F_Q(x) = \frac{\alpha}{\pi}\arctan\left(\frac{x + \cos\pi/\alpha}{\sin\pi/\alpha}\right) + 1 - \frac{\alpha}{2}. \tag{3.3.26}$$

The function $F_\alpha(x)$ has another representation in integral form which is connected with a number of interesting facts and to which we devote some attention below. The facts presented are taken from the two recent papers [248] and [249].

The second representation of $F_\alpha(x)$ is

$$F_\alpha(x) = \frac{1}{2} + \frac{1}{\alpha\pi^2}\int_0^\infty \log\left(\frac{1+|x+y|^\alpha}{1+|x-y|^\alpha}\right)\frac{dy}{y}. \tag{3.3.27}$$

This formula can be proved with the help of the following simple but useful lemma.

LEMMA 3.3.1. *Suppose that $F(x,y)$ and $f(u,v)$ are the distribution function and characteristic function, respectively, of a pair (U,V) of random variables such that*

$$\mathsf{E}\log^2(1+|U|)+\mathsf{E}\log^2(1+|V|)<\infty; \tag{3.3.28}$$

for any $\varepsilon>0$, and

$$\iint\limits_{D_\varepsilon} |f(u,v)|\frac{du\,dv}{|uv|}, \tag{3.3.29}$$

where $D_\varepsilon=\{(u,v)\colon |u|\ge\varepsilon, |v|\ge\varepsilon\}$. Then

$$\begin{aligned}F(0,0) = &-\tfrac14+\tfrac12(F(0,\infty)+F(\infty,0))\\ &-\frac{1}{4\pi^2}\int_0^\infty\int_0^\infty\{f(u,v)-f(-u,v)-f(u,-v)+f(-u,-v)\}\frac{du\,dv}{uv}.\end{aligned} \tag{3.3.30}$$

PROOF. It is not hard to verify that

$$\begin{aligned}F(0,0)=\frac14-\frac14\iint \operatorname{sgn} x\,dF(x,y)\\ -\frac14\iint\operatorname{sgn} y\,dF(x,y)+\frac14\iint\operatorname{sgn} x\operatorname{sgn} y\,dF(x,y).\end{aligned}$$

By the equality

$$\operatorname{sgn} x=\frac{1}{i\pi}\int_0^\infty(e^{iux}-e^{-iux})\frac{du}{u},$$

we get (3.3.30) from the last expression for $F(0,0)$ after changing the order of integration (which is allowed by the conditions (3.3.28) and (3.3.29) of the lemma).

COROLLARY. Let Y_1 and Y_2 be independent random variables with the same distribution as $Y(\alpha,0)$, $0<\alpha\le2$. Consider the variables $U=Y_1-xY_2$ and $V=Y_2$. According to the lemma,

$$\begin{aligned}F_\alpha(x)&=F(0,0)\\ &=\frac12+\frac{1}{\pi^2}\int_0^\infty\int_0^\infty\exp(-|u|^\alpha)[\exp(-|ux-v|^\alpha)-\exp(-|ux+v|^\alpha)]\frac{du\,dv}{uv}.\end{aligned}$$

But this integral occurs in tables (see, for example, (2.4.1.28) in [251]). It can be expressed by means of a single integral, and leads us to (3.3.27).

In addition to being connected in a simple way with the distribution $\hat F_\alpha(x)$ in (3.3.22) of a cutoff, the distribution $F_\alpha(x)$ also has a simple connection with the distribution $G_\alpha(x)$ of the random variable

$$\Delta=\min(|L|,|L|^{-1})=\min(\hat L,\hat L^{-1}). \tag{3.3.31}$$

Namely,

$$G_\alpha(x) = 4(F_\alpha(x) - \tfrac{1}{2}) = 2\hat{F}_\alpha(x), \qquad 0 \le x \le 1.$$

The distribution of Δ is used (as in [248]) to estimate the parameter α.

An interesting feature of the distribution $F_\alpha(x)$ (and also of the distributions $\hat{F}_\alpha(x)$ and $G_\alpha(x)$ connected with it) is the possibility of extending the domain of variation of the parameter α beyond the right-hand bound $\alpha = 2$. It turns out that $F_\alpha(x)$ remains a distribution function even when $\alpha > 2$.

LEMMA 3.3.2. *The function $F_\alpha(x)$ defined by (3.3.27) for $\alpha > 0$ is a distribution function.*

PROOF. The fact that $F_\alpha(x)$ is nondecreasing follows from the obvious fact that its derivative

$$F'_\alpha(x) = \frac{1}{\pi^2|x|}\int_0^\infty ((1+|x|^\alpha\,|1-y|^\alpha)^{-1} - (1+|x|^\alpha\,|1+y|^\alpha)^{-1})\frac{dy}{y}$$

is nonnegative. Replacing the variable y by yx and letting x go to ∞, we get that

$$F_\alpha(\infty) = \frac{1}{2} + \frac{1}{\pi^2}\int_0^\infty \log\left|\frac{1+y}{1-y}\right|\frac{dy}{y}.$$

The integral in the second term can be found in tables (see (2.6.14.24) in [251]). It is equal to $\frac{1}{2}$ (note the factor $1/\pi^2$).

Since

$$F_\alpha(-\infty) = \frac{1}{2} - \frac{1}{\pi^2}\int_0^\infty \log\left|\frac{1+y}{1-y}\right|\frac{dy}{y}$$

as $x \to -\infty$, it has thus been proved that $F_\alpha(-\infty) = 0$ and $F_\alpha(\infty) = 1$ for any $\alpha > 0$.

We note here a property of F_α in the case when $\alpha = 2n$ is an even number. Denote by R_n a random variable taking the n values

$$-\frac{n-1}{n}, -\frac{n-3}{n}, \dots, \frac{n-3}{n}, \frac{n-1}{n},$$

each with probability $1/n$. Then consider a random variable $Y_A(1,0,0,1)$ independent of R_n and form with its help the new random variable

$$Y_A(1,0,0,1)\cos(\pi R_n/2) + \sin(\pi R_n/2) \stackrel{d}{=} Y_C(1, R_n). \tag{3.3.32}$$

The last relation is a consequence of (2.3.5a). We present the promised property of $F_\alpha(x)$ as the following theorem, whose proof we omit, referring the interested reader to the original source [249].

THEOREM 3.3.5. *For any integer $n \geq 1$,*

$$F_{2n}(x) = P(Y_C(1, R_n) < x). \tag{3.3.33}$$

Thus, the distribution $F_{2n}(x)$ is a mixture of Cauchy distributions with linearly transformed argument (cf. (3.3.32)). Since $R_n \xrightarrow{d} R$ as $n \to \infty$, where R is a random variable uniformly distributed in the interval $(-1, 1)$, the limit distribution

$$F_\infty(x) = P(Y_C(1, R) < x)$$

exists and has density (see (2.3.5a))

$$P_\infty(x) = \frac{1}{\pi^2} \int_{-1}^{1} \frac{du}{1 - 2ux + u^2} = \frac{1}{\pi^2 x} \log \left| \frac{1+x}{1-x} \right|, \qquad x \neq 1.$$

Precisely the same relation holds between the density of the distribution $F_\alpha(x)$, $2 < \alpha < \infty$, and the symmetric trans-stable functions $\sigma(x, \alpha, 0)$ as is observed in the case when $0 < \alpha \leq 2$.

§3.4. Properties of extremal strictly stable distributions

The class $\mathfrak{W}_1$ which was the main subject of the preceding section has a subclass uniting the distributions with parameters (α, θ_α). As a matter of fact, the main role among them is played by the distributions with $\alpha < 1$, because $\theta_1 = 1$ corresponds to the distribution degenerate at the point $x = 1$, while the distributions with $\alpha > 1$ (more precisely, the part of them concentrated on the semi-axis $x > 0$) can, by the duality property, be expressed in terms of the distributions in the first group. The random variables $Y(\alpha, 1)$, $0 < \alpha < 1$, whose distributions form this group (and only they) are positive with probability 1; hence $Z(\alpha, 1) = Y(\alpha, 1)$. The abbreviated notation $Y(\alpha)$ will be used below for the variables $Y(\alpha, 1)$.

THEOREM 3.4.1. *Suppose that $\omega_1, \dots, \omega_n$ are numbers such that $0 < \omega_j < 1$. Then for any $n \geq 2$*

$$Y(\alpha_n) \stackrel{d}{=} Y(\omega_1) Y^{1/\omega_1}(\omega_2) Y^{1/\omega_1\omega_2}(\omega_3) \cdots Y^{1/\omega_1,\dots,\omega_{n-1}}(\omega_n), \tag{3.4.1}$$

where $\alpha_n = \omega_1\omega_2 \cdots \omega_n$.

REMARK. By relabeling the numbers ω_j in inverse order it is possible to give (3.3.23) the form

$$Y^{\alpha_n}(\alpha_n) \stackrel{d}{=} Y^{\omega_1}(\omega_1) Y^{\omega_1\omega_2}(\omega_2) \cdots Y^{\omega_1\cdots\omega_n}(\omega_n) \tag{3.4.2}$$

with the help of a simple transformation.

The proof of (3.4.1) is by induction. For $n = 2$, (3.4.1) coincides with (3.3.4). If it is true for a set of $n-1$ variables, then, by (3.3.4),

$$Y(\alpha_n) \stackrel{d}{=} Y(\alpha_{n-1}\omega_n) \stackrel{d}{=} Y(\alpha_{n-1}) Y^{1/\alpha_{n-1}}(\omega_n).$$

If we now replace $Y(\alpha_{n-1})$ by its expression according to (3.4.1), we get (3.4.1) for n random variables.

Despite the fact that (3.4.1) and (3.4.2) are simple consequences of (3.3.4), they serve as the source of a number of interesting connections between the distributions in the subclass of stable laws under consideration. Since infinite products of positive random variables will be encountered below, we agree to say that these products converge if the series of logarithms of their factors converge with probability 1.

The condition that the product $\mathsf{E}X_1^s \mathsf{E}X_2^s \cdots$ converge for some real value $s \neq 0$ is for us a convenient necessary and sufficient condition for the convergence of an infinite product $X_1 X_2 \cdots$ of independent and positive (with probability 1) random variables. This is not hard to establish if the question of convergence of a product of random variables is reduced to the question of convergence of a sum of random variables.

As before, E denotes a random variable with exponential distribution

$$\mathsf{P}(E > x) = \exp(-x), \qquad x \geq 0.$$

THEOREM 3.4.2. *Suppose that $\omega_1, \omega_2, \ldots$ is a sequence of numbers such that $0 < \omega_j \leq 1$, and $\alpha = \omega_1\omega_2 \cdots$ is their product. If $\alpha > 0$, then*

$$Y(\alpha) \stackrel{d}{=} Y(\omega_1) Y^{1/\omega_1}(\omega_2) Y^{1/\omega_1\omega_2}(\omega_3) \cdots, \tag{3.4.3}$$

$$Y^{\alpha}(\alpha) \stackrel{d}{=} Y^{\omega_1}(\omega_1) Y^{\omega_1\omega_2}(\omega_2) \cdots. \tag{3.4.4}$$

If $\alpha = 0$, then

$$Y^{\omega_1}(\omega_1) Y^{\omega_1\omega_2}(\omega_2) \cdots \stackrel{d}{=} 1/E; \tag{3.4.5}$$

in particular, for any $0 < \omega < 1$

$$Y^{\omega}(\omega) Y^{\omega^2}(\omega) Y^{\omega^3}(\omega) \cdots \stackrel{d}{=} 1/E. \tag{3.4.6}$$

The proof of (3.4.3) and (3.4.4) reduces to a passage to the limit as $n \to \infty$ in (3.4.1) and (3.4.2), respectively, with use of the Mellin transforms of both sides of the equalities and the criterion given above for convergence of infinite products of independent random variables.

For brevity let $Z(\alpha) = Z(\alpha, 1/\alpha)$, $1 \leq \alpha \leq 2$. The duality relation (3.2.18), namely, $Y(1/\alpha) = Z^{-\alpha}(\alpha)$ for $1 \leq \alpha \leq 2$, makes it possible to get relations of the type (3.4.1)–(3.4.4) and (3.4.6) for the random variables $Z(\alpha)$. We present them below.

If $\omega_1, \omega_2, \ldots$ is a finite or infinite sequence of numbers such that $\omega_j \geq 1$ and $\alpha = \omega_1\omega_2 \cdots \leq 2$, then

$$Z^{\alpha}(\alpha) \stackrel{d}{=} Z^{\omega_1}(\omega_1) Z^{\omega_1\omega_2}(\omega_2) \cdots, \tag{3.4.7}$$

$$Z(\alpha) \stackrel{d}{=} Z(\omega_1) Z^{1/\omega_1}(\omega_2) Z^{1/\omega_1\omega_2}(\omega_3) \cdots. \tag{3.4.8}$$

Application of the duality relation to (3.4.6) leads for any α, $1 < \alpha \leq 2$, to the relation

$$Z(\alpha) Z^{1/\alpha}(\alpha) Z^{1/\alpha^2}(\alpha) \cdots \stackrel{d}{=} E. \tag{3.4.9}$$

It is easy to see that

$$Z(2) \stackrel{d}{=} |Y(2,0)| \quad \text{and} \quad Y(2,0) \stackrel{d}{=} \sqrt{2}N \stackrel{d}{=} UZ(2),$$

where N is a $(0,1)$-normally distributed random variable, and U is a random variable independent of $Z(2)$ and taking values $+1$ and -1 with probability $\frac{1}{2}$ each. These considerations together with (3.4.7) and (3.4.8) make it possible to get the relations

$$\begin{aligned} N &\stackrel{d}{=} \frac{U}{\sqrt{2}} Z^{\omega_1/2}(\omega_1) Z^{\omega_1\omega_2/2}(\omega_2) \cdots \\ &\stackrel{d}{=} \frac{U}{\sqrt{2}} Z(\omega_1) Z^{1/\omega_1}(\omega_2) Z^{1/\omega_1\omega_2}(\omega_3) \cdots \end{aligned} \tag{3.4.10}$$

for any decomposition of the number 2 into a finite or infinite product $2 = \omega_1\omega_2 \cdots$, $\omega_j \geq 1$.

Similarly, using the duality relation again, we can claim that

$$\begin{aligned} N &\stackrel{d}{=} \frac{U}{\sqrt{2}} Y^{-1/2}(\omega_1) Y^{-1/2\omega_1}(\omega_2) Y^{-1/2\omega_1\omega_2}(\omega_3) \cdots \\ &\stackrel{d}{=} \frac{U}{\sqrt{2}} Y^{-\omega_1}(\omega_1) Y^{-\omega_1\omega_2}(\omega_2) \cdots \end{aligned} \tag{3.4.11}$$

for any decomposition $\frac{1}{2} = \omega_1\omega_2 \cdots$, $\omega_j \leq 1$ into a finite or infinite product.

We give three more relations in which the random variables N appear in an interesting way. Namely, as is clear from (3.3.19), for $\frac{1}{2} \leq \alpha < 1$,

$$\frac{Z(2\alpha)}{Y(\alpha)} \stackrel{d}{=} \left|\sqrt{2}N\right|^{1/\alpha}. \tag{3.4.12}$$

Further, if we choose $\omega_1 = \cdots = \omega_n = \frac{1}{2}$ in (3.4.1) and use the equality $Y(\frac{1}{2}) \stackrel{d}{=} N^{-2}/2$, which follows from the duality relation, then we arrive at the representation of $Y(1/2^n)$ as a product of powers of independent normally distributed random variables:

$$Y(1/2^n) \stackrel{d}{=} 2^{1-2^n} N^{-2} N^{-2^2} \cdots N^{-2^n}. \tag{3.4.13}$$

Finally, we get the third relation by taking the logarithm of (3.4.9) and setting $\alpha = 2$ that

$$\log E \stackrel{d}{=} \log 2 + \log|N| + 2^{-1}\log|N| + 2^{-2}\log|N| + \cdots. \tag{3.4.14}$$

However, the expansion of the random variable $\log E$ in a series of independent random variables is possible with the help of variables other than $\log|N|$. This is shown by (3.4.6) if the logarithm of both sides is taken. Namely,

$$\log E \stackrel{d}{=} -\{\alpha \log Y(\alpha) + \alpha^2 \log Y(\alpha) + \alpha^3 \log Y(\alpha) + \cdots\} \tag{3.4.15}$$

for any $0 < \alpha < 1$.

Both finite and infinite expansions of random variables in products of independent random variables can very simply be adapted for computing the distributions of, so to speak, "products of ratios". With this purpose we first compute the distribution of the ratio $Y(\alpha)/Y(\alpha)$. According to (3.3.18), for $\alpha < 1$

$$Y(\alpha)/Y(\alpha) \stackrel{d}{=} Z^{1/\alpha}(1, \alpha). \tag{3.4.16}$$

The distribution function $F_\alpha(x)$ of the random variable $Z^{1/\alpha}(1, \alpha/2)$, $0 < \alpha < 2$, has already appeared in the analysis of an example given in the last section. And we can use its expression (3.3.23) for computing the distribution function of the right-hand side of (3.4.16). After simple transformations we get that for any $0 < \alpha < 1$

$$\begin{aligned} R_\alpha(x^\alpha) &= \mathsf{P}(Y(\alpha)/Y(\alpha) < x) = \mathsf{P}(Z(1, \alpha) < x^\alpha) \\ &= \frac{1}{\pi\alpha} \arctan\left(\frac{x^\alpha + \cos\pi\alpha}{\sin\pi\alpha}\right) - \frac{1}{2\alpha} + 1. \end{aligned} \tag{3.4.17}$$

We know (see (2.9.2)) that $Y^\alpha(\alpha) \stackrel{d}{\to} E^{-1}$ as $\alpha \to 0$. The distribution $R_\alpha(x^\alpha)$ thus does not have a proper limit distribution as $\alpha \to 0$, but, on the other hand, $R_\alpha(x)$ does. It is not hard to compute that as $\alpha \to 0$

$$R_\alpha(x) \to R_0(x) = \mathsf{P}(E/E < x) = x/(1+x), \qquad x \geq 0.$$

We take two each of the complex of relations (3.4.1) and (3.4.2), with the random variables making up the left-hand (right-hand) sides mutually independent, and then we divide the corresponding parts of these relations one by the other. As a result,

$$\left(\frac{Y(\omega_1)}{Y(\omega_1)}\right)\left(\frac{Y(\omega_1)}{Y(\omega_1)}\right)^{1/\omega_1} \cdots \left(\frac{Y(\omega_1)}{Y(\omega_1)}\right)^{1/\omega_1\cdots\omega_{n-1}} \stackrel{d}{=} \frac{Y(\alpha_n)}{Y(\alpha_n)} \stackrel{d}{=} Z^{1/\alpha_n}(1, \alpha_n), \tag{3.4.18}$$

$$\left(\frac{Y(\omega_1)}{Y(\omega_1)}\right)^{\omega_1} \cdots \left(\frac{Y(\omega_1)}{Y(\omega_1)}\right)^{\omega_1\cdots\omega_n} \stackrel{d}{=} Z(1, \alpha_n). \tag{3.4.19}$$

Analogous considerations can be written for finite or infinite products of powers of variables $Z(\omega_j)/Z(\omega_j)$, based on (3.4.7), (3.4.8), and the fact (a consequence of (3.3.17)) that for $\alpha > 1$

$$Z(\alpha)/Z(\alpha) \stackrel{d}{=} Z(1, 1/\alpha).$$

It is clearly possible to take the limit as $n \to \infty$ in (3.4.18) and (3.4.19) if $\alpha_n \to \alpha > 0$. However, if $\alpha_n \to 0$, then the limit distribution in (3.4.18) does not exist, though it exists in (3.4.19), and we get that

$$\left(\frac{Y(\omega_1)}{Y(\omega_1)}\right)^{\omega_1} \left(\frac{Y(\omega_1)}{Y(\omega_1)}\right)^{\omega_1\omega_2} \cdots \stackrel{d}{=} \frac{E}{E}. \tag{3.4.20}$$

An analogous assertion follows from (3.4.9) for the random variables $Z(\alpha)$ with $1 < \alpha \leq 2$. Namely,

$$\left(\frac{Z(\alpha)}{Z(\alpha)}\right) \left(\frac{Z(\alpha)}{Z(\alpha)}\right)^{1/\alpha} \left(\frac{Z(\alpha)}{Z(\alpha)}\right)^{1/\alpha^2} \cdots \stackrel{d}{=} \frac{E}{E}. \tag{3.4.21}$$

The following relation stands rather alone. Denote by Γ_ν, $\nu > 0$, a random variable having Γ-distribution with parameter ν, i.e., a density of the form

$$p_\nu(x) = \frac{1}{\Gamma(\nu)} x^{\nu-1} \exp(-x), \qquad x \geq 0.$$

THEOREM 3.4.3. *For any integer $n \geq 2$*

$$1/Y(1/n) \stackrel{d}{=} \Gamma_{1/n}\Gamma_{2/n} \cdots \Gamma_{(n-1)/n} n^n. \tag{3.4.22}$$

COROLLARY. By the relation (3.3.1) with $\alpha = 1$ and $\alpha' = 1/n$, (3.4.22) implies that a random variable with the parameters $\alpha = 1/n$ and $|\theta| \leq 1$ has the representation

$$\begin{aligned} Y^{-1}(1/n, \theta) &\stackrel{d}{=} Y^{-1}(1, \theta)\Gamma_{1/n}\Gamma_{2/n} \cdots \Gamma_{(n-1)/n} \cdot n^n \\ &\stackrel{d}{=} Y(1, \theta)\Gamma_{1/n}\Gamma_{2/n} \cdots \Gamma_{(n-1)/n} \cdot n^n. \end{aligned}$$

PROOF OF THE THEOREM. The Mellin transform of the distribution of $1/Y(1/n)$ has, by (3.0.2), the form

$$M\left(-s, \frac{1}{n}, 1\right) = \frac{\Gamma(1+ns)}{\Gamma(1+s)}.$$

This fraction can be expanded in a product of ratios of Γ-functions by using the so-called multiplication theorem for the functions $\Gamma(x)$, $x > 0$:

$$\Gamma(x)\Gamma\left(x + \frac{1}{n}\right) \cdots \Gamma\left(x + \frac{n-1}{n}\right) = \frac{(2\pi)^{(n-1)/2}}{n^{(2nx-1)/2}} \Gamma(nx).$$

We have

$$\frac{\Gamma(1+ns)}{\Gamma(1+s)} = n^{ns}\frac{\Gamma(s+1/n)}{\Gamma(1/n)}\frac{\Gamma(s+2/n)}{\Gamma(2/n)}\cdots\frac{\Gamma(s+(n-1)/n)}{\Gamma((n-1)/n)}. \tag{3.4.23}$$

Since the ratio $\Gamma(s+\nu)/\Gamma(\nu)$ is the Mellin transform of the distribution of the variable Γ_ν, a comparison of the Mellin transforms on the left and right sides of (3.4.23) gives us (3.4.22) ([3.6]).

This concludes our getting acquainted with the "multiplicative" properties of strictly stable laws. Here something should be said about where these numerous and diverse relations between distributions (which for conciseness we have clothed as relations between random variables) might turn out to be useful.

First of all, such relations can be a convenient tool in computing distributions of various kinds of functions of independent random variables $Y(\alpha,\theta)$ and $Z(\alpha,\rho)$. If we supplement the relations of a multiplicative nature in this chapter by the additive relations of §2.1, we obtain a kind of distinctive "algebra" in the set of independent random variables of a special form. This is especially noticeable if we single out a set of independent random variables having distributions in $\mathfrak{W}_1$ with the parameter value $\theta=\theta\alpha$, $0<\alpha\le 2$.

To illustrate, we show how (3.4.13) and (3.4.22) help to get expressions of a new form for the densities $g(x,\frac{1}{4},1)$ and $g(x,\frac{1}{3},1)$ as integrals of positive functions. According to the indicated relations,

$$Y(\tfrac{1}{4})\stackrel{d}{=}Y(\tfrac{1}{2})Y^2(\tfrac{1}{2}), \qquad Y^{-1}(\tfrac{1}{3})\stackrel{d}{=}27\Gamma_{1/3}\Gamma_{2/3},$$

and hence

$$g\left(x,\frac{1}{4},1\right) = \int_0^\infty g\left(\frac{x}{y},\frac{1}{2},1\right)g\left(\sqrt{y},\frac{1}{2},1\right)\frac{dy}{2y\sqrt{y}}, \tag{3.4.24}$$

$$x^{-2}g\left(x^{-1},\frac{1}{3},1\right) = \int_0^\infty p_{1/3}\left(\frac{x}{y}\right)p_{2/3}\left(\frac{y}{27}\right)\frac{dy}{27y}. \tag{3.4.25}$$

The form of the densities $g(x,\frac{1}{2},1)$ and $p_\nu(x)$ is known. The first was given back in §2.2 (it is also easy to compute with the help of Theorem 3.4.3, since $Y^{-1}(\frac{1}{2})\stackrel{d}{=}4\Gamma_{1/2}$ according to (3.4.22)), and the second is well known. After substitution of these densities in (3.4.24) and (3.4.25) and a simplifying change of variable, we find that for $x>0$

$$g\left(x,\frac{1}{4},1\right) = \frac{1}{2\pi}x^{-4/3}\int_0^\infty \exp\left\{-\frac{1}{4}x^{-1/3}(y^4+y^{-2})\right\}dy, \tag{3.4.26}$$

$$g\left(x,\frac{1}{3},1\right) = \frac{1}{2\pi}x^{-3/2}\int_0^\infty \exp\left\{-\frac{1}{3\sqrt{3x}}(y^3+y^{-3})\right\}dy. \tag{3.4.27}$$

The second sphere of application of the "multiplicative" properties of stable laws is the construction of various kinds of useful statistics with explicitly computable distributions. Two examples of such statistics were analyzed at the end of the preceding section.

Finally, some of the relations turn out to be useful in stochastic modeling problems for generating sequences of random variables with stable distributions. For instance, given a sequence of random numbers having the standard normal distribution, it is very easy with the help of (3.4.13) to construct sequences of random variables distributed like $Y(1/2^n)$.

§3.5. M-infinite divisibility of the distributions of the variables $Y(\alpha,\theta)$ and $Z(\alpha,\rho)$

The classical theory of limit theorems for sums of independent random variables has been generalized in various directions. One such direction was a scheme with a generalized concept of sum: a certain operation with respect to which the set of possible values of the random elements being "added" forms a locally compact abelian group. By analogy with the addition scheme (the A-scheme), it is possible to introduce in this generalized scheme the concept of infinitely divisible laws (see [101]). The scheme for multiplying independent random variables (the M-scheme) is a direct generalization of the A-scheme. If in the former we single out the case when the random variables being multiplied are positive, then we obtain the A-scheme to within an isomorphism. The analogue of infinitely divisible distributions in the M-scheme can be defined as follows (this definition, used in [101], differs in form from the definition used in [64], but it singles out the very same class of distributions, which we agree to denote by $\mathfrak{M}$).

DEFINITION. The distribution F_X of a random variable X belongs to $\mathfrak{M}$ if there exists a sequence of integers $1 < n_1 < n_2 < \cdots$ such that for each number n_j the random variable X can be represented as a product of n_j independent and identically distributed random variables, i.e.,

$$X \stackrel{d}{=} X_1 X_2 \cdots X_{n_j}, \quad \text{where } X_k \stackrel{d}{=} X_l.$$

The distributions $F_X \in \mathfrak{M}$ can be described in terms of the corresponding characteristic transforms $W_X(t)$. It turns out that F_X is an M-infinitely divisible distribution if and only if the functions $w_k(t)_X$ have the form

$$w_0(t)_X = a_0 \mathfrak{f}_1(t)\mathfrak{f}_2(t), \qquad w_1(t)_X = a_1 \mathfrak{f}_1(t)/\mathfrak{f}_2(t),$$

where $\mathfrak{f}_1$ and $\mathfrak{f}_2$ are the characteristic functions of A-infinitely divisible laws, and

$$\mathfrak{f}_2(t) = \exp\left\{\int (e^{itx} - 1)\, dH_2(t)\right\},$$

where H_2 is a spectral function, and the real numbers a_0 and a_1 satisfy the conditions

$$0 \le a_0 \le 1, \qquad |a_1| \le \exp\left(-2\int dH_2(x)\right). \tag{3.5.1}$$

In the class $\mathfrak{M}$ we single out the special subclass $\mathfrak{M}'$ by the condition that among the sequences $n_1 < n_2 < \cdots$ there is a sequence consisting of even numbers. A distribution $F \in \mathfrak{M}$ is in the subclass $\mathfrak{M}'$ under the additional (necessary and sufficient) condition that $a_1 \ge 0$. A feature of the subclass $\mathfrak{M}'$ is that the random variables X with distributions in $\mathfrak{M}'$ can be decomposed into a product of $n \ge 1$ independent and identically distributed factors for any number n, even or odd. At the same time, an X with distribution in $\mathfrak{M}\backslash\mathfrak{M}'$ can only be decomposed into a product of an odd number of independent factors.

In connection with the random variables $Y(\alpha,\theta)$ and $Z(\alpha,\rho)$ of interest to us, the membership of their distributions in $\mathfrak{M}$ has to do with the corresponding representation of the characteristic transform $W(s,\alpha,\theta)$ and the Mellin transform $M(s,\alpha,\rho)$. Namely, $G(x,\alpha,\theta) \in \mathfrak{M}$ if and only if

$$\begin{aligned} w_0(s,\alpha,\theta) &= a_0\varphi_1(s)\varphi_2(s), \\ w_1(s,\alpha,\theta) &= a_1\varphi_1(s)/\varphi_2(s), \end{aligned} \tag{3.5.2}$$

where

$$\varphi_1(s) = \exp\left\{as + \int (e^{sx} - 1 - s\sin x)\, dH_1(x)\right\},$$

$$\varphi_2(s) = \exp\left\{\int (e^{sx} - 1)\, dH_2(x)\right\},$$

H_1 and H_2 are spectral functions, and a_0 and a_1 satisfy the requirements indicated above. Since $w_0(t)_X = w_1(t)_X$ for positive random variables X, the condition

$$M(s,\alpha,\rho) = \varphi_1(s) \tag{3.5.3}$$

ensures that $\hat{G}(x,\alpha,\rho)$ is in $\mathfrak{M}$.

It is not hard to see that if the distribution of a nonnegative random variable belongs to $\mathfrak{M}$, then it must be in $\mathfrak{M}'$.

THEOREM 3.5.1. *For any admissible values of the parameters (α,ρ) the distribution of a random variable $Z(\alpha,\rho)$ belongs to $\mathfrak{M}'$.*

PROOF. Let us show that $M(s,\alpha,\rho)$ is representable in the form (3.5.2). We have that

$$M(s,\alpha,\rho) = \frac{\sin\pi\rho s}{\rho\sin\pi s}\,\frac{\Gamma(1-s/\alpha)}{\Gamma(1-s)} = \frac{\Gamma(1-s)\Gamma(1+s)\Gamma(1-s/\alpha)}{\Gamma(1-\rho s)\Gamma(1+\rho s)\Gamma(1-s)}. \tag{3.5.4}$$

It is not hard to derive the following equalities from the theory of the functions Γ (for this we can use, for example, the integral representation of $\log\Gamma(z)$ (see 8.341 in [27])):

1. In the strip $|\operatorname{Re} s| < a/b,\ a > 0, b > 0$,

$$\log\left(\frac{\Gamma(a-bs)\Gamma(a+bs)}{\Gamma^2(a)}\right) = \int (e^{su} - 1 - su)\nu(u,a,b)\,du, \tag{3.5.5}$$

where

$$\nu(u,a,b) = \exp(-|u|a/b)(|u|(1-\exp(-|u|/b)))^{-1}.$$

2. In the half-plane $\operatorname{Re} s < 1/b,\ b > 0$,

$$\log\Gamma(1-bs) = -b\mathbb{C}s + \int_0^\infty (e^{su} - 1 - su)\nu(u,1,b)\,du, \tag{3.5.6}$$

where $\mathbb{C}$ is Euler's constant.

With the help of these equalities $\log M(s,\alpha,\rho)$ can be written in the following integral form:

$$\log M(s,\alpha,\rho) = (-1/\alpha+1)\mathbb{C}s + \int (e^{su} - 1 - su)H_1'(u)\,du,$$

where

$$H_1'(u) = \nu(u,1,1) - \nu(u,1,\rho) + \tfrac{1}{2}(1+\operatorname{sgn} u)(\nu(u,1,1/\alpha) - \nu(u,1,1)).$$

Since the function $\nu(u,a,b)$ is nondecreasing as b increases for fixed u and a, it follows that on the semi-axis $u < 0$

$$H_1'(u) = \nu(u,1,1) - \nu(u,1,\rho) \geq 0,$$

just as on the semi-axis $u > 0$

$$H_1'(u) = \nu(u,1,1/\alpha) - \nu(u,1,\rho) \geq 0.$$

Consequently, $H_1'(u)$ is the density of some measure. If we add the fact that

$$\int_{-1}^1 u^2 H_1'(u)\,du < \infty,$$

then the representation of $M(s,\alpha,\rho)$ in the form (3.5.3) becomes obvious.

REMARK. For symmetric random variables $\tilde{Z}(\alpha,\rho) = UZ(\alpha,\rho)$, where U is a random variable independent of $Z(\alpha,\rho)$ and taking the values $+1$ and -1 with probability $\frac{1}{2}$, the assertion of the theorem remains in force, because the characteristic transforms corresponding to them have the form

$$w_0(t) = M(it,\alpha,\rho), \qquad w_1(t) \equiv 0.$$

THEOREM 3.5.2. *The distribution of $Y(\alpha,\theta)$ belongs to $\mathfrak{M}$ if and only if either $\alpha \le 1$, or $\alpha > 1$ and $\theta = 0$. These distributions are in $\mathfrak{M}'$ if and only if, in addition, $\theta \ge 0$.*

PROOF. Let us transform the functions $w_k(s,\alpha,\theta)$ to the form (3.5.2) and determine when H_1 and H_2 are nondecreasing.

We begin with the case $\theta = 0$. Since here $w_1(s,\alpha,\theta) = 0$, it is necessary to transform only one function $w_0(s,\alpha,\theta)$, for nothing prevents us from taking $\varphi_2(s) \equiv 1$ and $a_1 = 0$ at once. We have that

$$w_0(s,\alpha,\theta) = \frac{1}{\cos\frac{\pi}{2}s}\frac{\Gamma(1-s\alpha)}{\Gamma(1-s)} = \frac{\Gamma(\frac{1}{2}-\frac{1}{2}s)\Gamma(\frac{1}{2}+\frac{1}{2}s)}{\Gamma^2(\frac{1}{2})}\frac{\Gamma(1-s\alpha)}{\Gamma(1-s)}.$$

According to (3.5.5) and (3.5.6), we can write $\log w_0(s,\alpha,\theta)$ in the form

$$\mathrm{C}(1-1/\alpha)s + \int (e^{su}-1-su)H_1'(u)\,du,$$

where

$$H_1'(u) = \nu(u,\tfrac{1}{2},\tfrac{1}{2}) + \tfrac{1}{2}(1+\operatorname{sgn} u)\{\nu(u,1,1/\alpha) - \nu(u,1,1)\}.$$

Obviously, $H_1'(u) = \nu(u,\frac{1}{2},\frac{1}{2}) \ge 0$ on the semi-axis $u < 0$. After uncomplicated transformations, $H_1'(u)$ takes the form $H_1'(u) = \nu(u,1,1/\alpha) - \nu(u,1,\frac{1}{2})$ on the semi-axis $u > 0$. This difference is nonnegative, because $\nu(u,1,b)$ is nondecreasing as b increases. Therefore, $G(x,\alpha,0) \in \mathfrak{M}$ (more precisely, it belongs to $\mathfrak{M}'$, since $a_1 = 0$).

Suppose now that $\theta \ne 0$. We have that

$$w_k(s,\alpha,0) = \frac{\cos\frac{\pi}{2}(k-\theta s)}{\cos\frac{\pi}{2}(k-s)}\frac{\Gamma(1-s/\alpha)}{\Gamma(1-s)}, \qquad k = 0,1.$$

Hence $a_0 = 1, a_1 = \theta$, and

$$\varphi_1^2 = w_0 w_1/\theta, \qquad \varphi_2^2 = \theta w_0/w_1.$$

Consequently,

$$\varphi_1^2(s) = \frac{\Gamma(1-s)\Gamma(1+s)}{\Gamma(1-\theta s)\Gamma(1+\theta s)}\frac{\Gamma^2(1-s/\alpha)}{\Gamma^2(1-s)},$$

$$\varphi_2^2(s) = \frac{\Gamma(\frac{1}{2}-\frac{1}{2}s)\Gamma(\frac{1}{2}+\frac{1}{2}s)\Gamma(1-\frac{\theta}{2}s)\Gamma(1+\frac{\theta}{2}s)}{\Gamma(\frac{1}{2}-\frac{\theta}{2}s)\Gamma(\frac{1}{2}+\frac{\theta}{2}s)\Gamma(1-\frac{1}{2}s)\Gamma(1+\frac{1}{2}s)}.$$

We begin with the function $\varphi_2^2(s)$. According to (3.5.5) and (3.5.6),

$$\log\varphi_2(s) = \int (e^{su}-1-su)H_2'(u)\,du,$$

where

$$H_2'(u) = \tfrac{1}{2}(\nu(u, \tfrac{1}{2}, \tfrac{1}{2}) - \nu(u, \tfrac{1}{2}, |\theta|/2) + \nu(u, 1, |\theta|/2) - \nu(u, 1, \tfrac{1}{2}))$$
$$= [(1+e^u)^{-1} - (1+e^{u/|\theta|})^{-1}]/2u \geq 0.$$

Moreover, as follows from 3.412 in [27],

$$\exp(-2\int H_2'(u)\,du) = \log|\theta|.$$

Finally, since $H_2'(u)$ is an even function,

$$\log\varphi_2(s) = \int (e^{su}-1)H_2'(u)\,du,$$

where $H_2(u)$ satisfies (3.5.1).

Analogous arguments involving the function $\log\varphi_1(s)$ enable us to write it in the form

$$\log\varphi_1(s) = \mathrm{C}(1-1/\alpha)s + \int (e^{su}-1-su)H_1'(u)\,du,$$

where

$$H_1'(u) = \tfrac{1}{2}[\nu(u,1,1) - \nu(u,1,|\theta|) + (1+\operatorname{sgn} u)(\nu(u,1,1/\alpha) - \nu(u,1,1))].$$

It is obvious that $H_1'(u) \geq 0$ on the semi-axis $u < 0$. On the semi-axis $u > 0$

$$H_1'(u) = \tfrac{1}{2}[2\nu(u,1,1/\alpha) - \nu(u,1,1) - \nu(u,1,|\theta|)].$$

It is completely obvious that $H_1'(u) \geq 0$ if $\alpha \leq 1$. Since

$$H_1'(u) \sim -\tfrac{1}{2}[\nu(u,1,1) + \nu(u,1,|\theta|)]$$

as $u \to \infty$ in the case when $\alpha > 1$, the inequality $H_1'(u) \geq 0$ is false for large u on the semi-axis $u > 0$, i.e., H_1 is not a measure. It was mentioned above that $a_1 = \theta$, and thus the condition that the distribution of $Y(\alpha,\theta)$ belong to $\mathfrak{M}'$ coincides with the condition that $\theta \geq 0$.

REMARK 1. We mention two particular cases corresponding to the variables $Y(2,0)$ (the normal distribution with parameters $(0,2)$) and $Y(1,0)$ (the Cauchy distribution). For them $w_1 = 0$ and

$$w_0(s,2,0) = \exp\left(-\frac{1}{2}\mathrm{C}s + \int_0^\infty (e^{-su}-1+su)\frac{du}{2u\sinh u}\right),$$
$$w_0(s,1,0) = \exp\left(\int (e^{su}-1+su)\frac{du}{2u\sinh u}\right).$$

REMARK 2. The random variables $|Y(\alpha,\theta)|$ and

$$Y^r(\alpha,\theta) = |Y(\alpha,\theta)|^r \operatorname{sgn} Y(\alpha,\theta)$$

are M-infinitely divisible together with $Y(\alpha,\theta)$.

§3.6. The logarithmic moments of $Y(\alpha,\theta)$ and $Z(\alpha,\rho)$

Although the random variables $|Y(\alpha,\theta)|$ and $Z(\alpha,\rho)$ have finite moments only of order less than α, their logarithms have moments of all orders, and this sometimes (for example, in certain statistics problems) turns out to be a very valuable property. It happens that moments of logarithmic type have a relatively simple form of expression. We occupy ourselves below with the computation of the following two quantities of similar type ($k = 0, 1$, and $n = 1, 2, \ldots$):

$$y_{kn}(\alpha,\theta) = \mathsf{E}(\log|Y(\alpha,\theta)|)^n (\operatorname{sgn} Y(\alpha,\theta))^k,$$
$$z_n(\alpha,\rho) = \mathsf{E}(\log Z(\alpha,\rho))^n.$$

It turns out that the logarithmic moments y_{kn} and z_n are polynomials of degree n in the variable $1/\alpha$ and polynomials of degree at most $n+1$ in the variable θ (or ρ). Explicit expressions for them can be written with the help of the Bell polynomials $C_n(u_1,\ldots,u_n)$ which were encountered in §2.5.

We introduce the quantities $q_1, q_2, \ldots$ by setting $q_1 = (1/\alpha - 1)\mathbb{C}$, where $\mathbb{C}$ is Euler's constant, and

$$q_k = A_k \pi^k |B_k|/k + (1/\alpha^k - 1)\Gamma(k)\varsigma(k) \quad \text{for } k \geq 2,$$

where the B_k are the Bernoulli numbers, $\varsigma(k)$ is the value at k of the Riemann ς-function, and the numbers A_k are defined for the different logarithmic moments as follows:

$$\begin{aligned}
A_k &= (2^k - 1)(1 - \theta^k) && \text{for the moments } y_{0n},\\
A_k &= 1 - \theta^k && \text{for the moments } y_{1n},\\
A_k &= 2^k(1 - \rho^k) && \text{for the moments } z_n.
\end{aligned}$$

THEOREM 3.6.1. *For any admissible values of the parameters (α,θ) or (α,ρ) the following equalities are valid $(n = 1, 2, \ldots)$:*

$$\begin{aligned}
y_{0n} &= C_n(q_1, q_2, \ldots, q_n),\\
y_{1n} &= \theta C_n(q_1, q_2, \ldots, q_n),\\
z_n &= C_n(q_1, q_2, \ldots, q_n).
\end{aligned} \tag{3.6.1}$$

PROOF. Since

$$M(s,\alpha,\rho) = \mathsf{E}Z^s(\alpha,\rho)$$

and

$$w_k(s,\alpha,\theta) = \mathsf{E}|Y(\alpha,\theta)|^s(\operatorname{sgn} Y(\alpha,\theta))^k,$$

it follows that

$$y_{kn} = (d/ds)^n w_k(s,\alpha,\theta)|_{s=0},$$
$$z_n = (d/ds)^n M(s,\alpha,\rho)|_{s=0}$$

are the coefficients in the series expansion of the functions w_k and M in powers of s. However, it is more convenient to expand these functions not directly but after first expanding their logarithms in series. We remark that in the case $\theta = 0$ we have that $w_1(s,\alpha,\theta) = 0$, so that the computation of y_{1n} should be carried out under the additional condition that $\theta \neq 0$. We have that

$$\begin{aligned}\log w_0(s,\alpha,\theta) = \; & \log\cos\tfrac{\pi}{2}\theta s - \log\cos\tfrac{\pi}{2}s \\ & + \log\Gamma(1-s/\alpha) - \log\Gamma(1-s).\end{aligned}$$

The function $\log\cos x$ can be expanded in a power series just like $\log\Gamma(1-x)$ (about this see 1.518 and 8.342 in [27]):

$$\log\cos x = -\sum_{k=0}^{\infty} \frac{2^k(2^k-1)|B_k|}{k\Gamma(k+1)} x^k,$$

$$\log\Gamma(1-x) = \mathrm{C}x + \sum_{k=2}^{\infty} \frac{\varsigma(k)}{k} x^k.$$

With the help of these formulas we get as a result that

$$\log w_0(s,\alpha,\theta) = \sum_{k=1}^{\infty} q_k \frac{s^k}{k!},$$

where the numbers q_k were defined above. If we now use Faa di Bruno's formula, we find that

$$w_0(s,\alpha,\theta) = \exp\left(\sum_{k=1}^{\infty} q_k \frac{s^k}{k!}\right) = 1 + \sum_{n=1}^{\infty} C_n(q_1,q_2,\ldots,q_n)\frac{s^n}{n!},$$

which implies the first equality in (3.6.1).

Considering $w_1(s,\alpha,\theta)$ with $\theta \neq 0$, it is convenient to write it in the form

$$w_1(s,\alpha,\theta) = \theta \frac{\sin\frac{\pi}{2}\theta_s}{\theta\sin\frac{\pi}{2}s}\,\frac{\Gamma(1-s/\alpha)}{\Gamma(1-s)} = \theta h(s)$$

and expand the function $h(s)$ in a series of powers of s. Of course, all the series expansion coefficients for w_1 are proportional to θ. The expansion of $h(s)$ is carried out just like that of w_0, with, however, the difference that the formula (1.518 in [27])

$$\log\left(\frac{x}{\sin x}\right) = \sum_{k=2}^{\infty} \frac{2^k|B_k|}{k\Gamma(k+1)} x^k$$

is used here. The same device can be used to expand $M(s,\alpha,\rho)$ in a power series in s.

Since the quantities q_n, $n \geq 2$, are polynomials of degree n with respect to the variables $1/\alpha$ and θ (or $1/\alpha$ and ρ), by returning to the explicit expression for the Bell polynomials (it was given in §2.5) it is not hard to establish that $C_n(q_1,\ldots,q_n)$ has the same property. In the case of the logarithmic moment y_{1n} the maximal power of θ is equal to $n+1$, because $y_{1n} = \theta C_n$.

We mention a computational detail which can simplify the expression for the quantities q_k in a number of cases. The values $\varsigma(k)$ do not have explicit expressions for odd k, though they rapidly approach 1 as k increases. For even k the values $\varsigma(k)$ can be expressed in terms of the Bernoulli numbers, which are rational fractions:

$$\varsigma(k) = 2^{k-1}\pi^k |B_k| / \Gamma(k+1).$$

In conclusion we give expressions for the logarithmic moments when $n=1$ and $n=2$, considering that $C_1(q_1) = q_1$ and $C_2(q_1,q_2) = q_1^2 + q_2$:

$$y_{01} = z_1 = (1/\alpha - 1)\mathrm{C}, \qquad y_{11} = \theta y_{01}, \tag{3.6.2}$$

$$y_{02} = (1/\alpha - 1)^2\mathrm{C}^2 + \pi^2(2/\alpha^2 - 3\theta^2 + 1)/12, \tag{3.6.3}$$

$$y_{12} = \theta[(1/\alpha - 1)^2\mathrm{C}^2 + \pi^2(2/\alpha^2 + \theta^2 - 1)/12], \tag{3.6.4}$$

$$z_2 = (1/\alpha - 1)^2\mathrm{C}^2 + \pi^2(1/\alpha^2 - 2\rho^2 + 1)/6. \tag{3.6.5}$$

CHAPTER 4

Estimators of the Parameters of Stable Distributions

§4.0. Introduction

It can be said without exaggeration that the problem of constructing statistical estimators of stable laws entered into mathematical statistics due to the work of Mandelbrot [118]–[128]. The economics models considered in those papers contained stable distributions whose parameters had to be determined empirically. Furthermore, it was discovered at once that mathematical statistics, while having at its disposal a large arsenal of methods, can be of little help in this case, since these methods are based mainly on such assumptions as the availability of an explicit form for the density, the existence of some particular number of moments, and so on, assumptions which are automatically not satisfied for distributions in $\mathfrak{S}$. In the best case they have only one moment of integral order (if $\alpha \neq 2$), and only in a few cases are there explicit expressions for the density that would enable us to concretize the algorithms for estimating the parameters (say, by the maximum likelihood method).

However, the problem had emerged, and the search began for solutions of it. This search was conducted in various directions and led to recommendations that satisfied the practical workers to various degrees. The investigations are still continuing, and thus far it is too early to speak of a complete solution of the problem if one has in mind the creation of algorithms meeting the requirements of the contemporary theory, i.e., estimators that are unbiased, efficient, etc., at least in the asymptotic sense. In this limited space we shall not even acquaint the reader with the achievements presently at hand, although more than a few of them are very interesting. The purpose of this chapter is not to throw light on the state of the problem as a whole. Even a brief survey would require space that could simply not be justified by the direction of the present monograph.

Below we propose a new approach to the problem of estimating the parameters of stable laws, based on the use of explicit expressions for the corresponding characteristic transforms and the method of sample moments well known in statistics. The latter, as we shall see, gives rise to a certain closeness between the proposed method and the classical method of estimating the parameters of a normal law. (4.1)

To the reader interested in the contemporary state of the general problem we can recommend the fairly comprehensive survey [31], which encompasses the period up to 1973. A list of later investigations is in part III of the Bibliography.

A feature of the proposed approach is that the problem of estimating parameters is solved for distributions in the class $\mathfrak{W}$, and then the general problem is reduced to the problem already solved by some special devices. The necessity of this division of the general problem is explained by the fact that explicit formulas for the characteristic transforms of stable laws exist only for distributions in the class $\mathfrak{W}$, and such formulas form the analytic basis of the method.

In working with the stable laws of this class the most convenient form of expression for their characteristic functions turns out to be the form (E) in the Introduction.

For convenience we reproduce here the expression for the logarithm of the characteristic functions $\mathfrak{g}(t,\nu,\theta,\tau)$ in this form:

$$\log \mathfrak{g}(t,\nu,\theta,\tau) = -\exp\{\nu^{-1/2}(\log|t| + \tau - i\tfrac{\pi}{2}\theta\,\mathrm{sgn}\,t) + \mathbb{C}(\nu^{-1/2}-1)\}, \quad (4.0.1)$$

where $\mathbb{C}$ is the Euler constant, and the parameters ν, θ, and τ vary within the bounds

$$\nu \geq \tfrac{1}{4}, \quad |\theta| \leq \min(1, 2\sqrt{\nu}-1), \quad |\tau| < \infty$$

and are connected with the parameters α, θ, and λ (of the form (C)) by the relations

$$\nu = \alpha^{-2}, \quad \theta_E = \theta_C, \quad \tau = \tfrac{1}{\alpha}\log\lambda + \mathbb{C}(\tfrac{1}{\alpha}-1).$$

Within the class $\mathfrak{W}$ the parameters α, β, γ, and λ of the form (B) (we shall also have to deal with this form in what follows) are connected with the parameters ν, θ, and τ by the relations

$$\nu = \alpha^{-2}, \qquad \theta = \begin{cases} \beta K(\alpha)/\alpha & \text{if } \alpha \neq 1, \\ \frac{2}{\pi}\arctan(\frac{2}{\pi}\gamma), & \text{if } \alpha = 1, \end{cases}$$

$$\tau = \begin{cases} \frac{1}{\alpha}\log\lambda + \mathbb{C}(\frac{1}{\alpha}-1) & \text{if } \alpha \neq 1, \\ \log\lambda + \frac{1}{2}\log(\gamma^2+\pi^2/4) & \text{if } \alpha = 1, \end{cases}$$

with inverse relations

$$\alpha = 1/\sqrt{\nu},$$
$$\beta = \theta \max(1, 1/(2\sqrt{\nu} - 1)) \operatorname{sgn}(1 - 1/\nu),$$
$$\gamma = \begin{cases} 0 & \text{if } \nu \neq 1, \\ \frac{\pi}{2}\tan(\pi\theta/2) & \text{if } \nu = 1, \end{cases}$$
$$\lambda = \begin{cases} \exp(\tau/\sqrt{\nu} - \mathbb{C}(1 - 1/\sqrt{\nu})) & \text{if } \nu \neq 1, \\ \exp(\tau + (\log\cos(\pi\theta/2) - \log(\pi/2)) & \text{if } \nu = 1. \end{cases}$$

A number of auxiliary facts will be needed in constructing estimators for the parameters of stable laws. These facts are in the next section, where a somewhat greater amount of material is collected than is used later.

§4.1. Auxiliary facts

To the form (E) there corresponds a distinctive form of expression for the characteristic transform $W(s, \nu, \theta, \tau)$, which is not hard to get from (2.6.26) by using the formulas for passing from the form (B) to (E). It is

$$w_k(s, \nu, \theta, \tau) = \exp\{\tau s + \mathbb{C}s(1 - \sqrt{\nu})\} \times \frac{\cos\frac{\pi}{2}(k - s\theta)}{\cos\frac{\pi}{2}(k - s)} \frac{\Gamma(1 - s\sqrt{\nu})}{\Gamma(1 - s)}, \tag{4.1.1}$$

where $k = 0$ or 1 and s varies in the strip $-1 < \operatorname{Re} s < 1/\sqrt{\nu}$.

LEMMA 4.1.1. *In the disk defined by $|s| < \min(1, 1/\sqrt{\nu})$ the functions $\log w_k(s, \nu, \theta, \tau)$ are analytic and are represented there by the following power series $(k = 0, 1)$:*

$$\log w_k(s, \nu, \theta, \tau) = k \log\theta + \tau s + \sum_{n=2}^{\infty} [a_{kn}(1 - \theta^n) + b_{kn}(\nu^{n/2} - 1)] \frac{s^n}{n!}, \tag{4.1.2}$$

where it is assumed that $0 \cdot \log\theta = 0$ for any values of θ and that $w_1(s, \nu, 0, \tau) = 0$ for any s, ν, and τ. The coefficients a_{kn} and b_{kn} have for $k = 0$ or 1 and $n \geq 2$ the form

$$a_{kn} = (\pi^2/u)(2^{n(1-k)} - (1 - k))|B_n|, \qquad b_{kn} = \Gamma(n)\varsigma(n).$$

Here $\varsigma(n)$ is the value of the Riemann ς-function at the point n.

REMARK. For odd $n \geq 3$ we have that $a_{kn} = 0$, since the numbers B_n are 0 for such n. If n is even, then $b_{kn} = (2\pi)^n|B_n|/(2n)$.

The lemma is proved on the basis of the power series expansions of $\log\cos x$, $\log(x/\sin x)$, and $\log\Gamma(1 - x)$ which we have already encountered in proving Theorem 3.6.1.

For simplicity we agree to omit the subscript E on the random variable $Y_E(\nu,\theta,\tau)$ and the characteristics associated with it for the duration of this chapter. There is no danger of confusing this notation with that used earlier for the same random variable $Y(\alpha,\theta,\lambda)$ since the letters ν and τ are reserved for the form (E).

We introduce the abbreviation notation

$$U = \operatorname{sgn} Y(\nu,\theta,\tau), \qquad V = \log|Y(\nu,\theta,\tau)|,$$
$$\overline{U} = U - \mathsf{E}U, \qquad \overline{V} = V - \mathsf{E}V.$$

With this notation the functions $w_k(s,\nu,\theta,\tau)$ are written as

$$w_k(s,\nu,\theta,\tau) = \mathsf{E}U^k \exp(sV), \qquad k = 0,1. \tag{4.1.3}$$

By following the arguments in Theorem 3.6.1 it is not hard to get from (4.1.2) explicit expressions for the logarithmic moments of the random variables $Y(\nu,\theta,\tau)$, i.e., the mixed moments of the variables U and V. Indeed, since (as is clear from (4.0.1))

$$\begin{aligned} Y(\nu,\theta,\tau) &= e^{\tau} Y(\nu,\theta,0) \\ &= \exp(\tau + \mathbb{C}(1-\sqrt{\nu}))Y_B(\alpha,\theta) \end{aligned} \tag{4.1.4}$$

for any admissible parameter values, the moments $\omega_{kn} = \mathsf{E}U^kV^n$ turn out to be connected with the logarithmic moments (in which $1/\alpha$ should be replaced by $\sqrt{\nu}$) by the following equalities ($k = 0$ or 1 and $n = 0, 1, \ldots$):

$$\begin{aligned} \omega_{kn} &= \mathsf{E}(\tau + \mathbb{C}(1-\sqrt{\nu}) + \log|Y_B(\alpha,\theta)|)^n (\operatorname{sgn} Y_B(\alpha,\theta))^k \\ &= \sum_{j=0}^{n} \binom{n}{j} y_{kj} (\tau + \mathbb{C}(1-\sqrt{\nu}))^{n-j}. \end{aligned} \tag{4.1.5}$$

In particular, the moments ω_{kn} up through order $n = 2$ have the following form:

$$\mathsf{E}U^0 = \mathsf{E}V^0 = \mathsf{E}U^2 = 1, \quad \mathsf{E}U = \theta, \quad \mathsf{E}V = \tau, \tag{4.1.6}$$

$$\mathsf{E}UV = \theta\tau = \mathsf{E}U\mathsf{E}V, \tag{4.1.7}$$

$$\mathsf{E}V^2 = \tau^2 + \pi^2(2\nu - 3\theta^2 + 1)/12, \tag{4.1.8}$$

$$\mathsf{E}UV^2 = \theta(\tau^2 + \pi^2(2\nu - \theta^2 - 1)/12). \tag{4.1.9}$$

REMARK. We point out an interesting fact. It was mentioned in §3.2 that the modulus $|Y_B(\alpha,\theta)|$ and the sign $\operatorname{sgn} Y_B(\alpha,\theta)$ of the random variable $Y_B(\alpha,\theta)$ are independent if and only if either $\theta = 0$, or $0 < \alpha < 1$ and $|\theta| = 1$. Because of (4.1.4), this implies that $|Y(\nu,\theta,\tau)|$ and $\operatorname{sgn} Y(\nu,\theta,\tau)$ are independent random variables if (and only if) either $\theta = 0$, or $\nu > 1$ and $|\theta| = 1$. It is clear that the same conclusion is preserved also in regard to the random variables V and U. As a supplement to this fact the equality (4.1.7) allows us to assert that U and V are always uncorrelated random variables.

The mixed central moments $\mu_{kn} = \mathsf{E}\overline{U}^k\overline{V}^n$ $(k, n = 0, 1, \ldots)$ can be expressed in terms of noncentral moments, and can thus also be written explicitly. For any integer $m \geq 0$

$$\mathsf{E}U^{2m}\overline{V}^n = \mathsf{E}\bar{V}^n, \qquad \mathsf{E}U^{2m+1}\overline{V}^n = \mathsf{E}U\overline{V}^n,$$

and so, since $\mathsf{E}U = \theta$,

$$\begin{aligned}\mu_{kn} &= \mathsf{E}(U-\theta)^k\overline{V}^n = \sum_{j=0}^{k}\binom{k}{j}(-\theta)^{k-j}\mathsf{E}U^j\overline{V}^n \\ &= (-1)^k\mathsf{E}\overline{V}^n\sum_0\binom{k}{j}\theta^{k-j} - (-1)^k\mathsf{E}U\overline{V}^n\sum_1\binom{k}{j}\theta^{k-j},\end{aligned}$$

where $\sum_0$ and $\sum_1$ denote summation over the respective even and odd values of j not exceeding k. Let $L_{0n} = \mu_{0n}$ and $L_{1n} = \mu_{1n}/\theta$, with the latter understood as the limit of L_{1n} as $\theta \to 0$ in the case when $\theta = 0$. It is not hard to compute L_{kn} with the help of (4.1.2) and (4.1.3). According to (4.1.6), $\mathsf{E}V = \tau$; hence

$$\begin{aligned}\sum_{n=0}^{\infty} L_{kn}\frac{s^n}{n!} &= \exp(-s\tau - k\log\theta)\mathsf{E}U^k\exp(sV) \\ &= \exp(-s\tau - k\log\theta)w_k(s,\nu,\theta,\tau) \\ &= \exp\left(\sum_{n=2}^{\infty} p_{kn}s^n/n!\right),\end{aligned}$$

where $p_{kn} = a_{kn}(1-\theta^n) + b_{kn}(\nu^{n/2} - 1)$. From this we get

$$L_{kn} = C_n(0, p_{k2}, p_{k3}, \ldots, p_{kn}), \tag{4.1.10}$$

where the C_n are the Bell polynomials which appeared in §§2.5 and 3.6. Thus, we have arrived at the following assertion.

LEMMA 4.1.2. *For any admissible values of the parameters ν, θ, and τ and any $r, n = 0, 1, \ldots$*

$$(-1)^r\mathsf{E}\overline{U}^r\overline{V}^n = L_{0n}\sum_0\binom{r}{j}\theta^{r-j} - L_{1n}\sum_1\binom{r}{j}\theta^{r-j+1}.$$

Let us see what some of the first moments μ_{kn} look like. Explicit expressions for them will be needed in what follows. In (4.1.10) we have $p_{k1} = 0$, so the first five Bell polynomials (see [70]) have the form

$$C_0 = 1, \quad C_1 = 0, \quad C_2 = p_{k2}, \quad C_3 = p_{k3}, \quad C_4 = p_{k4} + 3p_{k2}^2.$$

This implies that

$$\begin{aligned}
L_{00} &= L_{10} = 1, \qquad L_{01} = L_{11} = 0, \\
L_{02} &= \frac{\pi^2}{4}(1-\theta^2) + \frac{\pi^2}{6}(\nu - 1), \\
L_{12} &= \frac{\pi^2}{12}(1-\theta^2) + \frac{\pi^2}{6}(\nu - 1), \\
L_{03} &= L_{12} = 2\varsigma(3)(\nu^{3/2} - 1), \\
L_{04} &= \frac{7\pi^4}{120}(1-\theta^4) + \frac{\pi^4}{15}(\nu^2 - 1) + 3\pi^4\left(\frac{1}{4}(1-\theta^2) + \frac{1}{6}(\nu-1)\right)^2, \\
L_{14} &= \frac{\pi^4}{120}(1-\theta^4) + \frac{\pi^4}{15}(\nu^2 - 1) + \frac{\pi^4}{12}\left(\frac{1}{2}(1-\theta^2) + (\nu-1)\right)^2,
\end{aligned}$$

which, in turn, leads to the following expressions for the mixed moments:

$$\mathsf{E}\overline{U}^r = (-1)^r\left(\sum_0 \binom{r}{j}\theta^{r-j} - \sum_1 \binom{r}{j}\theta^{r-j+1}\right), \tag{4.1.11}$$

$$\mathsf{E}\overline{U}^r\overline{V} = 0, \tag{4.1.12}$$

$$\mathsf{E}\overline{U}^r\overline{V}^2 = \frac{\pi^2}{6}(1-\theta^2)\sum_0 \binom{r}{j}\theta^{r-j} + \frac{\pi^2}{12}(2\nu - \theta^2 - 1)(-1)^r\mathsf{E}\overline{U}^r, \tag{4.1.13}$$

$$\mathsf{E}\overline{U}^r\overline{V}^3 = 2\varsigma(3)(\nu^{3/2} - 1)(-1)^r\mathsf{E}\overline{U}^r, \tag{4.1.14}$$

$$\mathsf{E}\overline{U}^r\overline{V}^r = \frac{\pi^4}{20}(1-\theta^4)\sum_0 \binom{r}{j}\theta^{r-j} + \frac{\pi^4}{120}(8\nu^2 - \theta^4 - 7)(-1)^r\mathsf{E}\overline{U}^r. \tag{4.1.15}$$

LEMMA 4.1.3. *For any positive integers r and n,*

$$\begin{aligned}
\operatorname{cov}(\overline{U}^r, \overline{V}^n) &= (-1)^{r-1}\operatorname{cov}(\overline{U}, \overline{V}^n)\sum_1 \binom{r}{j}\theta^{r-j} \\
&= (-1)^r(L_{0_n} - L_{1_n})\sum_1 \binom{r}{j}\theta^{r-j+1}.
\end{aligned} \tag{4.1.16}$$

PROOF. Using the fact that $\mathsf{E}\overline{V}^n = L_{0n}$ and the equality (4.1.11), we have that

$$\begin{aligned}
\operatorname{cov}(\overline{U}^r, \overline{V}^n) &= \mathsf{E}\overline{U}^r\overline{V}^n - \mathsf{E}\overline{U}^r\overline{V}^n \\
&= (-1)^r(L_{0n} - L_{1n})\sum_1 \binom{r}{j}\theta^{r-j+1}.
\end{aligned}$$

In this equality we set $r = 1$ and obtain

$$\operatorname{cov}(\overline{U}, \overline{V}^n) = -\theta(L_{0n} - L_{1n}),$$

which implies the second part of (4.1.16).

COROLLARY. *For any positive integers r and n the random variables $\overline{U}^r$ and $\overline{V}^n$ are uncorrelated if and only if one of the following two conditions holds*:

1) $\theta = 0$;
2) $L_{0n} = L_{1n}$.

The case $\theta = 0$ is of little interest, since, as we have seen, the stronger assertion that $\overline{U}^r$ and $\overline{V}^n$ are independent is true for it.

The second condition involves nontrivial situations. The Bernoulli numbers B_n are 0 for all odd $n \geq 3$; consequently, $p_{0n} = p_{1n} = \Gamma(n)\varsigma(n)(\nu^{n/2} - 1)$. From this it follows that $L_{0n} = L_{1n}$, i.e., the random variables $\overline{U}^r$ and $\overline{V}^n$ are uncorrelated for such n. In the case $n = 1$ this conclusion follows from the fact that $L_{01} = L_{11} = 0$.

We consider a sample of size $n \geq 2$, i.e., a collection $X_1, \dots, X_n$ of independent and identically distributed random variables. Denote by

$$A_X = \frac{1}{n}\sum_{j=1}^{n} X_j \quad \text{and} \quad S_X^2 = \frac{1}{n-1}\sum_{j=1}^{n}(X_j - A_X)^2$$

the sample mean and sample variance.

LEMMA 4.1.4. *Assume that the sample random variables $X_1, \dots, X_n$ have finite fourth moment, and let $a = \mathsf{E}X_1$, $b_2 = \operatorname{Var} X_1$ and $c^4 = \mathsf{E}(X_1 - a)^2$. Then*

$$\mathsf{E}A_X = a, \quad \operatorname{Var} A_X = b^2/n, \quad \mathsf{E}S_X^2 = b^2, \tag{4.1.17}$$

$$\operatorname{Var} S_X^2 = (c^4 - b^4)/n + 2b^4/n(n-1). \tag{4.1.18}$$

We do not give proofs of these equalities, which are known in mathematical statistics. Concerning this, see [43].

Consider a pair (L, M) of uncorrelated random variables having zero means and finite fourth moments. Let $(L_1, M_1), \dots, (L_n, M_n)$ be a collection of mutually independent pairs of random variables, each distributed like the pair (L, M). From the n-tuples $(L_1, \dots, L_n)$ and $(M_1, \dots, M_n)$ we construct the sample variances S_L^2 and S_M^2.

LEMMA 4.1.5. *For any $n \geq 2$*

$$\operatorname{cov}(S_L^2, S_M^2) = \tfrac{1}{n}\operatorname{cov}(L^2, M^2). \tag{4.1.19}$$

PROOF. Using the last equality in (4.1.17), we have that

$$\begin{aligned} \operatorname{cov}(S_L^2, S_M^2) &= \mathsf{E}S_L^2 S_M^2 - \mathsf{E}S_L^2 \mathsf{E}S_M^2 \\ &= \mathsf{E}S_L^2 S_M^2 - \operatorname{Var} L \operatorname{Var} M. \end{aligned} \tag{4.1.20}$$

It is easy to see that the sample variance can be written in the form

$$S_X^2 = \frac{n}{n-1}(A_{X^2} - (A_X)^2),$$

where A_{X^2} is the sample mean of the squares of the random variables X_j. Therefore,

$$\begin{aligned} \mathsf{E}S_L^2 S_M^2 &= \left(\frac{n}{n-1}\right)^2 \mathsf{E}[A_{L^2}A_{M^2} - A_{L^2}(A_M)^2 - (A_L)^2 A_{M^2} + (A_L)^2(A_M)^2] \\ &= \left(\frac{n}{n-1}\right)^2 \mathsf{E}(I_{11} - I_{12} - I_{21} + I_{22}). \end{aligned} \tag{4.1.21}$$

Letting the indices i, j, r, and t vary independently from 1 to n, we can write

$$\begin{aligned} \mathsf{E}I_{12} &= n^{-3}\sum \mathsf{E}(L_i^2 M_j M_r) = n^{-3}\sum \mathsf{E}(L_i^2 M_j^2) = n^{-1}\mathsf{E}I_{11}, \\ \mathsf{E}I_{21} &= n^{-3}\sum \mathsf{E}(L_i L_j M_r^2) = n^{-3}\sum \mathsf{E}(L_i^2 M_j^2) = n^{-1}\mathsf{E}I_{11}, \\ \mathsf{E}I_{22} &= n^{-4}\sum \mathsf{E}(L_i L_j M_r M_t) = n^{-4}\sum \mathsf{E}(L_i^2 M_j^2) = n^{-2}\mathsf{E}I_{11}. \end{aligned}$$

Consequently,

$$\begin{aligned} \mathsf{E}S_L^2 S_M^2 &= \mathsf{E}I_{11} = n^{-2}\sum \mathsf{E}(L_i^2 M_j^2) \\ &= n^{-2}[n(n-1)\mathsf{E}L^2\mathsf{E}M^2 + n\mathsf{E}L^2M^2] \\ &= \operatorname{Var} L \operatorname{Var} M + n^{-1}(\mathsf{E}L^2M^2 - \mathsf{E}L^2\mathsf{E}M^2). \end{aligned}$$

Substitution of this expression into (4.1.21) gives us (4.1.19).

§4.2. Estimators of parameters of distributions in the class $\mathfrak{W}$

Let $V_1, \dots, V_n$ be independent random variables distributed like the random variable $Y = Y(\nu, \theta, \tau)$, i.e., with the distribution $G(x, \nu, \theta, \tau)$, about which we know only that it belongs to the class $\mathfrak{W}$. Our problem is the statistical estimation of the parameters of G.

From this sample we construct two collections of independent (within each collection) random variables $U_1 = \operatorname{sgn} Y_1, \dots, U_n = \operatorname{sgn} Y_n$ and $V_1 = \log|Y_1|, \dots, V_n = \log|Y_n|$, which are distributed like $U = \operatorname{sgn} Y(\nu, \theta, \tau)$ and $V = \log|Y(\nu, \theta, \tau)|$, respectively.

The idea for constructing estimators for the parameters ν, θ, and τ is based on the following three equalities:

$$\nu = \frac{6}{\pi^2}\operatorname{Var} V - \frac{3}{2}\operatorname{Var} U + 1, \quad \theta = \mathsf{E}U, \quad \tau = \mathsf{E}V. \tag{4.2.1}$$

The last two were noted in (4.1.6), and the validity of the first is not hard to verify by computing the variances of U and V. By (4.1.6)–(4.1.8),

$$\operatorname{Var} U = \mathsf{E}U^2 - (\mathsf{E}U)^2 = 1 - \theta^2,$$

$$\operatorname{Var} V = \mathsf{E}V^2 - (\mathsf{E}V)^2 = \frac{\pi^2}{12}(2\nu - 3\theta^2 + 1).$$

The needed relation is obtained by eliminating θ^2 from these equalities.

The idea itself is simple and not new in mathematical statistics. As a clarifying example we recall the classical problem of estimating the parameters of the normal distribution with density

$$p(x) = \frac{1}{\sigma\sqrt{2\pi}} \exp\left(-\frac{1}{2}\left(\frac{x-a}{\sigma}\right)^2\right). \tag{4.2.2}$$

Let $X_1, \ldots, X_n$ be independent random variables with the distribution of (4.2.2). Since $a = \mathsf{E}X_1$ and $\sigma^2 = \operatorname{Var} X_1$, the fact that

$$A_X \xrightarrow{P} a, \quad S_X^2 \xrightarrow{P} \sigma^2 \quad \text{(convergence in probability)}$$

as the sample size goes to infinity permits us to choose $\tilde{a} = A_X$ and $\tilde{\sigma}^2 = S_X^2$ as estimators $\tilde{a}$ and $\tilde{\sigma}^2$ of the parameters a and σ^2.

The method of sample moments is not very favored in mathematical statistics. It is regarded, not without reason, as an estimation method that is far from economical. However, in a number of cases when the distribution has sufficiently nice analytic properties (for example, the existence of moments of any order, etc.) the method of moments is capable of giving parameter estimators meeting contemporary requirements. And this category includes the case under consideration of the distributions of the random variables U and V, which have finite moments of any order.

On the basis of the collections $U_1, \ldots, U_n$ and $V_1, \ldots, V_n$ generated by the independent sample $Y_1, \ldots, Y_n$ we form the sample means A_U and A_V and take them as estimators of the parameters θ and τ, i.e.

$$\tilde{\theta} = A_U, \qquad \tilde{\tau} = A_V. \tag{4.2.3}$$

LEMMA 4.2.1. *The statistics $\tilde{\theta}$ and $\tilde{\tau}$ are unbiased consistent estimators of the parameters θ and τ with variances*

$$\sigma_\theta^2 = \operatorname{Var} \tilde{\theta} = (1 - \theta^2)n^{-1}, \tag{4.2.4}$$

$$\sigma_\tau^2 = \operatorname{Var} \tilde{\tau} = \frac{\pi^2}{12}(2\nu - 3\theta^2 + 1)n^{-1}. \tag{4.2.5}$$

PROOF. The fact that the estimators (4.2.3) are unbiased is a consequence of (4.2.1). The form of the variances is obtained from (4.1.17) and (4.1.6),

(4.1.8). The consistency of the estimators follows from the fact that the variances (4.2.4) and (4.2.5) of the estimators go to zero as $n \to \infty$.

Since the random variable U takes only the two values $+1$ and -1 with respective probabilities equal to $p = (1+\theta)/2$ and $1-p = (1-\theta)/2$, estimation of θ is equivalent to estimation of p, which is a well-known problem in statistics. It is known that $\tilde{p} = (1+\tilde{\theta})/2$ is an efficient estimator for the parameter p (see, for example, [84]).

Contrary to the case of the estimators (4.2.3), construction of an estimator for the parameter ν has its complexities. It might seem that the statistic

$$\hat{\nu} = \frac{6}{\pi^2} S_V^2 - \frac{3}{2} S_U^2 + 1 \tag{4.2.6}$$

could serve as such an estimator (consistent and unbiased), because, on the one hand,

$$S_U^2 = \frac{n}{n-1}(A_{U^2} - (A_V)^2) \xrightarrow{P} \mathsf{E}U^2 - (\mathsf{E}U)^2 = \operatorname{Var} U$$

as $n \to \infty$, and, similarly, $S_{V^2}^2 \xrightarrow{P} \operatorname{Var} V$; which implies that $\hat{\nu} \xrightarrow{P} \nu$. On the other hand, $\mathsf{E}\hat{\nu} = \nu$ by (4.1.18).

However, taking $\hat{\nu}$ as an estimator of ν is hindered by the fact that the domain $Q = \{(\nu, \theta) \colon |\theta| \le \min(1, 2\sqrt{\nu} - 1)\}$ of variation of the parameters ν and θ does not coincide with the domain $\hat{Q}_n$ of variation of the values of the pair $\hat{\nu}$, $\tilde{\theta}$, which has the form (see Figure 5) $\hat{Q}_n = \{(\hat{\nu}, \hat{\theta}) \colon |\tilde{\theta}| \le 1,\ \hat{\nu} \ge -(1+\eta_n)/2\}$, where $\eta_n = 3/(n-1)$ for even n and $\eta_n = 3/n$ for odd n. But this means that pairs of values of $\hat{\nu}, \tilde{\theta}$ can appear that do not correspond to any stable distributions. Consequently, we must alter the estimators $\hat{\nu}$ and $\tilde{\theta}$ in such a way that their new domain of variation coincides with Q. This can be done in different ways, for example, by drawing the normal from the point $(\hat{\nu}, \tilde{\theta})$ to the boundary of Q (if, of course, the point is outside Q) and taking the coordinates of the points of the normal on the boundary of Q as new estimators.

But we choose a simpler method when $\tilde{\theta}$ does not vary in general, but only $\hat{\nu}$ varies. Namely, let

$$\tilde{\nu} = \max(\hat{\nu}, (1 + |\tilde{\theta}|)^2/4). \tag{4.2.7}$$

With this definition of $\tilde{\nu}$ the domain of variation of the values of the pair $(\tilde{\nu}, \tilde{\theta})$ coincides with Q.

LEMMA 4.2.2. *For any $n \ge 2$*

$$\begin{aligned}\sigma_\nu^2 = \operatorname{Var}\hat{\nu} &= [\tfrac{22}{5}(\nu-1)^2 + \tfrac{6}{5}(9-5\theta^2)(\nu-1) + 3(1-\theta^2)(3+\theta^2)]\tfrac{1}{n} \\ &\quad + [2(\nu-1)^2 + 6(1-\theta^2)(\nu-1) + 9(1-\theta^2)^2]/n(n-1).\end{aligned} \tag{4.2.8}$$

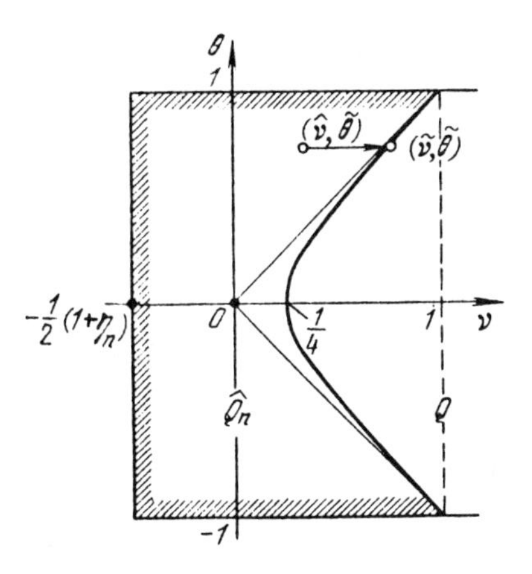

FIGURE 5

PROOF. According to (4.2.6),

$$\begin{aligned}\sigma_\nu^2 &= \operatorname{Var}\left(\frac{6}{\pi^2}S_V^2 - \frac{3}{2}S_U^2\right)\\ &= \frac{36}{\pi^4}\operatorname{Var} S_V^2 + \frac{9}{4}\operatorname{Var} S_U^2 - \frac{18}{\pi^2}\operatorname{cov}(S_U^2, S_V^2).\end{aligned}$$

Let us transform the right-hand side of this equality, using a property of the sample variance (4.1.18) and Lemma 4.1.5:

$$\begin{aligned}\sigma_\nu^2 = &\left\{\frac{36}{\pi^4}[\mathsf{E}\overline{V}^4 - (\operatorname{Var} V)^2] + \frac{9}{4}[\mathsf{E}\overline{U}^4 - (\operatorname{Var} U)^2] - \frac{18}{\pi^2}\operatorname{cov}(\overline{U}^2, \overline{V}^2)\right\}\frac{1}{n}\\ &+ \left[\frac{72}{\pi^4}(\operatorname{Var} V)^2 + \frac{9}{2}(\operatorname{Var} U)^2\right]\frac{1}{n(n-1)}.\end{aligned}$$

We have available explicit expressions for the mixed central moments (Lemma 4.1.2). With the help of (4.1.11), (4.1.13), and (4.1.15), along with the expression for $\operatorname{Var} U$ and $\operatorname{Var} V$ we get after simple transformations that

$$\mathsf{E}\overline{U}^4 = 1 + 2\theta^2 - 3\theta^4 = 4\operatorname{Var} U(1 - \operatorname{Var} U) + (\operatorname{Var} U)^2,$$

$$\mathsf{E}\overline{U}^2\overline{V}^2 = \operatorname{Var} U \operatorname{Var} V + \frac{\pi^2}{3}\operatorname{Var} DU(1 - \operatorname{Var} DU),$$

$$\begin{aligned}\mathsf{E}\overline{V}^4 = &(\operatorname{Var} V)^2 + 2[(\operatorname{Var} V)^2 - \frac{\pi^4}{16}(\operatorname{Var} U)^2] + \frac{12}{5}(\operatorname{Var} V - \frac{\pi^2}{4}\operatorname{Var} U)^2\\ &+ \frac{4}{5}\pi^2(\operatorname{Var} V - \frac{\pi^2}{4}\operatorname{Var} U) + \frac{\pi^4}{4}\operatorname{Var} U.\end{aligned}$$

This tells us that the first three terms in the representation of σ_ν^2 can be given the forms

$$\frac{36}{\pi^4}[\mathsf{E}\overline{V}^4 - (\operatorname{Var} V)^2] = \frac{22}{5}W^2 + 6\left(\operatorname{Var} U + \frac{4}{5}\right)W + 9\operatorname{Var} U,$$
$$\frac{9}{4}[\mathsf{E}\overline{U}^4 - (\operatorname{Var} U)^2] = 9\operatorname{Var} U(1 - \operatorname{Var} U),$$
$$\frac{18}{\pi^2}\operatorname{cov}(\overline{U}^2, \overline{V}^2) = 6\operatorname{Var} U(1 - \operatorname{Var} U),$$

where $W = 6\operatorname{Var} V/\pi^2 - 3\operatorname{Var} U/2$. Consequently,

$$\sigma_\nu^2 = \left[\frac{22}{5}W^2 + 6\left(\operatorname{Var} U + \frac{4}{5}\right)W + 3\operatorname{Var} U(4 - \operatorname{Var} U)\right]\frac{1}{n} + [2W^2 + 6W\operatorname{Var} U + 9(\operatorname{Var} U)^2]\frac{1}{n(n-1)}.$$

To obtain (4.2.8) it remains for us to substitute the values of the variables $W = \nu - 1$ and $\operatorname{Var} U = 1 - \theta^2$.

LEMMA 4.2.3.

$$(\mathsf{E}\tilde{\nu} - \nu)^2 \le \mathsf{E}(\tilde{\nu} - \nu)^2 \le \sigma_\nu^2 + \sigma_\theta^2. \tag{4.2.9}$$

PROOF. Since

$$|\theta| \le \min(1, 2\sqrt{\nu} - 1),$$

it follows that $\nu = \max(\nu, (1 + |\theta|)^2/4)$. Further, since

$$\max(a, b) - \max(a', b') \le \max(|a - a'|, |b - b'|)$$

for any real a, a', b, and b' and since $|\tilde{\theta}| \le 1$ and $|\theta| \le 1$, we have

$$\begin{aligned}\tilde{\nu} - \nu &= \max(\hat{\nu}, (1 + |\tilde{\theta}|)^2/4) - \max(\nu, (1 + |\theta|)^2/4) \\ &\le \max(|\hat{\nu} - \nu|, |\tilde{\theta} - \theta|).\end{aligned}$$

The inequality $\nu - \tilde{\nu} \le \max(|\nu - \hat{\nu}|, |\theta - \tilde{\theta}|)$, is obtained quite similar, i.e.

$$|\tilde{\nu} - \nu| \le \max(|\hat{\nu} - \nu|, |\tilde{\theta} - \theta|).$$

Hence,

$$(\tilde{\nu} - \nu)^2 \le \max((\hat{\nu} - \nu)^2, (\tilde{\theta} - \theta)^2) \le (\hat{\nu} - \nu)^2 + (\tilde{\theta} - \theta)^2.$$

Consequently, $\mathsf{E}(\tilde{\nu} - \nu)^2 \le \mathsf{E}(\hat{\nu} - \nu)^2 + \mathsf{E}(\tilde{\theta} - \theta)^2$.

We can now formulate the following assertion, which is a consequence of Lemmas 4.2.2 and 4.2.3:

THEOREM 4.2.1. *The statistic $\tilde{\nu}$ is an asymptotically unbiased and consistent estimator of the parameter ν, and the square of the bias and the square deviation of $\tilde{\nu}$ from the true value of ν does not exceed the sum $\sigma_\nu^2 + \sigma_\theta^2$, with order $O(1/n)$ as $n \to \infty$. The exact values of the terms in this sum are given by* (4.2.4) *and* (4.2.8).

§4.3. Estimators of parameters of distributions in the family $\mathfrak{S}$: the parameters α, β, and λ

Consider a distribution $G(x, \alpha, \beta, \gamma, \lambda) \in \mathfrak{S}$ and assume that we have an independent sample $Y_1, \dots, Y_{6n}$ with this distribution. The idea for constructing estimators for the parameters α, β, and λ amounts to transforming the sample in such a way that the result is another collection $Y_1', \dots, Y_m'$ of random variables with distribution in the class $\mathfrak{W}$ and with defining parameters related in a one-to-one manner to the parameter α, β, and λ. The size m of the new collection will be essentially smaller than the original size; however, this must be regarded as a necessary price for the distributions of the random variables Y_j' to have the properties we need (membership in the class $\mathfrak{W}$). As a transformation of the original sample we choose

$$Y_j^* = Y_{3j-2} - \xi Y_{3j-1} - (1-\xi) Y_{3j}, \qquad j = 1, 2, \dots, 2n,$$

where $0 < \xi \leq \frac{1}{2}$ is a fixed number.

The limit case $\xi = 0$ of this transformation is singled out in the transformation

$$Y_j^0 = Y_{2j-1} - Y_{2j}, \qquad j = 1, 2, \dots, 3n,$$

that stands by itself (the size $6n$ of the original sample was taken so that for partition into thirds and halves we do not need to concern ourselves with remainders).

LEMMA 4.3.1. *The random variables Y^0 are distributed like $Y(\alpha, 0, 0, 2\lambda)$, and the distribution of Y_j^* coincides with the distribution of $Y(\alpha, \beta^*, \gamma^*, \lambda^*)$, where the parameters β^*, γ^*, and λ^* are connected with α, β, γ, and λ (if all these correspond to the form (A)) by the equalities*

$$\beta^* = T_1(\xi)\beta, \quad \lambda* = T_2(\xi)\lambda, \quad \gamma^* = T_3(\xi)\beta, \tag{4.3.1}$$

where

$$T_2(\xi) = 1 + \xi^\alpha + (1-\xi)^\alpha,$$
$$T_1(\xi) = (1 - \xi^\alpha - (1-\xi)^\alpha)/T_2(\xi)$$

and

$$T_3(\xi) = \begin{cases} 0 & \text{if } \alpha \neq 1, \\ \frac{1}{\pi}(\xi \log \xi + (1-\xi)\log(1-\xi)) & \text{if } \alpha = 1. \end{cases}$$

Both the random variables Y_j^0 and the random variables Y_j^* have distributions in $\mathfrak{W}$.

PROOF. The statement concerning the form of the parameters β^*, λ^*, and γ^* reduces the use of Corollary 2.3 in this case. The fact that the distributions of Y_j^0 and Y_j^* are in $\mathfrak{W}$ is obvious, because for Y_j^* we have that $\gamma^* = 0$ if $\alpha \neq 1$, and $\beta^* = 0$ if $\alpha = 1$.

Although the distributions of Y_j^0 and Y_j^* belong to $\mathfrak{W}$, construction of estimators for the parameters of these distributions by the method of §4.2 is possible only after we pass from the form (A), to which the parameter expressions (4.3.1) correspond, to the form (E).

Since the transition to the random variables Y_j^* leads us to elimination of the parameter γ, the scheme of transformations of the original collection of parameters α, β, and λ to the corresponding collections of parameters of the other forms appears as follows:

$$\begin{aligned}(\alpha, \beta, \cdot, \lambda)_A &\leftrightarrow (\alpha, \beta^*, \gamma^*, \lambda^*)_A \leftrightarrow (\alpha, \beta^*, \gamma^*, \lambda^*)_B \\ &\leftrightarrow (\alpha, \theta^*, \lambda^*)_C \leftrightarrow (\nu^*, \theta^*, \tau^*)_E. \end{aligned} \tag{4.3.2}$$

All the transitions are invertible; therefore, after constructing estimators $\tilde{\nu}, \tilde{\theta}^*$, and $\tilde{\tau}^*$ of the parameters ν^*, θ^*, and τ^*, we can return by the same path, bringing with us estimators $\tilde{\alpha}, \tilde{\beta}$, and $\tilde{\lambda}$ of the parameters α, β, and λ.

An analogous situation arises in the case of transition from the original sample to the collection $\{Y_j^0\}$. In this case the parameter β is also eliminated in addition to γ. Thus, only the pair of parameters α and λ are to be estimated. The corresponding transition scheme has the form

$$\begin{aligned}(\alpha, \cdot, \cdot, \lambda)_A &\leftrightarrow (\alpha^0, 0, 0, \lambda^0)_A \leftrightarrow (\alpha^0, 0, \lambda^0)_B \\ &\leftrightarrow (\alpha^0, 0, \lambda^0)_C \leftrightarrow (\nu^0, 0, \tau^0)_E. \end{aligned} \tag{4.3.3}$$

There is no trouble involved in finding the connection between the original parameters α, β, and λ in the form (A) and their equivalents ν^*, θ^*, and τ^*, or between α, λ (also in the form (A)) and their equivalents ν^0, τ^0 if we use the transition formulas from one form to the other according to the indicated schemes.

We shall not give details of these simple but cumbersome computations here, but simply write out the final result:

$$\begin{aligned}
\nu^* &= \nu = \alpha^{-2}, \\
\theta^* &= \theta^*(\alpha,\beta) = \frac{2}{\pi\alpha} \arctan\left(\beta T_1(\xi)\tan\frac{\pi}{2}\alpha\right) \quad \text{if } \alpha \neq 1, \\
\theta^* - \theta^*(1,\beta) &= \lim_{\alpha-1} \theta^*(\alpha,\beta) = \frac{2}{\pi}\arctan(\beta T_3(\xi)), \qquad (4.3.4) \\
\tau^* &= \tau^*(\alpha,\beta,\lambda) \\
&= \frac{1}{\alpha}\Big\{\log\lambda + \mathrm{C}(1-\alpha) \\
&\qquad + \frac{1}{2}\log\left[T_2^2(\xi) + \beta^2 T_1^2(\xi) t_2^2(\xi)\tan^2\left(\frac{\pi}{2}\alpha\right)\right]\Big\} \quad \text{if } \alpha \neq 1, \\
\tau^* &= \tau^*(1,\beta,\lambda) = \lim_{\alpha\to 1}\tau^*(\alpha,\beta,\lambda) = \log(2\lambda) + \frac{1}{2}\log[1+\beta^2 T_3^2(\xi)].
\end{aligned}$$

Let us note at once two features of these formulas.

The first is that the sign of the asymmetry parameter changes in passing from β to θ^*. It is not hard to avoid this effect if desired by considering the collection of random variables $-Y_j^*$ instead of the collection of Y_j^*.

The second consists in a narrowing of the domain of variation of the parameters ν, θ^*, and τ^* in comparison with the domain of variation of the parameters in the form (E). This happens because in the case $\alpha \neq 1$

$$\max_{\xi} |T_1(\xi)| = |T_1(\tfrac{1}{2})| = |1 - 2^{1-\alpha}|/(1 + 2^{1-\alpha}) \leq \tfrac{1}{3}, \qquad (4.3.5)$$

while in the case $\alpha = 1$

$$\max_{\xi} |T_3(\xi)| = |T_3(\tfrac{1}{2})| = \frac{\log 2}{\pi} = 0.22\ldots.$$

As a result, θ^* varies not from $-\min(1, 2\sqrt{\nu}-1)$ to $\min(1, 2\sqrt{\nu}-1)$, but within narrower limits. We shall see that the change in the domain of variation of θ^* introduces additional complications in the construction of estimators for the parameters ν, θ^*, and τ^* and for the parameters α, β, and λ connected with them. It is in our interest to reduce the narrowing of the domain of variation of θ^* to a minimum, which is equivalent to choosing a value of ξ such that the functions $|T_1(\xi)|$ and $|T_3(\xi)|$ attain a maximal value, i.e., $\xi = \frac{1}{2}$. According to these considerations, a transformation of the first type (transition to the collection of random variables Y_j^*) with the value $\xi = \frac{1}{2}$ is used in what follows. An expression for the corresponding parameters ν, θ^*, and τ^* is obtained from (4.3.4) after substitution of the indicated value of ξ.

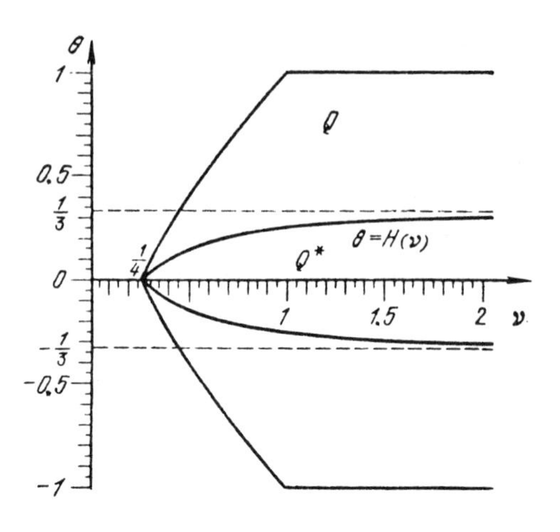

FIGURE 6

The inverse relations, i.e., those expressing the parameters α, β, and λ in terms of ν, θ^*, and τ^*, are

$$\alpha = \frac{1}{\sqrt{\nu}},$$

$$\beta = \beta(\nu, \theta^*)$$

$$= \frac{1 + \exp[(1 - 1/\sqrt{\nu}) \log 2] \tan\left(\frac{\pi}{2}\theta^*/\sqrt{\nu}\right)}{1 - \exp[(1 - 1/\sqrt{\nu}) \log 2] \tan\ \left(\frac{\pi}{2}/\sqrt{\nu}\right)} \quad \text{if } \nu \neq 1,$$

$$\beta = \beta(1, \theta^*) = \lim_{\nu \to 1} \beta(\nu, \theta^*) = -\frac{\pi}{\log 2} \tan\left(\frac{\pi}{2}\theta^*\right),$$

$$\lambda = \lambda(\nu, \theta^*, \tau^*) = \frac{\cos(\frac{\pi}{2}\theta^*/\sqrt{\nu})}{1 + \exp[(1 - 1/\sqrt{\nu}) \log 2]}$$

$$\times \exp\left[\frac{\tau^*}{\sqrt{\nu}} - \mathbb{C}\left(1 - \frac{1}{\sqrt{\nu}}\right)\right] \quad \text{if } \nu \neq 1,$$

$$\lambda = \lambda(1, \theta^*, \tau^*) = \lim_{\nu \to 1} \lambda(\nu, \theta^*, \tau^*) = \frac{1}{2} e^{\tau^*} \cos\left(\frac{\pi}{2}\theta^*\right). \tag{4.3.6}$$

For the second type of transformation (i.e., transition to the Y_j^0),

$$\nu^0 = \nu = \alpha^{-2}, \quad \theta^0 = 0, \quad \tau^0 = [\log(2\lambda) + \mathbb{C}(1 - \alpha)]/\alpha \tag{4.3.7}$$

The relations inverse to them have the form

$$\alpha = 1/\sqrt{\nu}, \quad \lambda = \tfrac{1}{2} \exp[(\tau^0 + \mathbb{C})\nu^{-1/2} - \mathbb{C}]. \tag{4.3.8}$$

As is easy to see from the connection formulas (4.3.4), the domain Q^* of variation of the parameters (ν, θ^*) is essentially smaller than the domain Q of variation of the parameters (ν, θ) which together with τ determine the distributions in $\mathfrak{W}$ (the domain of variation of τ^* remains the same as the domain of variation of τ: the whole real axis). Namely (Figure 6),

$$Q^* = \{(\nu, \theta^*)\colon \nu \geq \tfrac{1}{4}, |\theta^*| \leq H(\nu)\},$$

where

$$H(t) = \frac{2}{\pi}\sqrt{t} \arctan\left\{\frac{\exp[(1 - 1/\sqrt{t})\log 2] - 1}{\exp[(1 - 1/\sqrt{t})\log 2] + 1} \tan\left(\frac{\pi}{2\sqrt{t}}\right)\right\}.$$

The value of $H(t)$ at the point $t = 1$ is determined by continuity:

$$H(1) = \lim_{t\to 1} H(t) = \frac{2}{\pi}\arctan\left(\frac{\log 2}{\pi}\right).$$

We construct estimators $\tilde{\theta}^*$ and $\tilde{\tau}^*$ of the parameters θ^* and τ^* (by (4.2.3)) and an estimator $\tilde{\nu}$ of the parameter ν (according to (4.2.6) and (4.2.7)) from the transformed sample $Y_1^*, \ldots, Y_{2n}^*$ by the methods of the preceding section. The parameter τ^* has the whole real axis as domain of values, and the estimator $\tilde{\tau}^*$ of it has the same domain of values. But the estimators $\tilde{\nu}$ and $\tilde{\theta}^*$, which vary in Q, have a larger domain of values than the domain Q^* of values of the parameters ν and θ^* being estimated by them. This means that the estimators $\tilde{\nu}$ and $\tilde{\theta}^*$ should be changed in such a way that their domain of values coincides with Q^*. One possible variant of such a change is the following:

$$\bar{\nu} = \tilde{\nu}, \quad \bar{\theta}^* = \min(H(\tilde{\nu}), |\tilde{\theta}^*|)\operatorname{sgn}\tilde{\theta}^*, \quad \bar{\tau}^* = \tilde{\tau}^*.$$

Let us investigate the properties of the estimators $\bar{\nu}, \bar{\theta}^*$, and $\overline{\tau^*}$. First of all, the estimators $\bar{\nu}$ and $\bar{\tau}^*$, which are not to be changed, fall entirely under Theorem 4.2.1 and the corresponding part of Lemma 4.2.1. Hence, only the properties of the estimator θ^* must be analyzed.

LEMMA 4.3.2. *As* $n \to \infty$

$$\begin{aligned}\mathsf{E}(\bar{\theta}^* - \theta^*)^2 &\leq \tfrac{4}{9}\mathsf{E}(\tilde{\nu} - \nu)^2 + \mathsf{E}(\tilde{\theta}^* - \theta^*)^2 \\ &\leq \tfrac{4}{9}\sigma_\nu^2 + \tfrac{13}{9}\sigma_{\theta^*}^2 = O(1/n),\end{aligned} \tag{4.3.9}$$

where σ_ν^2 *and* $\sigma_{\theta^*}^2$ *are computed by* (4.2.8) *and* (4.2.4), *with account taken of the fact that the sample size is* $2n$.

PROOF. Since $|\theta^*| \leq H(\nu)$, it follows that

$$\bar{\theta}^* - \theta^* = \min(H(\tilde{\nu}), |\tilde{\theta}^*|)\operatorname{sgn}\tilde{\theta}^* - \min(H(\nu), |\theta^*|)\operatorname{sgn}\theta^*.$$

Elementary computations involving the various possible cases show that for any positive numbers a_1 and a_2 and any real numbers b_1 and b_2

$$|\min(a_1, |b_1|)\operatorname{sgn} b_1 - \min(a_2, |b_2|)\operatorname{sgn} b_2| \leq \max(|a_1 - a_2|, |b_1 - b_2|).$$

Therefore,

$$\mathsf{E}(\bar{\theta}^* - \theta^*)^2 \le \mathsf{E}\max\{(H(\tilde{\nu}) - H(\nu))^2, (\tilde{\theta}^* - \theta^*)^2\}$$
$$\le \mathsf{E}(H(\tilde{\nu}) - H(\nu))^2 + \mathsf{E}(\tilde{\theta}^* - \theta^*)^2. \qquad (4.3.10)$$

The function $H(t)$ is monotonically increasing on the semi-axis $t \ge 0$, where it varies from 0 to $\frac{1}{3}$ and is concave. Since $H(t) \sim 2t/3$ as $t \to 0$, and $H'(t)$ is monotonically decreasing as t increases, it follows that

$$|H(\tilde{\nu}) - H(\nu)| \le H'(+0)|\tilde{\nu} - \nu| = \tfrac{2}{3}|\tilde{\nu} - \nu|.$$

From this we get by using (4.2.9) that

$$\mathsf{E}(H(\tilde{\nu}) - H(\nu))^2 \le \tfrac{4}{9}\mathsf{E}(\tilde{\nu} - \nu)^2 \le \tfrac{4}{9}(\sigma_\nu^2 + \sigma_{\theta^*}^2).$$

This together with (4.3.10) yields (4.3.9).

We can thus make the following assertion.

THEOREM 4.3.1. *The statistics $\bar{\nu}, \bar{\theta}^*$, and $\bar{\tau}^*$ are consistent asymptotically unbiased estimators for the parameters ν, θ^*, and τ^*, with mean-square deviation from the estimated parameter and with magnitude of bias of order $1/\sqrt{n}$.*

Estimators of the original parameters α, β, and λ can be derived from (4.3.6) by replacing the parameters ν, θ^*, and τ^* by their estimators, i.e.,

$$\bar{\alpha} = (\bar{\nu})^{-1/2}, \quad \bar{\beta} = \beta(\bar{\nu}, \bar{\theta}^*), \quad \bar{\lambda} = \lambda(\bar{\nu}, \bar{\theta}^*, \bar{\tau}^*).$$

Since the quantities α, β, and λ depend smoothly on the parameters ν, θ^*, and τ^*, i.e., all the functions in (4.3.6) are jointly differentiable with respect to their variables (only one-sided derivatives apply on the boundary of the domain Q^* of variation of the parameters), it follows that the assertion of Theorem 4.3.1 extends also to the estimators $\bar{\alpha}, \bar{\beta}$, and $\bar{\lambda}$. However, the explicit formulas (4.3.6) and (4.3.4) connecting them with the parameters α, β, and λ being estimated turn out to be very unwieldy, and this complicates the determination of concise estimates of the mean-square deviations and biases of the statistics $\bar{\alpha}, \bar{\beta}$, and $\bar{\lambda}$ from the corresponding parameters α, β, and λ. We do not give these extensive computations, but refer the interested reader to [166]. Here we analyze a simpler case involving estimators of the parameters α and λ on the basis of the transformed sample $Y_1^0, \dots, Y_{3n}^0$.

Let the sample mean A_0 and the sample variance S_0^2 be constructed from the collection of variables $V_{j0} = \log|Y_j^0|$, $j = 1, \dots, 3n$, and define $\hat{\nu}_0 = 6S_0^2/\pi^2 - \frac{1}{2}$. We consider the statistics

$$\tilde{\nu}_0 = \max(\tfrac{1}{4}, \hat{\nu}_0) \quad \text{and} \quad \tilde{\tau}_0 = A_0 \qquad (4.3.11)$$

as estimators of the parameters $\nu^0 = \nu, \tau^0$ connected with the parameters α and λ by (4.3.7) and (4.3.8).

The great advantage of the estimators $\tilde{\nu}_0$ and $\tilde{\tau}_0$ over the analogous estimators $\bar{\nu}$ and $\bar{\tau}^*$ is the simplicity of the relations between the parameters they estimate and the parameters α and λ. As a result the estimators $\tilde{\alpha}$ and $\tilde{\lambda}$ corresponding to them have the following simple form:

$$\tilde{\alpha} = \tilde{\nu}_0^{-1/2}, \qquad \tilde{\lambda} = \tfrac{1}{2}\exp[(\tilde{\tau}_0 + \mathbb{C})\tilde{\nu}_0^{-1/2} - \mathbb{C}]. \tag{4.3.12}$$

How well the statistics $\tilde{\alpha}$ and $\tilde{\lambda}$ play the roles of estimators for the parameters of the original distribution can be judged by the next assertion.

THEOREM 4.3.2. *For any pairs of admissible values of the parameters α and λ (in the form (A)), the estimators $\tilde{\alpha}$ and $\tilde{\lambda}$ of them constructed according to* (4.3.11) *and* (4.3.12) *from the transformed sample $Y_1^0, \dots, Y_{3n}^0$ satisfy the inequalities*

$$\mathsf{E}(\tilde{\alpha} - \alpha)^2 \le \frac{16}{5}(22\nu^2 + 10\nu + 1)\frac{1}{3n} + 8(2\nu + 1)^2\frac{1}{3n(3n-1)} \quad \text{for } n \ge 1, \tag{4.3.13}$$

and

$$\mathsf{E}(\tilde{\lambda} - \lambda)^2 \le 2\left[\sqrt{P_2(\nu)} + 27(\log 2\lambda + \mathbb{C})^2\sqrt{P_8(\sqrt{\nu})}\right] \times \left[\max(1, 4\lambda^2) + \exp\left(2\left(3 - \frac{2}{\alpha}\right)\mathbb{C}\right)\Delta_n^{3n/2}\right]\frac{1}{3n} \quad \textit{for } 3n > \frac{8}{\alpha}, \tag{4.3.14}$$

where $P_2(\nu) = \pi^4(36\nu^2 + 20\nu + 19)/60$, $P_8(\nu)$ is a polynomial of degree eight coinciding with the value of L_{08} if we set $\theta = 0$ in it, and the value of Λ_n is given by (4.3.20).

PROOF. We have that

$$\tilde{\alpha} - \alpha = \tilde{\nu}_0^{-1/2} - \nu^{-1/2} = (\nu - \tilde{\nu}_0)/(\nu\tilde{\nu}_0^{-1/2} + \tilde{\nu}_0\nu^{-1/2}).$$

The quantities ν and $\tilde{\nu}_0$ are not less than $\frac{1}{4}$; therefore, the right-hand side of the equality is at most $4|\nu - \tilde{\nu}_0|$ in absolute value. At the same time,

$$|\tilde{\nu}_0 - \nu| = |\max(\tfrac{1}{4}, \nu_0) - \max(\tfrac{1}{4}, \nu)| \le |\hat{\nu}_0 - \nu|.$$

Taking into account (4.1.17), (4.1.18), and the fact that the sample size is $3n$, we get that

$$\begin{aligned}\mathsf{E}(\tilde{\nu}_0 - \nu)^2 \le \mathsf{E}(\hat{\nu}_0 - \nu)^2 &= \frac{36}{\pi^4}\operatorname{Var} S_0^2 \\ &= \frac{12}{\pi^4}[\mathsf{E}(\overline{V}_0)^4 - (\operatorname{Var} V_0)^2]\frac{1}{n} + \frac{24}{\pi^4}(\operatorname{Var} V_0)^2\frac{1}{n(3n-1)},\end{aligned}$$

where $V^0 \stackrel{d}{=} V_{j0}$ and $\overline{V}_0 = V_0 - \mathsf{E}V_0$. According to (4.1.15) and (4.1.13),

$$\mathsf{E}(\overline{V}_0)^4 = \frac{\pi^4}{240}(36\nu^2 + 20\nu + 3), \qquad (\operatorname{Var} V_0)^2 = \frac{\pi^4}{144}(2\nu+1)^2,$$

from which it follows that

$$\frac{12}{\pi^4}[\mathsf{E}\overline{V}_0)^4 - (\operatorname{Var} V_0)^2] = \frac{1}{15}(22\nu^2 + 10\nu + 1).$$

The estimate (4.3.13) is now obtained without trouble from the above inequalities and equalities.

Further, the obvious inequality

$$|e^x - e^y| \le |x - y| \max(e^x, e^y), \qquad x, y \in R^1,$$

gives us that

$$\begin{aligned} |\tilde{\lambda} - \lambda| &\le |\log \tilde{\lambda} - \log \lambda| \max(\tilde{\lambda}, \lambda) \\ &= \left| (\tilde{\tau}_0 - \tau_0)\tilde{\nu}^{-1/2} + (\log 2\lambda + \mathbb{C}) \frac{\nu - \tilde{\nu}_0}{\tilde{\nu}_0 + \sqrt{\nu\tilde{\nu}_0}} \right| \max(\tilde{\lambda}, \lambda) \\ &\le (|\tilde{\tau}_0 - \tau_0| + |\log 2\lambda + \mathbb{C}|\, |\tilde{\nu}_0 - \nu|) \max(2\tilde{\lambda}, 2\lambda). \end{aligned}$$

Moreover, it follows from (4.3.12) that

$$\max(2\tilde{\lambda}, 2\lambda) \le \max(1, 2\lambda, \exp(2\tilde{\tau}_0 + \mathbb{C})).$$

By Hölder's inequality, these inequalities imply the estimate

$$\begin{aligned} \mathsf{E}(\tilde{\lambda} - \lambda)^2 &\le [\mathsf{E}(|\tilde{\tau}_0 - \tau_0| + |\log 2\lambda + \mathbb{C}|\, |\tilde{\nu}_0 - \nu|)^4]^{1/2} \\ &\quad \times [\mathsf{E}\max(1, (2\tilde{\lambda})^4, (2\lambda)^4)]^{1/2} \\ &\le 2[(\mathsf{E}(\tilde{\tau}_0 - \tau_0)^4)^{1/2} + (\log 2\lambda + \mathbb{C})^2(\mathsf{E}(\tilde{\nu}_0 - \nu)^4)^{1/2}] \\ &\quad \times [\max(1, 4\lambda^2) + (\mathsf{E}\exp(4\mathbb{C} + 8\tilde{\tau}_0))^{1/2}]. \end{aligned} \tag{4.3.15}$$

Since $|\tilde{\nu}_0 - \nu| \le |\hat{\nu}_0 - \nu|$, we have from Lemma 4.1.2 that

$$\mathsf{E}(\tilde{\nu}_0 - \nu)^4 \le \mathsf{E}(\hat{\nu}_0 - \nu)^4 = \frac{6^4}{\pi^8}\mathsf{E}(S_0^2 - \operatorname{Var}\overline{V}_0)^4.$$

The sample size is $3n$; hence, the last expression can be given the form

$$\begin{aligned} &\frac{6^4}{\pi^8}\left(\frac{3n}{3n-1}\right)^4 \mathsf{E}\left\{\frac{1}{3n}\sum_{j=1}^{3n}(\overline{V}_0^2 - \operatorname{Var} V_0) - \left(\frac{1}{3n}\sum_{j=1}^{3n}\overline{V}_0\right)^2\right\}^4 \\ &\le \frac{4 \cdot 6^4}{\pi^8}\left(\frac{3n}{3n-1}\right)^4 \left[\mathsf{E}\left(\frac{1}{3n}\sum_{j=1}^{3n}(\overline{V}_0^2 - \operatorname{Var} V_0)\right)^4 + \mathsf{E}\left(\frac{1}{3n}\sum_{j=1}^{3n}\overline{V}_0\right)^8\right]. \end{aligned}$$

To estimate the last sum we use the following easily verified inequalities.

If $X_1, \dots, X_N$, $N \geq 2$, are independent and identically distributed random variables with mean value a, then

$$\mathsf{E}\left(\frac{1}{N}\sum_{j=1}^{N}(X_j - a)\right)^4 \leq 4\mathsf{E}(X_1 - a)^4 N^{-2},$$

$$\mathsf{E}\left(\frac{1}{N}\sum_{j=1}^{N}(X_j - a)\right)^8 \leq 15\mathsf{E}(X_1 - a)^8 N^{-2}.$$

Moreover, simple computations show that

$$\mathsf{E}(\overline{V}_0^2 - \operatorname{Var} V_0)^4 \leq \mathsf{E}\overline{V}_0^8 = L_{08}^0 = C_8(0, p_{02}, \dots, p_{08})$$

(here the superscript 0 reminds us that we are dealing with the parameters $\nu_0 = \nu$ and θ_0, and the function L_{08}^0 depends only on them).

Consequently,

$$\begin{aligned}\mathsf{E}(\tilde{\nu}_0 - \nu)^4 &\leq \frac{4 \cdot 6^4}{\pi^8}\left(\frac{3n}{3n-1}\right)^4 (4L_{08}^0 + 15L_{08}^0)(3n)^{-2} \\ &= \frac{2^6 \cdot 3^4 \cdot 19}{\pi^8}\left(\frac{3n}{3n-1}\right)^4 C_8(0, p_{02}, \dots, p_{08})(3n)^{-2}.\end{aligned}$$

The polynomial C_8 has a fairly complicated expression, but in our case (when $p_{01} = 0$) it becomes essentially simpler and takes the form

$$\begin{aligned}C_8 = p_{08} &+ 28p_{06}p_{02} + 56p_{05}p_{03} + 35p_{04}^2 \\ &+ 210p_{04}p_{02}^2 + 280p_{03}^2p_{02} + 105p_{02}^4,\end{aligned} \tag{4.3.17}$$

where (since $\theta_0 = 0$) the quantities p_{0k} are functions of the single parameter ν:

$$\begin{aligned}p_{02} &= \tfrac{\pi^2}{12}(2\nu + 1), & p_{05} &= 24\varsigma(5)(\nu^{5/2} - 1), \\ p_{03} &= 2\varsigma(3)(\nu^{3/2} - 1), & p_{06} &= \tfrac{\pi^6}{252}(32\nu^3 + 31), \\ p_{04} &= \tfrac{\pi^4}{120}(8\nu^2 + 7), & p_{08} &= \tfrac{\pi^3}{240}(128\nu^4 + 127).\end{aligned}$$

Consequently, $L_{08}^0 = P_8(\sqrt{\nu})$ is a polynomial of degree eight in the variable $\sqrt{\nu}$. If desired, it can be written explicitly by using (4.3.17) and the explicit expressions for p_{0k}. Thus,

$$\mathsf{E}(\tilde{\nu}_0 - \nu)^4 \leq \frac{4 \cdot 19 \cdot 6^4}{\pi^8}\left(\frac{3n}{3n-1}\right)^4 P_8(\sqrt{\nu})(3n)^{-2} \tag{4.3.18}$$

(note that $4 \cdot 19 \cdot 6^4/\pi^8 < 180$). Further,

$$\tilde{\tau}_0 - \tau = \frac{1}{3n}\sum_{j=1}^{3n}\overline{V}_0,$$

which, by (4.1.15) and (4.3.16), implies the estimator

$$\mathsf{E}(\tilde{\tau}_0 - \tau_0)^4 \le 4\mathsf{E}(\overline{V}_0)^4(3n)^{-2} = P_2(\nu)(3n)^{-2}, \tag{4.3.19}$$

where

$$P_2(\nu) = \frac{\pi^4}{60}(36\nu^2 + 20\nu + 19).$$

To conclude the construction of the estimate (4.3.15) it is necessary to estimate the quantity

$$(\mathsf{E}\exp(4\mathbb{C} + 8\tilde{\tau}_0))^{1/2} = \exp(2\mathbb{C})(\mathsf{E}\exp(8\tilde{\tau}_0))^{1/2}.$$

We have that

$$\begin{aligned}\mathsf{E}\exp(8\tilde{\tau}_0) &= \mathsf{E}\exp\left(\frac{8}{3n}\sum_{j=1}^{3n}(V_{j0} - \tau_0)\right) \\ &= \exp\left(-\frac{8}{3n}\tau_0\right)\mathsf{E}\prod_{j=1}^{3n}|Y_j^0|^{8/3n} \\ &= \exp\left(-\frac{8}{3n}\tau_0\right)\prod_{j=1}^{3n}\mathsf{E}|Y_j^0|^{8/3n}.\end{aligned}$$

Here it should be recalled that in the case $8/3n < \alpha$ the expectation $\mathsf{E}|Y_j^0|^{8/3n}$ exists and that, by (2.6.26), it can be given the form

$$\mathsf{E}|Y_j^0|^{8/3n} = w_0(8/3n, \alpha, 0, 0, 2\lambda) = (2\lambda)^{8/3n}\Delta_n,$$

where

$$\Delta_n = \left(\cos\frac{4\pi}{3n}\right)^{-1}\Gamma\left(1 - \frac{8}{3\alpha n}\right)\Big/\Gamma\left(1 - \frac{8}{3n}\right). \tag{4.3.20}$$

Consequently,

$$\begin{aligned}\exp(2\mathbb{C})(\mathsf{E}\exp(8\tilde{\tau}_0))^{1/2} &= \exp\left\{2\left(\frac{2}{\alpha}\log(2\lambda) - 2\tau_0 + \mathbb{C}\right)\right\}\Delta_n^{3n/2} \\ &= \exp\left(2\left(3 - \frac{2}{\alpha}\right)\mathbb{C}\right)\Delta^{3n/2}.\end{aligned} \tag{4.3.21}$$

It is not hard to see that $\Delta_n^{3n/2} \sim \exp(4\mathbb{C}(1/\alpha - 1))$ as $n \to \infty$. Therefore, for large n,

$$(\mathsf{E}\exp(4\mathbb{C} + 8\tilde{\tau}_0))^{1/2} \sim \exp(2\mathbb{C}).$$

Inequality (4.3.14) is now obtained by combining the estimates (4.3.18)–(4.3.20) with inequality (4.3.15).

COROLLARY. *The statistics* $\tilde{\alpha}$ *and* $\tilde{\lambda}$ *formed according to* (4.3.11) *and* (4.3.12) *are consistent and asymptotically unbiased estimators of the parameters* α *and* λ, *with the bias and the mean-square deviation of the estimated parameters of order* $O(1/\sqrt{n})$.

REMARK. In this whole section the estimators of parameters of a stable distribution have been associated with an original sample size of $6n$, $n \geq 1$. This assumption enabled us to form the transformed samples $Y_1^*, \dots, Y_{2n}^*$ and $Y_1^0, \dots, Y_{3n}^0$ without anything left over, and thus the corresponding estimators were associated with samples of sizes $2n$ and $3n$.

All the inequalities given in Theorems 4.3.1 and 4.3.2 remain true, of course, if instead of the sequences $\{2n\}$ and $\{3n\}$ in them we consider the sequence of integers $n = 2, 3, \dots$. This corresponds to original samples of various sizes ($3n$ in the first case, and $2n$ in the second).

§4.4. Estimators of the parameter γ

The parameter γ occupies a special position in the scheme of the above approach to solving the general problem of statistical estimation of parameters of stable laws. The device used to construct estimators of the parameters α, β, and λ amounted to transforming the original sample into a new sample of smaller size which was associated in turn with distributions in the class $\mathfrak{W}$, and within this class the logarithmic moments have comparatively simple expressions in terms of the parameters being estimated. An analogous universal transformation of the original sample that would enable us to estimate the parameter γ in the same way has not been found, and it seems that such a transformation does not exist in general. Therefore, new ways must be found to construct estimators of γ.

In the case $\alpha > 1$ the distribution $G_A(x, \alpha, \beta, \gamma, \lambda)$ has finite mean value equal to $\lambda\gamma$, and this makes it possible in principle to use the sample mean for an estimator of it. However, these distributions do not have finite variances for $\alpha < 2$, only finite moments of order less than α. The scatter of the values of the sample mean with respect to the variable $\lambda\gamma$ being estimated thus turns out to be large, and it is all the layer, the closer α is to 1.

The main idea for constructing an estimator for the parameter γ from an independent sample $Y_1, \dots, Y_n$ of sufficiently large size reduces to the following. Let α, β, γ, and λ be the parameters of the stable law (in the form (A)) which is the distribution of the random variables Y_j. We consider a simplified variant of the problem, assuming that we know the parameters α, β, and λ, and that only the single parameter γ must be estimated. In the general situation we must first estimate α, β, and λ, and then take into account the effect of substituting estimators of these parameters for their exact values. A problem of this kind is significantly more complicated, and we do not treat the general case, recalling that the purpose of this chapter is rather to illustrate than to be independent.

It is assumed below that $Y_1', Y_2', \dots$ is a given sequence of random numbers distributed according to the stable law $G_A(x, \alpha, \beta, 0, \lambda)$ and that the values of the parameters α, β, and λ are known. Of course, the problem of generating such a sequence of numbers can itself turn out not to be a simple matter, since, generally speaking, we have at our disposal only some fairly complicated integral representations of the function $G_A(x, \alpha, \beta, 0, \lambda)$, or series representations of it. But these difficulties are of a computational rather than of a fundamental theoretic nature.

The original sample $Y_1, \dots, Y_n$ is transformed with the help of the random numbers $Y_1', \dots, Y_n'$ into a new collection of variables $\tilde{Y}_1, \dots, \tilde{Y}_n$ by the rule

$$\tilde{Y}_j = \lambda^{-1}(Y_j - Y_j'), \qquad j = 1, 2, \dots, n.$$

The variables $\tilde{Y}_j$ can be interpreted as an independent sample from a collection subordinate to the stable distribution. According to Properties 2.3b and 2.1,

$$\tilde{Y}_j \stackrel{d}{=} \lambda^{-1} Y_A(\alpha, 0, \gamma/2, 2\lambda) \stackrel{d}{=} \lambda^{-1} Y_A(\alpha, 0, 0, 2\lambda) + \gamma,$$

from which it is clear that the $\tilde{Y}_j$ have a symmetric stable law biased by the quantity γ. More precisely,

$$\tilde{F}(x - \gamma) = \mathsf{P}(\tilde{Y}_j < x) = G_A(\lambda(x - \gamma), \alpha, 0, 0, 2\lambda).$$

Consequently, γ coincides with the median of the distribution $\tilde{F}(x - \gamma)$, and this, in turn, allows us to use the sample median method well known in statistics for estimating γ (see, for example, [84]).

Let us arrange the observations $\tilde{Y}_j$ in increasing order and denote the terms of the resulting sequence by $w_i\colon w_1 < w_2 < \dots < w_n$. The sample median μ_n is defined by

$$\mu_n = \begin{cases} w_{(n+1)/2} & \text{if } n \text{ is odd} \\ \frac{1}{2}(w_{(n+2)/2} + w_{n/2}) & \text{if } n \text{ is even.} \end{cases}$$

We take this statistic as an estimator of the unknown parameter γ, i.e., we let $\tilde{\gamma} = \mu_n$. The cases of even and odd n do not differ in principle, and the corresponding asymptotic analysis (as $n \to \infty$) of the properties of the estimator $\tilde{\gamma}$ leads in both cases to the same conclusions, but in a technical respect the case of odd n is somewhat simpler, and for this reason we confine ourselves to an analysis of the case when $n = 2m + 1$.

Let $F_n(x) = \mathsf{P}\{\tilde{\gamma} - \gamma < x\}$. It is not hard to compute (see §17 in [84]) that

$$F_n(x) = \sum_{k=m+1}^{n} \binom{n}{k} \tilde{F}^k(x)(1 - \tilde{F}(x))^{n-k}. \tag{4.4.1}$$

Since the distribution function $\tilde{F}(x)$ has a density

$$\tilde{p}(x) = \lambda g_A(x\lambda, \alpha, 0, 0, 2\lambda),$$

$F_n(x)$ also has a density, and, as we see by differentiating (4.4.1),

$$\begin{aligned} p_n(x) = F_n'(x) &= n\binom{2m}{m}[\tilde{F}(x)(1 - \tilde{F}(x))]^m \tilde{p}(x) \\ &= a_n \exp(-m\psi(x))\tilde{p}(x), \end{aligned} \tag{4.4.2}$$

where $\psi(x) = -\log \tilde{F}(x) - \log(1 - \tilde{F}(x))$ and $a_n = n\binom{2m}{m}$.

It follows from (4.4.2) that the bias $\mathsf{E}(\tilde{\gamma} - \gamma)$ and the mean-square error $\mathsf{E}(\tilde{\gamma} - \gamma)^2$ of the estimator $\tilde{\gamma}$ can be written as the integrals

$$\mathsf{E}(\tilde{\gamma} - \gamma) = a_n \int x\tilde{p}(x) \exp(-m\psi(x))\, dx, \tag{4.4.3}$$

$$\mathsf{E}(\tilde{\gamma} - \gamma)^2 = a_n \int x^2\tilde{p}(x) \exp(-m\psi(x))\, dx. \tag{4.4.4}$$

The distribution $\tilde{F}(x)$ is symmetric, i.e., $1 - \tilde{F}(x) = \tilde{F}(-x)$ and $\tilde{p}(x) = \tilde{p}(-x)$ for all x. This implies, first, that the functions $\psi(x)$ and $p_n(x)$ are even, and, second, that $p_n(x) \sim \operatorname{const} x^{-\alpha(m+1)-1}$ as $x \to \infty$ (by 2.4.48), (2.5.4), and (2.5.23)). Consequently, the integrals (4.4.3) and (4.4.4) exist if $n > 4/\alpha - 1$ (the first integral exists if $n > 2/\alpha - 1$), and, moreover, the integral (4.4.3) is equal to zero, i.e., the estimator $\tilde{\gamma}$ is unbiased.

Let us now turn to the determination of the asymptotic behavior of the integral (4.4.4). Since the distribution $\tilde{F}(x)$ is symmetric,

$$\begin{aligned} &\psi(0) = 2\log 2, \qquad \psi'(0) = 0, \\ &\psi''(0) = 8\tilde{p}^2(0) = 8\lambda^2 g_A^2(0, \alpha, 0, 0, 2\lambda). \end{aligned} \tag{4.4.5}$$

The value of the density of a symmetric stable distribution at zero is known (see (2.2.11)). Therefore, by (2.1.2),

$$\psi''(0) = 2[\tfrac{1}{\pi}\Gamma(1 + 1/\alpha)(2\lambda)^{1-1/\alpha}]^2 > 0. \tag{4.4.6}$$

Properties (4.4.5) and (4.4.6) enable us to use the Laplace method to obtain an asymptotic representation of the integral (4.4.4) as $n \to \infty$; we already encountered this method under analogous circumstances in §2.5 (see [20]).

According to this method, standard arguments give us that

$$\begin{aligned} I_n &= a_n \int x^2 \tilde{p}(x) \exp(-m\psi(x))\, dx \\ &= a_n \exp(-m\psi(0)) \int x^2 \tilde{p}(x) \exp\left(-m\psi''(0)\frac{x^2}{2} + \cdots\right) dx. \end{aligned}$$

Since $\sigma_n^2 = m\psi''(0) \to \infty$ as $n \to \infty$, we get the following asymptotic formula after the change of variable $\sum_n x = t$ and corresponding estimates of the "tails" of the integral:

$$\begin{aligned} I_n &\sim a_n \sigma_n^{-3} \tilde{p}(0) \exp(-m\psi(0)) \int x^2 \exp(-x^2/2)\, dx \\ &= (2m+1)\binom{2m}{m} 2^{-2m} (8m)^{-3/2} \sqrt{2\pi} (\tilde{p}(0))^{-3} \sim cn^{-1}, \end{aligned}$$

where

$$c = (\pi(2\lambda)^{(1-\alpha)/\alpha} / \Gamma(1 + 1/\alpha))^2. \tag{4.4.7}$$

These arguments can now be summarized as the next statement.

THEOREM 4.4.1. *The statistic $\tilde{\gamma} = \mu_n$ is a consistent unbiased estimator of the parameter γ for all odd $n > (2-\alpha)/\alpha$, with mean-square error*

$$\mathsf{E}(\tilde{\gamma} - \gamma)^2 \sim cn^{-1} \quad \textit{as } n \to \infty,$$

where the constant c is given by (4.4.7).

REMARK 1. For even n the statistic $\tilde{\gamma} = \mu_n$ is a consistent and asymptotically unbiased estimator of γ.

REMARK 2. We indicate one more variant of an estimator for γ under the assumption that the values of the remaining parameters are known. Formally, this variant has more advantages than the one considered above, because it does not require a transformation of the original sample. The following considerations are taken as a basis for the estimator. According to (2.1.2),

$$G(x, \alpha, \beta, \gamma, \lambda) = G((x - l)\lambda^{-1/\alpha}, \alpha, \beta, 0, 1),$$

where $l = \lambda(\gamma + b_0)$, and b_0 is uniquely determined by the parameters α, β, and λ. From this it can be concluded that the median $m(\alpha, \beta, \gamma, \lambda)$ of the distribution $G(x, \alpha, \beta, \gamma, \lambda)$ is connected with the median $m(\alpha, \beta) = m(\alpha, \beta, 0, 1)$ by the relation

$$\lambda^{-1} m(\alpha, \beta, \gamma, \lambda) = \lambda^{1/\alpha - 1} m(\alpha, \beta) + b_0 + \gamma. \tag{4.4.8}$$

We then consider the sample median $\tilde{\mu}$ constructed from the original sample $Y_1, \ldots, Y_n$ with distribution $G(x, \alpha, \beta, \gamma, \lambda)$. Because of the nice analytic properties of stable laws, the statistic μ_n turns out to have an asymptotically normal distribution with mean value $\mu = m(\alpha, \beta, \gamma, \lambda)$ and variance

$$\sigma^2 \sim g^{-2}(\mu, \alpha, \beta, \gamma, \lambda)(4n)^{-1}, \qquad n \to \infty$$

(see, for example, [84]).

Consequently, replacing the median μ in (4.4.8) by the sample median $\tilde{\mu}$, we get the estimator of γ

$$\tilde{\gamma} = \lambda^{-1}\tilde{\mu} - b_0 - c\lambda^{1/\alpha-1}, \tag{4.4.9}$$

where $c = m(\alpha, \beta)$ is the unique solution of the equation $G(x, \alpha, \beta) = \frac{1}{2}$.

Like the estimator in Theorem 4.4.1, $\tilde{\gamma}$ is an asymptotically unbiased and consistent (more precisely, $1/\sqrt{n}$-consistent estimator of the parameter γ.

§4.5. Discussion of the estimators

The estimators given in §§4.2–4.4 for the parameters of stable laws are not best possible estimators even in the asymptotic sense as the sample size n increases to infinity. This is due, first of all, to the fact that all the estimators were constructed on the basis of the sample moments and the sample median—statistics of a simple structure that do not, as a rule, have the greatest efficiency. Also, in constructing estimators in the general situation we used a transformation of the original sample which led either to a reduction in its size by a factor of two or three, or to an increase of the scatter of the random variables making up the sample in isolated cases. At the same time the estimators we found have a number of merits that not only enable us to regard these estimators as a convenient tool for solving practical problems, but also make them a good basis for constructing best possible (in a definite sense) estimators of the corresponding parameters.

If for the time being we agree to let $\tilde{\mu}$ stand for an estimator of the parameter μ (which can be any of the parameters considered in §§4.2–4.4), then in respect to its properties we can assume that it:

1) has an algorithmically simple structure (this applies, first of all, to the estimators of the parameters α, θ, and τ corresponding to the form (E));

2) is asymptotically normal; and

3) has mean-square deviation $\mathsf{E}(\tilde{\mu} - \mu)^2$ of order $1/n$.

Property 2) has not been proved for any of the estimators in §§4.2–4.4, but it can be established without difficulty with the help of the central limit theorem by using the "locally additive" structure of the estimators. By the obvious inequalities

$$(\mathsf{E}\tilde{\mu} - \mu)^2 \le \mathsf{E}(\tilde{\mu} - \mu)^2, \qquad \operatorname{Var}\tilde{\mu} \le 4\mathsf{E}(\tilde{\mu} - \mu)^2,$$
$$\mathsf{P}(\sqrt{n}|\tilde{\mu} - \mu| \ge T) \le n\mathsf{E}(\tilde{\mu} - \mu)^2/T^2, \qquad T > 0,$$

property 3) implies that $\tilde{\mu}$ is asymptotically unbiased with bias of order at most $1/\sqrt{n}$, that the variance of the estimator has order at most $1/n$, and, finally, that $\tilde{\mu}$ is a $1/\sqrt{n}$-consistent estimator of the parameter μ.

We mention another circumstance that should be pointed out. The constructions of estimators for the parameters of stable distributions were carried out within two groups. One included the parameters α, β, and λ, and the other the parameter γ. These groups differed first of all in the form of the transformations of the original sample in constructing the estimators. The estimators of the first group can be regarded in total as the coordinates of a three-dimensional random vector which is obviously a $1/\sqrt{n}$-consistent estimator of the corresponding triple of parameters.

Contemporary mathematical statistics has available several methods for improving $1/\sqrt{n}$-consistent estimators, and they make it possible in principle to construct asymptotically efficient estimators of the parameters of stable laws, at least within the indicated groups, on the basis of the estimators we obtained. These methods can be divided into the following two categories.

The first contains methods that do not use information about the analytic expression for the distribution density of the sample units. Here we cite the comparatively recent paper [143], which contains, in particular, a brief survey of other publications in the same direction.

The second category includes methods in which knowledge of the distribution density of the sample units is assumed. Among the large number of papers connected with the use of such methods we recall [18], whose results can form a convenient basis for solving the general problem of asymptotically efficient estimators for stable distributions.

The preferability of the methods in the first category is obvious, especially if it is considered that the regularity conditions they use for the distributions of the sample units are weaker than the conditions usually present in the methods of the second category. Unfortunately, they have been considerably less developed than the latter methods and have been connected only with the problem of estimating the scalar shift parameter of a distribution, which allows us to use them only for estimating the parameter γ and, in some cases, λ. Therefore, in solving the problem of constructing asymptotically efficient estimators in the general situation we must resort to methods of the second category. Furthermore, it can be said at once that in constructing asymptotically efficient estimators of the first group we do not encounter fundamental difficulties on the indicated path. The difficulties begin if we want to construct similar estimators for the whole collection of parameters, since the problem of constructing a $1/\sqrt{n}$-consistent estimator for the parameter γ has not been solved when the remaining parameters are unknown.

In the framework of this discussion we cannot acquaint the reader in detail with the substance of the general results in the paper [18] cited above; all the more so, we cannot verify that these results can be used in the problem of estimating the parameters of stable distributions. We give only an exposition of the main idea of the method in the simplest situation when only one parameter μ is to be estimated.

In the role of μ we can take any of the stable law parameters for which we have a $1/\sqrt{n}$-consistent estimator. At the same time, it should be mentioned that the possibilities of this method are much broader, and that it enables us to obtain asymptotically efficient estimators for the parameter collection $\mu = (\alpha, \beta, \lambda)$, since, as we have seen, there are $1/\sqrt{n}$-consistent estimators for the vector-valued parameter μ.

Let $p(x, \mu)$ be the density of the distribution of independent random variables $X_1, \ldots, X_n$ forming a sample from which we want to construct an asymptotically efficient estimator of the parameter μ, given that we have a $1/\sqrt{n}$-consistent estimator $\tilde{\mu}$ of this parameter. Let

$$L(X \mid \mu) = \sum_{j=1}^{n} \log p(X_j, \mu)$$

be the likelihood function, and let $L'(X \mid \mu)$ and $L''(X \mid \mu)$ be its first and second derivatives with respect to μ.

It turns out that the statistic

$$\hat{\mu} = \tilde{\mu} - L'(X \mid \tilde{\mu})/L''(X \mid \tilde{\mu}) \tag{4.5.1}$$

is an asymptotically efficient estimator of μ under certain regularity conditions for the density $p(x, \mu)$, for example, the existence and continuity of the second derivative of $p(x, \mu)$ with respect to μ, etc. In the cases when the Fisher information

$$I(\mu) = \int \left(\frac{\partial}{\partial \mu} \log p(x, \mu) \right)^2 p(x, \mu)\, dx$$

associated with the distribution corresponding to $p(x, \mu)$ can be computed it is possible to use the statistic

$$\mu^* = \tilde{\mu} + L'(X \mid \tilde{\mu})/(nI(\mu)) \tag{4.5.2}$$

instead of (4.5.1).

The fact that an explicit expression for the density is used in forming the statistics $\hat{\mu}$ and μ^* creates certain inconveniences in the case of estimating the parameters of stable laws, because with some exceptions we have available only complicated forms of expressions for the corresponding densities as series or integrals. However, these are not the difficulties that arise in constructing the maximum likelihood estimator μ_0, where one must solve the transcendental equation $L'(X \mid \mu) = 0$ for the variable μ (μ_0 is the solution of this equation).

Suppose, for example, that we are looking for the value of only the single parameter μ of a stable law under the condition that we know the remaining parameters. To use the estimator (4.5.1) we obviously must at least have available some tables of values of the functions

$$\frac{\partial}{\partial \mu} \log p(x, \mu) \quad \text{and} \quad \frac{\partial^2}{\partial \mu^2} \log p(x, \mu).$$

The tabulation of these functions is not a very complicated task. The attempt at compiling such tables for the stable distributions themselves can serve as an affirmation of this (see [6] and [31]).

In constructing analogous estimators for several parameters (i.e., for estimators of vectors) the problem becomes significantly more complicated on the computational level, since it is required to tabulate all the first and second mixed derivatives of the logarithm of the density with respect to the parameters to be estimated.

As we know, among the stable laws there are those whose densities can be expressed with the help of elementary function. The parameter sets ($\alpha = \frac{1}{2}, \beta = 1, \gamma, \lambda$), ($\alpha = 1, \beta = 0, \gamma, \lambda$), and ($\alpha = 2, \beta, \gamma, \lambda$) correspond to them. The last set, which corresponds to the normal distributions, is well enough known that there is no need to comment on the associated problem of estimating parameters. The remaining two cases in this scheme are less known. It is thus useful, in our opinion, to analyze these cases by way of illustrating the considerations given above. We consider the simplest problem of estimating one of the parameters γ or λ under the condition that the value of the second is known.

In the first case,

$$g(x, \frac{1}{2}, 1, \gamma, \lambda) = \frac{\lambda}{2\sqrt{\pi}} (x - \gamma\lambda)^{-3/2} \exp\left(-\frac{\lambda^2}{4(x - \gamma\lambda)}\right), \quad x > \gamma\lambda. \quad (4.5.3)$$

Assume that the value λ is known, while the value of γ is unknown and must be estimated. We form the likelihood function $L(Y \mid \gamma)$. Its derivative with respect to γ has the form

$$L'(Y|\gamma) = \frac{3}{2}\lambda \sum_j (Y_j - \gamma\lambda)^{-1} - \frac{\lambda^3}{4} \sum_j (Y_j - \gamma\lambda)^{-2}.$$

The likelihood equation $L'(Y \mid \gamma) = 0$, although it reduces to an algebraic equation, does not allows us to write out the maximum likelihood estimator. Therefore, we look for an asymptotically efficient estimator $\hat{\gamma}$ of the parameter γ by following the general hints given above. First of all, the explicit form (4.5.3) of the density enables us to compute the corresponding Fisher information $I(\gamma)$. An uncomplicated computation shows that $I(\gamma) = 42/\lambda^2$. This permits us to construct an estimator $\hat{\gamma}$ of the form (4.5.2).

At our disposal are two $1/\sqrt{n}$-consistent estimators of the parameter γ. One was given in Theorem 4.4.1, and the second in the remark accompanying this theorem. The most convenient estimator in this case turns out to be (4.4.9). Namely,

$$\tilde{\gamma} = \lambda^{-1}\tilde{\mu} - c\lambda,$$

where $\tilde{\mu}$ is the sample median and $c = 1.08\ldots$ is the solution of the equation

$$G(x, \tfrac{1}{2}, 1) = 2[1 - \Phi(1/\sqrt{2x})] = \tfrac{1}{2},$$

where Φ is the distribution function of the standard normal law.

In summary, with the help of (4.5.2) we obtain the statistic

$$\hat{\gamma} = \tilde{\gamma} + \frac{\lambda^2}{42}L'(Y \mid \tilde{\gamma})n^{-1} \tag{4.5.4}$$

as an asymptotically efficient estimator of γ.

Suppose now that the value of γ is known, but the value of λ must be estimated. The course of the arguments is as before. We construct the likelihood function $L(Y \mid \lambda)$ and then find its derivative with respect to λ:

$$L'(Y \mid \lambda) = n\lambda^{-1} + \left(\frac{3}{2}\gamma - \lambda\right)\sum_j (Y_j - \gamma\lambda)^{-1} - \frac{\lambda^2\gamma}{4}\sum_j (Y_j - \gamma\lambda)^{-2}.$$

Of course, we should not try to solve the likelihood equation $L'(Y \mid \lambda) = 0$, and we must construct an asymptotically efficient estimator $\hat{\lambda}$ of λ according to the indicated rule. However, contrary to the preceding case, the Fisher information $I(\lambda)$, though it is computed, turns out to depend on λ. Therefore, $\hat{\lambda}$ should be constructed by the rule (4.5.1). The needed $1/\sqrt{n}$-consistent estimator $\tilde{\lambda}$ of λ can be found in §4.3 (equality (4.3.12)). Let us compute the second derivative of $L(Y \mid \lambda)$ with respect to λ:

$$L''(Y \mid \lambda) = n\lambda^{-2} - \sum_j (Y_j - \gamma\lambda)^{-1} + \frac{1}{2}\gamma(3\gamma - \lambda)\sum_j (Y_j - \gamma\lambda)^{-2} - \frac{1}{2}\lambda^2\gamma^2\sum_j (Y_j - \gamma\lambda)^{-3}.$$

According to the rule (4.5.1), the asymptotically efficient estimator $\hat{\lambda}$ has the form

$$\hat{\lambda} = \tilde{\lambda} - L'(Y \mid \tilde{\lambda})/L''(Y \mid \hat{\lambda}).$$

As we see, the estimator of λ has turned out to be more complicated than the estimator of γ. This has to do in the first place with the choice of the parametrization system. If instead of the system $(\alpha, \beta, \gamma, \lambda)$ we had considered the system $(\alpha, \beta, \gamma', \lambda)$, where $\gamma' = \gamma\lambda$, then the complexity of the estimator of γ' in the case when $\alpha = \frac{1}{2}$ and $\beta = 1$ would have stayed the same, while the estimator of λ given that γ' is known would have been essentially simpler to obtain, since the likelihood equation is fairly simple to solve. It turns out that the resulting maximum likelihood estimator is a sufficient statistic (see [166]).

Asymptotically efficient estimators of one of the parameters γ or λ in the case when $\alpha = 1$ and $\beta = 0$ can be constructed in the same way with the help of the explicit expression

$$g(x, 1, 0, \gamma, \lambda) = \frac{\lambda}{\pi}[(x - \gamma\lambda)^2 + \lambda^2]^{-1}$$

for the density. For the parameter system $(\alpha, \beta, \gamma', \lambda)$ mentioned above, the problem of constructing an asymptotically efficient estimator of γ' for known λ can be found in [84], but there it is solved with the help of an obsolete iteration procedure.

We conclude the discussion of the problem of estimating the parameters of stable laws by another remark. It is well known how important in practice it is to construct confidence intervals for parameters being estimated. In general, we have not treated this question, although the estimators given for the parameters ν, θ, and τ, due to their simple structure, enable us to construct confidence intervals for them in a fairly simple manner. For example, the estimator for θ is reduced to an estimator for the parameter $p = (1 + \theta)/2$ in the binomial scheme (see the explanation after Lemma 4.2.1), and the problem of constructing confidence intervals for this estimator has become classical (see [84] or [43]).

§4.6. Simulation of sequences of stable random variables

In creating diverse methods for estimating the parameters of stable laws, theoreticians need empirical tests of the effectiveness of these methods. The problem thus arises of creating algorithms for constructing numerical sequences that could be interpreted as sequences of observations of independent random variables with a common stable law.

This problem has been considered by several authors in [40], [178], [246], and [250]. The complexity, which is based on the relation

$$Y(\alpha, \beta, \gamma, \lambda) \stackrel{d}{=} G^{-1}(\omega, \alpha, \beta, \gamma, \lambda)$$

where G^{-1} is the function inverse to the distribution function G and ω is a random variable uniformly distributed on $(0, 1)$, turns out not to be very suitable in the general case. The reason is that there are no analytic expressions for the function G^{-1}. The only exception here is the case when $\alpha = 1$ and $\beta = 0$, for which

$$Y(1, 0, \gamma, \lambda) \stackrel{d}{=} \lambda \tan[\pi(\omega - \tfrac{1}{2})] + \lambda\gamma. \tag{4.6.1}$$

The case $\alpha = 2$ is in a special position, of course, since the problem of simulating sequences of normally distributed random numbers has long had the attention of specialists, and considerable progress has been made toward its solution (see [252]).

The multiplicative relations considered in §3.4 (the relations (3.4.13) and those in Theorem 3.4.3) help us to obtain a partial solution of the problem.

Indeed, having the possibility of simulating sequences of normally distributed random numbers, we can use (3.4.13) to generate sequences of random numbers having stable distributions with parameters $\alpha = 1/2^n$, $\beta = 1$, $\gamma = 0$, and $\lambda = 1$ for $n = 1, 2, \ldots$, and then with arbitrary parameters γ and λ (due to Property 2.1).

In precisely the same way, having created an algorithm for simulating sequences of random numbers distributed like $\Gamma_{k/n}$ (see Theorem 3.4.3), we have the possibility of constructing sequences of random numbers with distribution $G(x, 1/n, \beta, 0, 1)$ (and then also with distribution $G(x, 1, /n, \beta\gamma, \lambda))$ for $n = 2, 3, \ldots$.

The solution of the problem in the general situation was found by a path started in 1973 in the article [40], in which the particular case of the representation (2.2.18) with $0 < \alpha < 1$ and $\beta = 1$ is used to obtain the relation

$$Y(\alpha, 1) \stackrel{d}{=} (A(\pi\omega)/E)^{(1-\alpha)/\alpha}, \tag{4.6.2}$$

where

$$A(\varphi) = \frac{\sin(1-\alpha)\varphi}{\sin\alpha\varphi}\left(\frac{\sin\alpha\varphi}{\sin\varphi}\right)^{1/(1-\alpha)},$$

and the random variable E, which is independent of ω, has the standard exponential distribution (see the paragraph before Theorem 3.4.2 in §3.4).

The idea for obtaining (4.6.2) is based on the fact that the distribution function of the random variable $(Y(\alpha, 1))^{\alpha/(1-\alpha)}$, $0 < \alpha < 1$, is

$$\frac{1}{\pi}\int_0^\pi \exp(-xA(\varphi))\, d\varphi.$$

The general case was investigated in the same way in 1976 in [246], where the representations (2.2.27) and (2.2.28) of the distribution functions of stable laws were obtained as a basis for the author's arguments. The following relations were found there:

If $\alpha \neq 1$, then

$$Y(\alpha, \beta) \stackrel{d}{=} B_\alpha(\omega - 1/2)E^{1-1/\alpha}, \tag{4.6.3}$$

where ω and E, as in (4.6.2), are independent and have a uniform and an exponential distribution, respectively,

$$B_\alpha(\varphi) = \frac{\sin\frac{\pi}{2}\alpha(\varphi+\theta)}{\cos\frac{\pi}{2}[\varphi-\alpha(\varphi+\theta)]}\left(\frac{\cos\frac{\pi}{2}[\varphi-\alpha(\varphi+\theta)]}{\sin\frac{\pi}{2}\varphi}\right)^{1/\alpha}, \qquad |\varphi| < \frac{1}{2},$$

and $\theta = \beta K(\alpha)/\alpha$.

If $\alpha = 1$, then

$$Y(1,\beta) \stackrel{d}{=} B_1(\omega - \tfrac{1}{2}) - \tfrac{2}{\pi}\beta \log E, \tag{4.6.4}$$

where ω and E are the same as before, and

$$B_1(\varphi) = \frac{2}{\pi}\beta \log\left(\frac{1+\beta\varphi}{\cos\frac{\pi}{2}\varphi}\right) + (1+\beta\varphi)\tan\frac{\pi}{2}\varphi, \qquad |\varphi| < \frac{1}{2}.$$

The relations (4.6.3) and (4.6.4) turned out to be not only very convenient for constructing sequences of random numbers, but also fairly reliable with regard to the quality of the random sequences generated with their help. This has been verified in the process of a comparative analysis of different methods for estimating the parameters of stable laws.

In addition to what was said about the second way of solving the random number generation problem, we direct attention to the end of §2.2, where mention is made of the possibility of constructing representations that are analogous to (2.2.18) and (2.2.19) but at the same time differ essentially from them. It is not excluded that such variants of representations will enable us to obtain new relations of the type (4.6.3) and (4.6.4) and to exploit them in constructing sequences of random numbers (of course, the advantages of these variants look very problematical so far).

Comments

In our desire to lighten the burden on the main text as much as possible we have gathered here facts of a historical nature, along with facts which, in our view, are not of paramount importance; in particular, information about who is credited with various results and when they were first published. The absence in these Comments of an explanation and reference for some theorem means that the author does not have such a reference, and possibly that the corresponding statement is being published for the first time.

Introduction

I.1. Pólya introduced in probability theory a certain class of characteristic functions which has come to bear his name. This class is formed by the real-valued functions $\mathfrak{f}(t)$ defined on the whole t-axis and having the following properties:

$$\mathfrak{f}(0) = 1, \quad \mathfrak{f}(-t) = \mathfrak{f}(t), \qquad \mathfrak{f}'(t) \leq 0, \quad f''(t) \geq 0, \quad t > 0.$$

The functions $f_\lambda^{(\alpha)}$, $0 < \alpha \leq 1$, are easily seen to be in the Pólya class. See [68] for details about the Pólya class.

I.2. By definition, the class of infinitely divisible laws consists of the distributions G that can be represented as compositions of n identical distributions G_n for any integer $n \geq 2$. In 1934 Lévy obtained a description of the distributions G in this class in terms of the corresponding characteristic functions $\mathfrak{g}(t)$.

The so-called Lévy canonical form of expression for the functions $\mathfrak{g}(t)$ is given by (I.3).

Thus, Theorem A is a rephrasing of the Khintchine theorem asserting that the class $\mathfrak{G}$ introduced coincides with the class of infinitely divisible distributions. This result was first published by Khintchine in the paper *Zur Theorie der unbeschränkt teilbaren Verteilungsgesetze*, Mat. Sb. **2(44)** (1937), 79–117.

In his investigations Khintchine employed another form of expression for the functions $\mathfrak{g}(t)$, now called the Lévy-Khintchine canonical form. Theorem B, which was proved by Gnedenko, corresponds in the original paper to the Lévy-Khintchine canonical form, and not to the Lévy canonical form, with which the formulation we have given is associated.

I.3. The variants of expressions in the literature for the functions $\mathfrak{g}(t)$ have the form

$$\mathfrak{g}(t) = \exp(it\gamma - \lambda\psi(t, \alpha, \beta)).$$

In the forms of expression we used for the function $\log \mathfrak{g}(t)$ the parameter γ is replaced by $\lambda\gamma$, and as a result the function becomes proportional to λ. It turns out that this form of expression has analytic advantages, though at the same time it excludes degenerate distributions from the description of the family $\mathfrak{S}$. It is perhaps surprising that until now no one has wanted to write the characteristic functions of stable laws and the characteristic functions $\mathfrak{g}(t, \lambda)$ of homogeneous stable processes $X(\lambda)$ in a coordinated way. The fact of the matter is that (see [77])

$$\log \mathfrak{g}(t, \lambda) = \lambda \log \mathfrak{g}(t, 1),$$

where $\lambda > 0$ signifies a time parameter.

The form (A) of expression for $\log \mathfrak{g}(t)$, which is the best known form and until recently has been the one most often used, was proposed by Lévy in [51]. In the case $\alpha = 1$ this formula was given independently of Lévy by Khintchine in [44]. In Lévy's monograph [50] the form (A) was preceded by the following modification of it, obtained for strictly stable laws:

$$\log \mathfrak{g}(t) = \lambda\Gamma(-\alpha)(\cos(\pi\alpha/2) - i\beta \sin(\pi\alpha/2))t^{\alpha},$$

where $0 < \alpha \leq 2$, $|\beta| \leq 1$, $\lambda > 0$, and $t > 0$.

The forms (B), (M), (C), and (E) apparently appeared for the first time in the author's papers [96], [97], and [167]. The merits of the form (B) and (C) were pointed out long ago. For example, the form (C) can be encountered in Feller's monograph [22], while systematic use is made of the form (B) in the monographs of Ibragimov and Linnik [35] and Lukacs [54].

I.4. The following equalities establish a connection between the constants C_1 and C_2 in the expression (I.10) for the spectral function $H(x)$ of a stable law given in the form (A) by the parameters α, β, γ, and λ:

$$C_1 = \lambda(1+\beta)\pi^{-1}\Gamma(\alpha)\sin(\pi\alpha/2),$$
$$C_2 = \lambda(1-\beta)\pi^{-1}\Gamma(\alpha)\sin(\pi\alpha/2).$$

I.5. Criteria 2 and 3 stem from work of Lévy.

I.6. If (I.27) is replaced by the equality

$$c_1X_1 + c_2X_2 + \cdots + c_kX_k \stackrel{d}{=} X_1 + a,$$

where a is a constant, then we obtain a criterion for a nondegenerate distribution to belong to $\mathfrak{S}$.

I.7. There is one particular error which is repeated in fairly many papers (connected in one way or another with stable laws). Namely, to describe the family $\mathfrak{S}$ they use form (A) for $\log \mathfrak{g}(t)$ with the sign in front of $it\omega_A(t, \alpha, \beta)$ chosen to be "minus" in the case $\alpha \neq 1$. Along with this it is commonly assumed that the value $\beta = 1$ corresponds to the stable laws appearing in the scheme (I.6) as limit distributions of the normalized sums Z_n with positive terms. But this assumption is incompatible with the choice of "minus" in front of $it\omega_A$ in the form (A).

The error evidently became widespread because it found its way into the well-known monograph [26]. Hall [29] devoted a special note to a discussion of this error, calling it (with what seems unnecessary pretentiousness) a "comedy of errors".

Though he is undoubtedly right on the whole, in our view he exaggerates unnecessarily, presenting the matter as if the mistake he observed is almost universal. In reality this defect in [26] was noticed long ago. For example, special remarks on this were made in the papers [95] of Zolotarev and [79] of Skorokhod. And in general there are more than a few papers and books whose authors were sufficiently attentive and did not lapse into this "sin". For example, we can mention Linnik's paper [53] and Feller's book [22].

Chapter 1

1.1. In the late 1950s and early 1960s a probability seminar met at Moscow State University. At one session the leader, E. B. Dynkin, related the following curious problem, which I reproduce here as I remember it.

Consider in the (x, y)-plane a Brownian motion of a particle beginning at $(0, 0)$ and subject to a constant drift in the direction of the vector $a = (a_1, a_2)$ with $a_1 \geq 0$ and $a_2 \leq 0$. The line $y = -1$ is an absorbing barrier. For such a drift the particle is absorbed with probability 1. Let $(X_0, -1)$ be the absorption point. It turns out that in the case when $a_1 > 0$ and $a_2 < 0$ the random variable X_0 has a normal distribution on the line $y = -1$, while in the case when $a_1 > 0$ and $a_2 = 0$ it has a stable distribution with parameters $\alpha = 1$ and $\beta = 1$ (here the parameters can be associated both with the form (A) and with the form (B)). But if $a_1 = 0$ and $a_2 = 0$, then X_0 has a Cauchy distribution (i.e., $\alpha = 1$ and $\beta = \gamma = 0$).

1.2. A model of point sources of influence based on the use of a Poisson distribution for the number of points in a finite region is visible as far back as the short note of Lifshits [117], which dealt with the solution of a certain problem in physics (it is discussed in more detail in the second example). Somewhat later a particular case of this model became the subject of an independent investigation in Good's paper [112]. It is possible that interest in the model itself was stimulated by the paper of Holtsmark [114] (see Example 1) or by a result of that paper in Chandrasekhar's book [103].

But neither Holtsmark nor Chandrasekhar relied on a model in which the number of particles falling in a bounded volume is random and has a Poisson distribution. As a result their method looks more unwieldy and less universal.

A particular case of the model of point sources of influence lay at the basis of the solution of a certain problem in solid-state theory considered by Zolotarev and Strunin in [141].

1.3. The Holtsmark distribution has become widely known in probability theory because of the books of Chandrasekhar [103] and Feller [22]. There does not exist an explicit expression for the distribution density of ν in elementary functions. However, a relatively simple expression is known (see [168]) for the distribution function of the length of ν:

$$\mathsf{P}(|\nu\lambda^{-2/3}| < r) = 1 - \frac{2}{\pi}\int_0^{\pi/2} (1 + 3ar^3)\exp(-ar^3)\,d\varphi, \qquad r > 0,$$

where $a = (\cos\varphi)^2\cos(\varphi/2)(\sin(3\varphi/2))^{-3}$.

1.4. The example is taken from an article of Ovseevich and Yaglom [132]. This contains an error which essentially affects the content of the whole paper. After observing that the time function $F(t,\lambda)$ of the types of lines considered can be represented as a composition of any number n of identical time functions, $F(t,\lambda) = F(t,\lambda/n) * \cdots * F(t,\lambda/n)$, the authors conclude that $F(t,\lambda)$ must be the distribution function of a stable law. However, the property they noticed actually characterizes a broader class of distributions (see Comment I.2).

1.5. Only after the manuscript was prepared for printing did I become aware of an interesting instance of stable laws in the hereditary theory of elasticity. It is contained in the book *The mathematical theory of wave propagation in media with a memory* by A. A. Lokshin and Yu. V. Suvorov (Moscow State University Press, Moscow, 1982).

The general equation of motion of a homogeneous hereditary-elastic rod of unit density performing longitudinal oscillations under the action of an external load $f(t,x)$ has the form

$$Vu = f, \quad \text{where } V = \frac{\partial^2}{\partial t^2} - \frac{\partial^2}{\partial x^2} + R(t) * \frac{\partial^2}{\partial x^2}, \tag{D}$$

and where t is the time, x is the coordinate of a point of the rod, and $R(t)$ is the so-called relaxation kernel, which is connected with another characteristic of the rod—the creep kernel $K(t)$—by the relation

$$R(t) = K(t) - K(t) * K(t) + \cdots.$$

In the case when the creep kernel has the form $K = \varphi + \frac{1}{4}\varphi * \varphi$ on the interval $0 < t < t_0$, where $\varphi(t) = ht^{-\alpha}E(t)/\Gamma(1-\alpha)$, $h > 0$, $0 < \alpha < 1$, and $E(t)$ is the unit distribution, the fundamental solution $\mathcal{E}(t,x)$ of (D) is nonzero in the angle $t \geq |x|$ and can be written with the help of the functions $G(x,\alpha,1)$ in the region $|x| \leq t < |x| + t_0$:

$$\mathcal{E}(t,x) = \frac{1}{2}G\left((t-|x|)\Big/\left(\frac{1}{2}h|x|\right)^{1/\alpha}, \alpha, 1\right)$$
$$+ \frac{h}{4}\int_{|x|}^{t} \frac{(t-y)^{-\alpha}}{\Gamma(1-\alpha)} G\left((y-|x|)\Big/\left(\frac{1}{2}h|x|\right)^{1/\alpha}, \alpha, 1\right) dy.$$

Stable laws appear in other problems in the theory of hereditary elasticity in connection with the fact that the creep kernel $K(t) = \mathfrak{Э}_{-a}(-b,t)$ of Rabotnov, which is popular in this theory, is related to the Mittag-Leffler function $E_\sigma(x)$:

$$\mathfrak{Э}_a(b,t) = (1+a)t^a E'_{1+a}(bt^{1+a}).$$

Very recently I became acquainted with the paper "Stable measures and processes in statistical physics" by A. Weron and K. Weron, and with the survey paper "Stable processes and measures: A survey" by A. Weron, both published as preprints in the Center for Stochastic Processes at the University of North Carolina.* These papers contain a number of interesting examples of the appearance of stable processes in problems in contemporary theoretical physics.

Chapter 2

2.1. The reference to material in Chapter 3 is given only for the purpose of a general orientation in evaluating the possibilities of using Lemma 2.1.1, presented below.

* *Translator's note.* The second of these papers has since been published in Probability Theory on Vector Spaces. III (Proceedings, Lublin, 1983), Lecture Notes in Math., vol. 1080, Springer-Verlag, 1984, pp. 306–364.

2.2. There are several cases in which the densities of stable laws can be expressed by means of special functions. Corresponding formulas are given in §2.8.

2.3. The idea of using the representations (2.2.8)–(2.2.10) in analyzing the properties of stable distributions stems from Linnik [53], who considered formally only the case $\alpha < 1$. The equality (2.2.11) was given in the author's paper [97].

2.4. The representation (2.2.13) was obtained by Skorokhod [79].

2.5. Theorem 2.2.3 was proved in a somewhat modified form in the author's paper [98].

2.6. The fact that the stable distributions with $\alpha < 1, \beta = 1$, and $\gamma = 0$ are entirely concentrated on the semi-axis $x > 0$ was observed independently by several authors: Bergström [2], Linnik [53], and Ovseevich and Yaglom [132]. However, Lévy ((113) in [50]) was the first to direct attention to it.

2.7. Theorem 2.3.1 (more precisely, the equality (2.3.3) corresponding to the form (A)) is the main result in the author's article [95].

2.8. In [78] and [80] Skorokhod proved an assertion equivalent to the assertion of Theorem 2.4.1 in the part corresponding to the cases $\alpha < 1$ and $\alpha = 1$. The fact that the density $g(x, \alpha, \beta)$ is an entire analytic function when $\alpha > 1$ was mentioned in the book [26] of Gnedenko and Kolmogorov, with a reference to A. I. Lapin.

2.9. The representations (2.4.6) and (2.4.8) were found independently by Bergström [2], Feller [21], and Chao Chung-Jeh [9]. These series expansions for the densities $g(x, \alpha, 0)$ were also obtained by Wintner [89]. For the case $\beta = 1$ the equality (2.4.8) is mentioned in a paper of Pollard [67]. In the cited paper Bergström refers to a paper of Humbert in 1945, where the formal expansion (2.4.8) is given for the function $g(x, \alpha, 1)$ with $\alpha < 1$.

2.10. After the power series expansions of the entire functions $q(x, \alpha, \beta)$ with $\alpha \neq 1$ were found, computation of their order σ and type δ was not difficult, so the values of σ and δ were known to anyone interested in the properties of stable distributions. Theorem 2.4.3 first appeared in the book [54] of Lukacs in a somewhat modified formulation. However, the proof there of the fact that $\sigma = \infty$ when $\alpha = 1$ and $\beta \neq 0$ is erroneous. Our version of the proof was suggested by Ostrovskiĭ, who also pointed out the precise equality

$$\limsup_{r \to \infty} r^{-1} \log\log M(r) = 1/\beta.$$

2.11. N. V. Smirnov knew of the possibility of expanding the function $q(x, \alpha, 1)$, $\alpha < 1$, in Laguerre polynomials. I had occasion to hear about this at a seminar of Kolmogorov (who was working at Moscow State University during 1954–1955).

2.12. Kolmogorov was the first to point out the asymptotic relation (2.5.18), without a rigorous proof, at the seminar mentioned in Comment 2.11. The first terms in the asymptotic expansion (2.5.17) were obtained in the case $\alpha < 1$ by Linnik [53] and in the case $\alpha \geq 1$ by Skorokhod [79]. The asymptotic expansion (2.5.17) appeared in full scope, in a somewhat modified form, in the book [35] of Ibragimov and Linnik. It should be mentioned that the formulas presented there contain misprints.

2.13. The asymptotic expansion (2.5.25) was found by Skorokhod [79] (without explicit expressions for the coefficients d_n).

2.14. The equality (2.6.10) was obtained in the case $\alpha < 1$ by Pollard [67], and in the case $\alpha \geq 1$ by the author [97].

2.15. The equality (2.6.11) was proved by the author in [97].

2.16. The relation (2.6.15) is equivalent to (3.3.10).

2.17. Theorem 2.6.3 was proved by the author [97].

2.18. The equality (2.6.26) was proved by the author in [97].

2.19. The class L is a subset of $\mathfrak{G}$. It can be defined as the set of all possible complete limits* of distributions of increasing sums of independent random variables $(X_1 + \cdots + X_n)B_n^{-1} - A_n$ with the property that

$$\max(X_1, \ldots, X_n)B_n^{-1} \xrightarrow{d} 0 \quad \text{as } n \to \infty,$$

or as the set of infinitely divisible distributions with absolutely continuous spectral functions H such that the product $xH'(x)$ is nonincreasing on the semi-axes $x < 0$ and $x > 0$.

2.20. Theorem 2.7.2 is due to Khintchine [44].

2.21. Theorem 2.7.3 is a probabilistic interpretation of a known unimodality criterion (given in Corollary 2) of Khintchine. An equivalent statement in terms of the characteristic transforms (2.7.5) was presented by the author in [101]. It is mentioned in Feller's book [22] that Shepp also pointed out this interpretation.

---

*We have in mind "complete" convergence.

2.22. Lemma 2.7.2 was used by Wintner to prove Theorem 2.7.4 in [86]. We present a new proof of this lemma.

2.23. Theorem 2.7.5 was proved by Ibragimov and Chernin [34].

2.24. The system of operators used by us, which can be interpreted as fractional integration and fractional differentiation, is not unique. For example, we can use a system of operators of the form $(0 < r < 1)$

$$I^{-r}h(x) = \frac{1}{\Gamma(r)} \int_{-\infty}^{x} (x-t)^{r-1} h(t)\,dt,$$

$$I^{r}h(x) = \frac{r}{\Gamma(1-r)} \int_{-\infty}^{x} [h(x) - h(t)](x-t)^{-r-1}\,dt.$$

There are also other types of operators with analogous properties. For further information we refer the reader to Feller's paper [21]. Wolfe's paper [92] is one of the few articles which uses fractional differentiation and integration operators in probability theory problems.

2.25. Related in structure to (2.8.12) is an integrodifferential equation in the book [35] of Ibragimov and Linnik (Theorem 2.3.2) for the densities of stable laws; however, the equation contains an essential error. The error arises because the complex factor $\exp(-i\pi r)$ is absent in the definition of the fractional integration operator (which is analogous to that used by us). As a result, the operator inverse to the one defined by them cannot be regarded as a generalization of the concept of differentiation.

In the same place it is asserted that an integrodifferential equation of the form (2.8.12) can be transformed into the differential equation (2.8.25) in the case when α is rational. Actually, this equation is obtained as a consequence of an integrodifferential equation of the form (2.8.20).

2.26. In considering the densities $g(x, \alpha, \beta, 0, \lambda)$ of unbiased strictly stable laws, Medgyessy [62] derived for them a partial differential equation in the case when $\alpha = m/n$, a rational number with relatively prime m and n:

$$\sum_{j=1}^{3} K_j \frac{\partial^{a_j+b_j}}{\partial x^{a_j} \partial \lambda^{b_j}} g = 0,$$

where a_j, b_j, and K_j are constant numbers dependent on m, n, β, and a certain free parameter whose variation gives a family of equations. If we take account of the equality

$$g = \lambda^{-1/\alpha} g(x\lambda^{-1/\alpha}, \alpha, \beta, 0, 1),$$

then the connection between this equation and (2.8.24) becomes obvious.

2.27. The equation (2.8.26) is contained in Linnik's paper [53].

2.28. It is not hard to transform the right-hand side of (2.8.29) to the form

$$\frac{1}{\pi}\mathrm{Re}\{\sqrt{\pi}\varsigma\exp(-\varsigma^2)-2i\varsigma w(\varsigma)\},$$

where $\varsigma=-iz/2$ and $w(\varsigma)=\exp(-\varsigma^2)\int_0^\varsigma\exp(t^2)\,dt$. The function $w(\varsigma)$ is tabulated for complex ς in [42].

2.29. The equation (2.8.33) was established by Pollard [67].

2.30. Theorem 2.9.1 was proved by the author [97].

The paper [13] of Cressie studies the properties of the distribution

$$R_\alpha(x)=\mathsf{P}(|Y(\alpha,\beta,0,\lambda)^\alpha<x),\qquad \alpha<1,$$

and also solves problems such as the representation of the density $R'_\alpha(x)$ by a series in inverse powers of x, and the limit behavior of $R_\alpha(x)$ as $\alpha\to 0$ (we use our own notation here). It is easily seen that

$$R_\alpha(x)=G(x^{1/\alpha},\alpha,\beta,0,\lambda)-G(x^{1/\alpha},\alpha,-\beta,0,\lambda);$$

hence both of these problems get an obvious solution if one uses the elementary property $Y(\alpha,\beta,0,\lambda)=\lambda^{1/\alpha}Y(\alpha,0)$ and invokes (2.4.8) and the limit relation (2.9.1).

2.31. The asymptotic expansion (2.9.4) was obtained by the author in [97]. It is proved in the paper [7] of Brockwell and Brown that the asymptotic expansion (2.9.4) is a convergent series for all $\alpha<1$ in the case when $\theta=1$ and $x>0$. The structure of the asymptotic expansion (2.9.4) shows that its convergence for any admissible $\alpha<1$ and θ is a consequence of the result mentioned.

2.32. The asymptotic expansion (2.9.8) was presented in the author's paper [97]. However, the expression given there for the function H_1 is erroneous.

2.33. Theorem 2.10.1 is a result of the author [96].

2.34. The connections (2.10.9) and (2.10.12) between the Mittag-Leffler functions and stable laws were pointed out by the author in [96] and [97].

2.35. In some of his papers Studnev considered a generalization of the concept of infinitely divisible distributions within the limits of the set of functions $V(x)$ of bounded variation that satisfy the conditions $V(-\infty)=0$ and $V(\infty)=1$. The 1967 paper *On some generalizations of limit theorems in probability theory* (Teor. Veroyatnost. i Primenen. **12** (1967), 729–734=Theor. Probab. Appl. **12** (1967), 668–672) contains a remark about the class of "generalized stable laws" and gives a form for the corresponding Fourier-Stieltjes transforms. Although this class contains the Airy function, the question of whether it is a natural extension of the family of stable distributions has not been considered so far. For a criterion for "naturalness" we can take, in particular, the duality law, which is satisfied by the Airy function, as (2.10.15) shows.

Chapter 3

3.1. With the exception of Theorem 3.4.3 and the information in §3.5, the material in this chapter is based on results of the author in [97] (mainly the second part of it). There are several publications of other authors whose results are directly or indirectly connected with the facts presented below. These cases will be mentioned and commented on separately.

We note one important detail.

In the case when the random variable X takes both positive and negative values with positive probability, the use of the proposed notation for the generalized power can give rise to an ambiguous situation, since, for example, X^2 can be understood as $|X|^2$ or as $|X|^2 \operatorname{sgn}(X)$. Therefore, it would be more correct to give the generalized power its own notation, say $X^{(s)} = |X|^s \operatorname{sgn} X$. We did not do this, in order to achieve a certain simplification in the formulas. This will not cause any confusion in this chapter if we agree always to understand a power of a normally distributed random variable N in the usual sense, i.e., $N^{2k} = |N|^{2k}$, when k is an integer.

3.2. The statistical interpretation of the concept of a cutoff is as follows. If $X_1, \ldots, X_n$ is an independent sample subordinate to the same distribution as the random variable X, then by discarding among the X_j only those are positive, we obtain a sample $X_{i_1}, \ldots, X_{i_T}$ (of random size T subordinate to the binomial distribution) in which the elements have the same distribution as a cutoff $\hat{X}$.

3.3. The fact that the random variable $1/X$ has a Cauchy distribution if X itself does has apparently been known for a long time, but the author has not been able to find any references prior to his paper [97]. Menon [63] obtained the following interesting characterization of distributions of Cauchy type. Suppose that the random variables X and $1/X$ are stable, and $1/X \overset{d}{=} h(X)$, where $h(x) = Ax + O(1)$, A is a constant, and $h'(x) = A + O(|x|^{-\varepsilon})$, $\varepsilon > 0$, as $|x| \to \infty$. Then X has a distribution $G_A(x, 1, 0, \gamma, \lambda)$.

3.4. The following interesting equality can be derived from (3.3.4) if we choose $\alpha = \frac{1}{2}$ and use the explicit expression for the density $g(x, \frac{1}{2}, 1)$:

$$\int_0^\infty e^{-su} g(\sqrt{u}, \alpha', 1)\, du = \frac{\sqrt{\pi}}{2} s^{-3/2} g\left(\frac{1}{4s}, \frac{\alpha'}{2}, 1\right).$$

The equality (3.4.26) is a particular case of it (if we choose $\alpha' = \frac{1}{2}$).

3.5. The simplest case of the distribution of the statistic L is obtained when $\alpha = 2$. This follows from (3.3.21) and (3.3.17):

$$L \stackrel{d}{=} \frac{Z(2,1/2)}{Z(2,1/2)} U \stackrel{d}{=} Z(1,\tfrac{1}{2}) U \stackrel{d}{=} Y(1,0),$$

i.e, the statistic L, which in this case represents the ratio N/N of two independent random variables distributed according to the standard normal law, has a Cauchy distribution, a fact well known in probability theory. The following relation (a consequence of (3.3.17)) is analytically related to it:

$$\frac{Z(\alpha,1/\alpha)}{Z(\alpha,1/\alpha)} \stackrel{d}{=} Z\left(1, \frac{1}{\alpha}\right), \qquad 1 < \alpha \leq 2.$$

3.6. Theorem 3.4.3 and its corollary are due to Williams [85]. In his proof the form of the Mellin transform $M(-s, 1/n, 1)$ is found independently with the help of a clever argument which we regard as useful to present. If $\alpha = 1/n$, then for any $s > 0$

$$\exp(-s^\alpha) = \mathsf{E}\exp(-sY(\alpha)) = \mathsf{P}(sY(\alpha) < E),$$

where E is a random variable independent of $Y(\alpha)$ and having an exponential distribution. Since

$$\mathsf{P}((E/Y(\alpha))^\alpha > s^\alpha) = \exp(-s^\alpha),$$

it follows that

$$(E/Y(\alpha))^\alpha \stackrel{d}{=} E \quad \text{and} \quad Y(\alpha)/E \stackrel{d}{=} E^{-1/\alpha}.$$

Consequently, for any $r > -\alpha$

$$\mathsf{E}(Y^{-r}(\alpha)E^r) = \mathsf{E}(E^{r/\alpha}) = \Gamma(1 + r/\alpha)$$

and, moreover, since $Y(\alpha)$ and E are independent,

$$\mathsf{E}(Y^{-r}(\alpha)E^r) = \mathsf{E}Y^r(\alpha)\mathsf{E}(E^r) = \mathsf{E}Y^{-r}(\alpha)\Gamma(1+r).$$

A comparison of the above equalities shows that

$$M(-r, \alpha, 1) = \mathsf{E}Y^{-r}(\alpha) = \Gamma(1 + r/\alpha)/\Gamma(1+r).$$

Chapter 4

4.1. The material in this chapter is for the most part taken from the author's paper [167]. Theorem 4.3 exploits a technique of Januškevičienė [166], where, in particular, an analogous problem is solved.

Bibliography

I. References of a general-theoretic nature

1. Harold Bergström, *On the theory of the stable distribution functions*, Proc. Twelfth Scandinavian Math. Congr. (Lund, 1953, Håkan Ohlssons, Lund, 1954, pp. 12–13.

2. ——, *On some expansions of stable distribution functions*, Ark. Mat. **2** (1952/53), 375–378.

3. ——, *Eine Theorie der stabilen Verteilungsfunktionen*, Arch. Math. (Basel) **4** (1953), 380–391.

4. Ludwig Bieberbach, *Lehrbuch der Funktionentheorie*. Vol. 2, 2nd ed., Teubner, Leipzig, 1931; reprint, Johnson Reprint Corp., New York, 1968.

5. S. Bochner, *Stable laws of probability and completely monotone functions*, Duke Math. J. **3** (1937), 726–728.

6. L. N. Bol′shev et al., *Tables of stable one-sided distributions*, Teor. Veroyatnost. i Primenen. **15** (1970), 309–319; English transl. in Theor. Probab. Appl. 15 (1970).

7. P. J. Brockwell and B. M. Brown, *Expansions for the positive stable laws*, Z. Wahrsch. Verw. Gebiete **45** (1978), 213–224.

8. Augustin Cauchy, *Sur les résultats moyens d'observations de même nature, et sur les résultats les plus probables*, and *Sur la probabilité des erreurs qui affectant des résultats moyens d'observations de même nature*, C. R. Acad. Sci. Paris **37** (1853), 198–206, 264–272; reprinted in his *Oeuvres completes*, Ser. 1, Vol. 12, Gauthier-Villars, Paris, 1900, pp. 94–104, 104–114.

9. Chao Chung-Jeh, *Explicit formula for the stable law of distribution*, Acta Mat. Sinica **3** (1953), 177–185. (Chinese; English summary)

10. K. L. Chung, *Sur les lois de probabilité unimodales*, C. R. Acad. Sci. Paris **236** (1953), 583–584.

11. Harold Cramér, *On the approximation to a stable probability distribution*, Studies in Math. Anal. and Related Topics (Essays in Honor of George Pólya) Stanford Univ. Press, Stanford, Calif., 1962, pp. 70–76.

12. ——, *Mathematical methods of statistics*, Princeton Univ. Press, Princeton, N. J., 1946.

13. Noel Cressie, *A note on the behaviour of the stable distributions for small index* α, Z. Wahrsch. Verw. Gebiete **33** (1975/76), 61–64.

14. D. A. Darling, *The maximum of sums of stable random variables*, Trans. Amer. Math. Soc. **83** (1956), 164–169.

15. V. A. Ditkin and P. I. Kuznetsov, *Handbook of operational calculus*, GITTL, Moscow, 1951. (Russian)

16. R. L. Dobrushin and Yu. M. Sukhov, *Time asymptotics for some degenerate models of the evolution of systems with infinitely many particles*, Itogi Nauki: Sovremennye Problemy Mat., vol. 14, VINITI, Moscow, 1979, pp. 147–254; English transl. in J. Soviet Math. **16** (1981), no. 4.

17. Daniel Dugué, *Variables scalaires attachées à deux matrices de Wilks. Comparaison de deux matrices de Wilks en analyse des données*, C. R. Acad. Sci. Paris Sér. A-B **284** (1977), A899–A901.

18. K. O. Dzhaparidze, *On simplified estimators of unknown parameters with good asymptotic properties*, Teor. Veroyatnost. i Primenen. **19** (1974), 355–366; English transl. in Theor. Probab. Appl. **19** (1974).

19. A. Erdélyi et al., *Tables of integral transforms.* Vol. 1, McGraw-Hill, 1954.

20. M. A. Evgrafov, *Asymptotic estimates and entire functions*, GITTL, Moscow, 1957; English transl., Gordon and Breach, New York, 1961.

21. William Feller, *On a generalization of Marcel Riesz' potentials and the semi-groups generated by them*, Comm. Sém. Math. Univ. Lund=Medd., Lunds Univ. Mat., Sem., Tome Supplémentaire Dédié à Marcel Riesz, Gauthier-Villars, Paris, 1952, pp. 74–81.

22. ——, *An introduction to probability theory and its applications*, Vols. 1 (2nd ed.), 2, Wiley, 1957, 1966.

23. S. G. Ghurye, *A remark on stable laws*, Skand. Aktuarietidskr. **1958** 68–70 (1959).

24. V. L. Girko, *Theory of random determinants*, "Vishcha Shkola" (Izdat. Kiev. Univ.), Kiev, 1980. (Russian)

25. B. V. Gnedenko, *The theory of limit theorems for sums of independent random variables*, Izv. Akad. Nauk SSSR Ser. Mat. **3** (1939), 181–232. (Russian; English summary)

26. B. V. Gnedenko and A. N. Kolmogorov, *Limit distributions for sums of independent random variables*, GITTL, Moscow, 1949; English transl., Addison-Wesley, 1954.

27. I. S. Gradshteĭn and I. M. Ryzhik, *Tables of integrals, series, and products*, 4th rev. ed., Fizmatgiz, Moscow, 1963; English transl., Academic Press, 1965.

28. Beniamino Gulotta, *Leggi di probabilità condizionatamente stabili*, Giorn. Ist. Ital. Attuari **19** (1956), 22–30.

29. Peter Hall, *A comedy of errors: the canonical form for a stable characteristic function*, Bull. London Math. Soc. **13** (1981), 23–27.

30. G. H. Hardy, J. E. Littlewood and G. Pólya *Inequalities*, Cambridge Univ. Press, 1934.

31. Donald R. Holt and Edwin L. Crow, *Tables and graphs of the stable probability density functions*, J. Res. Nat. Bur. Standards Sect. B **77** (1973), 143–198.

32. Pierre Humbert, *Quelques résultats relatifs à la fonction de Mittag-Leffler*, C. R. Acad. Sci. Paris **236** (1953), 1467–1468.

33. I. A. Ibragimov, *On the composition of unimodal distributions*, Teor. Veroyatnost. i Primenen. **1** (1956), 283–288; English transl. in Theor. Probab. Appl. **1** (1956).

34. I. A. Ibragimov and K. E. Chernin, *On the unimodality of stable laws*, Teor. Veroyatnost. i Primenen. **4** (1959), 453–456; English transl. in Theor. Probab. Appl. **4** (1959).

35. I. A. Ibragimov and Yu. V. Linnik, *Independent and stationary sequences of random variables*, "Nauka", Moscow, 1965; English transl., Noordhooff, 1971.

36. Bernard Jesiak, *An uniqueness theorem for stable laws*, Math. Nachr. **92** (1979), 243–246.

37. H.-J. Rossberg and B. Jesiak, *On the unique determination of stable distribution functions*, Math. Nachr. **82** (1978), 297–308.

38. A. M. Kagan, Yu. V. Linnik and S. Radhakrishna Rao, *Characterization problems in mathematical statistics*, "Nauka", Moscow, 1972; English transl., Wiley, 1973.

39. E. Kamke, *Differentialgleichungen. Lösungsmethoden und Lösungen. Teil 1: Gewöhnliche Differentialgleichungen*, 6th ed., Akademische Verlag, Geest & Portig, Leipzig, 1959.

40. Marek Kanter, *Stable densities under change of scale and total variation inequalities*, Ann. Probab. **3** (1975), 697–707.

41. ———, *On the unimodality of stable densities*, Ann. Probab. **4** (1976), 1006–1008.

42. K. A. Karpov, *Tables of the function* $w(z) = e^{-z^2} \int_0^z e^{x^2}\,dx$ *in the complex domain*, Izdat. Akad. Nauk SSSR, Moscow, 1954; English transl., Pergamon Press, Oxford, and Macmillan, New York, 1965.

43. Maurice G. Kendall and Alan Stuart, *The advanced theory of statistics. Vol. 2: Inference and relationship*, 2nd ed., Hafner, New York, 1967.

44. A. Ya. Khinchin [Khintchine], *Limit laws for sums of independent random variables*, ONTI, Moscow, 1938. (Russian)

45. A. Khintchine and Paul Lévy, *Sur les lois stables*, C. R. Acad. Sci. Paris **202** (1937), 374–376.

46. A. N. Kolmogorov and B. A. Sevast′yanov, *Computation of final probabilities for random branching processes*, Dokl. Akad. Nauk SSSR **56** (1947), 783–786. (Russian)

47. V. M. Kruglov, *A remark on the theory of infinitely divisible laws*, Teor. Veroyatnost. i Primenen. **15** (1970), 330–336; English transl. in Theor. Probabl. Appl. **15** (1970).

48. R. G. Laha, *On a characterization of the stable law with finite expectation*, Ann. Math. Statist. **27** (1956), 187–195.

49. ——, *On the laws of Cauchy and Gauss*, Ann. Math. Statist. **30** (1959), 1165–1174.

50. Paul Lévy, *Calcul des probabilités*, Gauthier-Villars, Paris, 1925.

51. ——, *Théorie de l'addition des variables aléatoires*, 2nd ed., Gauthier-Villars, 1954.

52. ——, *Remarques sur un problème relatif aux lois stables*, Studies in Math. Anal. and Related Topics (Essays in Honor of George Pólya), Stanford Univ. Press, Stanford, Calif., 1962, pp. 211–218.

53. Yu. V. Linnik, *On stable probability laws with exponent less than one*, Dokl. Akad. Nauk SSSR **94** (1954), 619–621. (Russian)

54. Eugene Lukacs, *Characteristic functions*, 2nd ed., Hafner, New York, 1970.

55. ——, *On some properties of symmetric stable distributions* Analytic Function Methods in Probability Theory (Proc. Colloq., Debrecen, 1977; B. Gyires, editor), Colloq. Math. Soc. János Bolyai, vol. 21, North-Holland, 1979, pp. 227–241.

56. ——, *Stable distributions and their characteristic functions*, Janresber. Deutsch. Math. Verein **71** (1969), no. 2, 84–114.

57. ——, *Some properties of stable frequency functions*, Bull. Inst. Internat. Statist.=Bull. Internat. Statist. Inst. **42** (1969), 1213–1224.

58. ——, *Sur quelqués propriétés des lois stables et symétriques*, C. R. Acad. Sci. Paris Ser. A-B **286** (1978), A1213–A1214.

59. A. Marchand, *Sur les dérivées et sur les différences des fonctions de variables réelles*, J. Math. Pures Appl. (9) **6** (1927), 337–425.

60. P′al Medgyessy, *Partial integro-differential equations for stable density functions and their applications*, Publ. Math. Debrecen **5** (1958), 288–293.

61. ——, *Partial differential equations for stable density functions, and their applications*, Magyar Tud. Akad. Mat. Kutató Int. Közl. **1** (1956), 489–514. (Hungarian)

62. ——, *Partial differential equations for stable density functions, and their applications*, Magyar Tud. Akad. Mat. Kutató Int. Közl. **1** (1956), 516–518. (English summary of [61])

63. M. V. Menon, *A characterization of the Cauchy distribution*, Ann. Math. Statist. **33** (1962), 1267–1271.

64. K. R. Parthasarathy, R. Ranga Rao and S. R. S. Varadhan, *Probability distributions on locally compact abelian groups*, Illinois J. Math. **7** (1963), 337–369.

65. V. V. Petrov, *Sums of independent random variables*, "Nauka", Moscow, 1972; English transl., Springer-Verlag, 1975.

66. Harry Pollard, *The completely monotonic character of the Mittag-Leffler function* $E_a(-x)$, Bull. Amer. Math. Soc. **54** (1948), 1115–1116.

67. ——, *The representation of* e^{-x^λ} *as a Laplace integral*, Bull. Amer. Math. Soc. **52** (1946), 908–910.

68. Georg Pólya, *Herleitung des Guassschen Fehlergesetzes aus einer Funktionalgleichung*, Math. Z. **18** (1923), 96–108.

69. B. Ramachandran, *On characteristic functions and moments*, Sankhyā Ser. A **31** (1969), 1–12.

70. John Riordan, *An introduction to combinatorial analysis*, Wiley, New York, and Chapman & Hall, London, 1958.

71. Ken-iti Sato and Makoto Yamazato, *On distribution functions of class,* L, Z. Wahrsch. Verw. Gebiete **43** (1978), 273–308.

72. L. J. Savage, *A geometrical approach to the special stable distributions*, Zastos. Mat. **10** (1969), 43–46.

73. Yoichi R. Shimizu, *On the decomposition of stable characteristic functions*, Ann. Inst. Statist. Math. **24** (1972), 347–353.

74. B. A. Sevast′yanov, *Final probabilities for branching stochastic processes*, Teor. Veroyatnost. i Primenen. **2** (1957), 140–141; English transl. in Theor. Probab. Appl. **2** (1957).

75. D. N. Shanbhag, D. Pestana and M. Sreehari, *Some further results in infinite divisibility*, Math. Proc. Cambridge Philos. Soc. **82** (1977), 289–295.

76. I. S. Shiganov, *A metric approach to the investigation of stability of Pólya's theorem on characterizing the normal distribution*, Problems of Stability of Stochastic Models (Proc. Fifth All-Union Sem., Panevezhis, 1980; V. M. Zolotarev and V. V. Kalashnikov, editors), Vsesoyuz. Nauchno-Issled. Inst. Sistem. Issled., Moscow, 1981, pp. 145–154. (Russian)

77. A. V. Skorokhod, *Random processes with independent increments*, "Nauka", Moscow, 1964; English transls., Clearing House Federal Sci. Tech. Information, Springfield, Va., 1966, and *Theory of random processes*, Nat. Lending Library Sci. Tech., Boston Spa, Yorkshire, England, 1971.

78. ——, *On a theorem concerning stable distributions*, Uspekhi Mat. Nauk **9** (1954), no. 2 (60), 189–190; English transl. in Selected Transl. Math. Statist. and Probab., vol. 1, Amer. Math. Soc., Providence, R. I., 1961.

79. ——, *Asymptotic formulas for stable distribution laws*, Dokl. Akad. Nauk SSSR **98** (1954), 731–734; English transl. in Selected Transl. Math. Statist. and Probab., vol. 1, Amer. Math. Soc., Providence, R. I., 1961.

80. ——, *Analytic properties of stable probability distributions*, in: Student Scientific Papers of Kiev University, vyp. 16, Izdat. Kiev. Univ., Kiev, 1955, pp. 159–164. (Russian) R. Zh. Mat. **1957** #1623.

81. A. N. Tikhonov and A. A. Samarskiĭ, *The equations of mathematical physics*, GITTL, Moscow, 1951; English transls. of 2nd ed., Pergamon Press, Oxford, and Macmillan, New York, 1963, and Vols. I, II, Holden-Day, San Francisco, Calif., 1964, 1967.

82. E. C. Titchmarsh, *Introduction to the theory of Fourier integrals*, Clarendon Press, Oxford, 1937.

83. ——, *The theory of functions*, 2nd ed., Oxford Univ. Press, 1939.

84. B. L. van der Waerden, *Mathematische Statistik*, Springer-Verlag, 1957; English ed., 1969.

85. E. J. Williams, *Some representations of stable random variables as products*, Biometrika **64** (1977), 167–169.

86. Aurel Wintner, *On the stable distribution laws*, Amer. J. Math. **55** (1933), 335–339.

87. ——, *On a class of Fourier transforms*, Amer. J. Math. **58** (1936), 45–90.

88. ——, *The singularities of Cauchy's distributions*, Duke Math. J. **8** (1941), 678–681.

89. ——, *Cauchy's stable distributions and an "explicit formula" of Mellin*, Amer. J. Math. **78** (1956), 819–861.

90. ——, *Stable distributions and Laplace transforms*, Ann. Scuola Norm. Sup. Pisa Sci. Fis. Mat. (3) **10** (1956), 127–134.

91. ____, *Stable distributions and the transforms of Stieltjes and Le Roy*, Boll. Un. Mat. Ital. (3) **13** (1958), 24–33.

92. Stephen J. Wolfe, *On moments of probability distribution functions*, Functional Calculus and Its Applications (Proc. Internat. Conf., West Haven, Conn., 1974), Lecture Notes in Math., Vol. 457, Springer-Verlag, 1975, pp. 306–316.

93. Graham J. Worsdale, *Tables of cumulative distribution functions for symmetric stable distributions*, J. Roy Statist. Soc. Ser. C: Appl. Statist. **24** (1975), 123–131.

94. Makoto Yamazato, *Unimodality of infinitely divisible distribution functions of class L*, Ann. Probab. **6** (1978), 523–531.

95. V. M. Zolotarev, *Expression of the density of a stable distribution with exponent α greater than one in terms of a density with exponent* $1/\alpha$, Dokl. Akad. Nauk SSSR **98** (1954), 735–738; English transl. in Selected Transl. Math. Statist. and Probab., vol. 1, Amer. Math. Soc., Providence, R. I., 1961.

96. ____, *On analytic properties of stable distribution laws*, Vestnik Leningrad. Univ. **1956**, no. 1 (Ser. Mat. Mekh. Astr. vyp. 1), 49–52; English transl. in Selected Transl. Math. Statist. and Probab., vol. 1, Amer. Math. Soc., Providence, R. I., 1961.

97. ____, *The Mellin-Stieltjes transformation in probability theory*, Teor. Veroyatnost. i Primenen. **2** (1957), 444–469; English transl. in Theor. Probab. Appl. **2** (1957).

98. ____, *On representation of stable laws by integrals*, Trudy. Mat. Inst. Steklov. **71** (1964), 46–50; English transl. in Selected Transl. Math. Statist. and Probab, vol. 6, Amer. Math. Soc., Providence, R. I., 1966.

99. ____, *On the M-divisibility of stable laws*, Teor. Veroyatnost. i Primenen. **12** (1967), 559–562; English transl. in Teor. Probab. Appl. **12** (1967).

100. ____, *The analytic structure of infinitely divisible laws of class L*, Litovsk. Mat. Sb. **3** (1963), 123–140; English transl. in Selected Transl. Math. Statist. and Probab., vol. 15, Amer. Math. Soc., Providence, R. I., 1981.

101. ____, *On a general theory of multiplication of independent random variables*, Dokl. Akad. Nauk SSSR **142** (1962), 788–791; English transl. in Soviet Math. Dokl. **3** (1962).

II. References of an applied nature

102. B. Berndtsson and Peter Jagers, *Exponential growth of a branching process usually implies stable age distribution*, J. Appl. Probab. **16** (1979), 651–656.

103. S. Chandrasekhar, *Stochastic problems in physics and astronomy*, Rev. Modern Phys. **15** (1943), 1–89.

104. A. S. Davydov, *Quantum mechanics*, Fizmatgiz, Moscow, 1963; English transl. of 2nd ed., Pergamon Press, 1976.

105. R. L. Dobrushin, *A statistical problem, leading to stable laws, in the theory of signal detection on a background of noise in a multichannel system*, Teor. Veroyatnost. i Primenen. **3** (1958), 173–185; English transl. in Theory Probab. Appl. **3** (1958).

106. William H. DuMouchel and Richard A. Olshen, *On the distributions of claim costs*, Credibility: Theory and Applications (Proc. Berkeley Actuarial Res. Conf., 1974, dedicated to E. A. Lew), Academic Press, 1975, pp. 23–50, 409–414.

107. Katherine Dusak, *Futures trading and investo returns: an investigation of commodity market risk premiums*, J. Political Economy **81** (1973), 1387–1406.

108. J. D. Eshelby, *The continuum theory of dislocations*, IL, Moscow, 1963. (Russian)*

109. Eugene F. Fama, *Mandelbrot and the stable Paretian hypothesis*, J. Business **36** (1963), 420–429.

110. ——, *The behavior of stock-market prices*, J. Business **38** (1965), 34–105.

111. B. D. Fielitz and E. W. Smith, *Asymmetric stable distributions of stock price changes*, J. Amer. Statist. Assoc. **67** (1972), 813–814.

112. I. J. Good, *The real stable characteristic functions and chaotic acceleration*, J. Roy, Statist. Soc. Ser. B **23** (1961), 180–183.

113. Clive W. J. Granger and Daniel Orr, "*Infinite variance*" *and research strategy in time series analysis*, J. Amer. Statist. Assoc. **67** (1972), 275–285.

114. J. Holtsmark, *Über die Verbreiterung von Spektrallinien*, Ann. Physik (4) **58** (**363**) (1919), 577–630.

115. L. A. Khalfin, *Contribution to the decay theory of a quasistationary state*, Zh. Èksper. Teoret. Fiz. **33** (1957), 1371–1382; English transl. in Soviet Phys. JETP **6** (1958).

116. N. S. Krylov and V. A. Fok, *On the two main interpretations of the uncertainty relation for energy and time*, Zh. Èksper. Teoret. Fiz. **17** (1947), 93–107. (Russian)

117. I. M. Lifshits, *On temperature flashes in a medium under the action of nuclear radiation*, Dokl. Akad. Nauk SSSR **109** (1956), 1109–1111. (Russian)

**Editor's note.* This book is a collection of Russian translations of articles originally published in English.

118. Benoit Mandelbrot, *La distribution de Willis-Yule, relative aux nombres d'espèces dans les genres biologique*, C. R. Acad. Sci. Paris **242** (1956), 2223–2226.

119. ——, *Variables et processus stochastiques de Pareto-Lévy, et la répartition des revenus*, C. R. Acad. Sci. Paris **249** (1959), 613–615, 2153–2155.

120. ——, *The Pareto-Lévy law and the distribution of income*, Internat. Econ. Rev. **1** (1960), 79–106.

121. ——, *Stable Paretian random functions and the multiplicative variation of income*, Econometrica **29** (1961), 517-543.

122. ——, *The stable Paretian income distribution when the apparent exponent is near two*, Internat. Econ. Rev. **4** (1963), 111–115.

123. ——, *New methods in statistical economics*, J. Polit. Economy **71** (1963), 421–440.

124. ——, *The variation of certain speculative prices*, J. Business **36** (1963), 394–419.

125. ——, *Some noises with $1/f$ spectrum, a bridge between direct current and white noise*, IEEE Trans. Information Theory **IT-13** (1967), 289–298.

126. ——, *The variation of some other speculative prices*, J. Business Univ. Chicago **40** (1967), 393–413.

127. George L. Gerstein and Benoit Mandelbrot, *Random walk models for the spike activity of a single neuron*, Biophysical J. **4** (1964), 41–68.

128. Benoit Mandelbrot and Howard M. Taylor, *On the distribution of stock price differences*, Operations Res. **15** (1967), 1057–1062.

129. J. Huston McCulloch, *Continuous time processes with stable increments*, J. Business **51** (1978), 601–619.

130. P. Medgyessy, *On the characterization of the form of the graphs of distribution and density functions.* I, II, Magyar Tud. Akad. Mat. Fiz. Oszt. Közl. **14** (1964), 279–292; **17** (1967), 101–108. (Hungarian)

131. R. R. Officer, *The distribution of stock returns*, J. Amer. Statist. Assoc. **67** (1972), 807–812.

132. I. A. Ovseevich and A. M. Yaglom, *Monotone transfer processes in homogeneous long lines*, Izv. Akad. Nauk SSSR Otdel. Tekhn. Nauk **1954**, no. 7, 13–20. (Russian)

133. Richard Roll, *The behavior of interest rates. An application of the efficient market model to U. S. Treasury bills*, Basic Books, New York, 1970.

134. Hilary L. Seal, *Stochastic theory of a risk business*, Wiley, 1969.

135. V. I. Siforov, *On noise buildup and fadeouts in main radio relay communications lines*, Èlektrosvyaz′ **1956**, no. 5, 6–17. (Russian)

136. Yu. B. Sindler, *The accumulation of noise in FM radio relay communications lines due to fading of the signal*, Radiotekhn. i Èlektron. **1** (1956), 627–637. (Russian)

137. B. W. Stuck and B. Kleiner, *A statistical analysis of telephone noise*, Bell. System Tech. J. **53** (1974), 1263–1320.

138. I. G. Vitenzon, *On the relative motion of a material point with variable mass*, Khar′kov. Gos. Univ. Uchen. Zap. **24** = Zap. Nauchno-Issled. Inst. Mat. Mekh. i Khar′kov. Mat. Obshch. (4) **21** (1949), 87–99.

139. L. J. Walpole, *The elastic field of an inclusion in an anisotropic medium*, Proc. Roy. Soc. London Ser. A **300** (1967), 270–289.

140. Janice Moulton Westerfield, *An examination of foreign exchange risk under fixed and floating rate regimes*, J. Internat. Economics **7** (1977), 181–200.

141. V. M. Zolotarev and B. M. Strunin, *On the distribution of internal stresses under a random arrangement of point defects*, Fiz. Tverd. Tela **13** (1971), 594–596; English transl. in Soviet Phys. Solid State **13** (1971/72).

III. Estimators of the parameters of stable laws

142. Ruth W. Arad, *Parameter estimation for symmetric stable distributions*, Internat. Econ. Rev. **21** (1980), 209–220.

143. Rudolf Beran, *Asymptotically efficient adaptive rank estimates in location models*, Ann. Statist. **2** (1974), 63–74.

144. Robert C. Blattberg and Nicholas J. Gonedes, *A comparison of the stable and Student distributions as statistical models for stock prices*, J. Business **47** (1974), 244–280.

145. Robert Blattberg and Thomas Sargent, *Regression with non-Gaussian stable disturbances: some sampling results*, Econometrica **39** (1971), 501–510.

146. P. J. Brockwell and B. M. Brown, *High-efficiency estimation for the positive stable laws*, J. Amer. Statist. Assoc. **76** (1981), 626–631.

147. J. L. Bryant and A. S. Paulson, *Some comments on characteristic function-based estimators*, Sankhyā Ser. A **41** (1979), 109–116.

148. William H. DuMouchel, *On the asymptotic normality of the maximum likelihood estimate when sampling from a stable distribution*, Ann. Statist. **1** (1973), 948–957.

149. ____, *Stable distributions in statistical inference*. I: *Symmetric stable distributions compared to other symmetric long-tailed distributions*, J. Amer. Statist. Assoc. **68** (1973), 469–477.

150. ____, *Stable distributions in statistical inference*. II: *Information from stably distributed samples*, J. Amer. Statist. Assoc. **70** (1975), 386–393.

151. Eugene F. Fama and Richard Roll, *Parameter estimates for symmetric stable distributions*, J. Amer. Statist. Assoc. **66** (1971), 331–338.

152. Alan Paul Fenech, *Asymptotically efficient of location for a symmetric stable law*, Ann. Statist. **4** (1976), 1088–1100.

153. Andrey Feuerverger and Philip McDunnough, *On some Fourier methods for inference*, J. Amer. Statist. Assoc. **76** (1981), 379–387.

154. ———, *On the efficiency of empirical characteristic function procedures*, J. Roy. Statist. Soc. Ser. B **43** (1981), 20–27.

155. J. L. Hodges, Jr., and E. L. Lehmann, *Estimates of location based on rank tests*, Ann. Math. Statist. **34** (1963), 598–611.

156. I. A. Ibragimov and R. Z. Khas′minskiĭ, *Estimation of a distribution density*, Zap. Nauchn. Sem. Leningrad. Otdel. Mat. Inst. Steklov (LOMI) **98** (1980), 61–85; English transl. in J. Soviet Math. **21** (1983), no. 1.

157. I. Sh. Ibramkhalilov, *Estimation of parameters of distributions*, Izv. Akad. Nauk Azerbaĭdzhan. SSR Ser. Fiz.-Tekhn. i Mat. Nauk **1964**, no. 2, 31–41.

158. ———, *On estimates of functionally related parameters*, Theor. Probab. Math. Statist. No. 6 (1975), 59–72.*

159. R. A. Leitch and A. S. Paulson, *Estimation of stable law parameters: stock price behavior application*, J. Amer. Statist. Assoc. **70** (1975), 690–697.

160. B. F. Logan et al., *Limit distributions of self-normalized sums*, Ann. Probab. **1** (1973), 788–809.

161. A. S. Paulson, E. W. Holcomb and R. A. Leitch, *The estimation of the parameters of the stable laws*, Biometrika **62** (1975), 163–170.

162. S. James Press, *Estimation in invariate and multivariate stable distributions*, J. Amer. Statist. Assoc. **67** (1972), 842–846.

163. C. P. Quesenberry and H. A. David, *Some tests for outliers*, Biometrika **48** (1961), 379–390.

164. J. C. Thornton and A. S. Paulson, *Asymptotic distribution of characteristic function-based estimators for the stable laws*, Sankhyā Ser. A **39** (1977), 341–354.

165. G. J. Worsdale, *The estimation of the symmetric stable distribution parameters*, COMPSTAT 1976 (Proc. Second Sympos. Comput. Statist., West Berlin), Physica-Verlag, Vienna, 1976, pp. 55–63.

166. O. L. Yanushkyavichene [Januškevičiene], *Investigation of certain parameter estimates for stable distributions*, Litovsk. Mat. Sb. **21** (1981), no. 4, 195–209; English transl. in Lithuanian Math. J. **21** (1981).

---

* *Editor's note.* The Russian text cites specifically the English translation of this paper, *not* the Russian original.

167. V. M. Zolotarev, *Statistical estimates of the parameters of stable laws*, Mathematical Statistics (R. Bartoszyński et al., editors), Banach Center Publ. No. 6, PWN, Warsaw, 1980, pp. 359–376.

168. ____, *Integral transformations of distributions and estimates of parameters of multidimensional spherically symmetric stable laws*, Contributions to Probability (Collection Dedicated to Eugene Lukacs), Academic Press, 1981, pp. 283–305.

IV. Limit theorems

169. A. Aleshkyavichene [Aleškevičienė], *A local limit theorem for sums of random variables connected in a homogeneous Markov chain in the case of a stable limit distribution*, Litovsk. Mat. Sb. **1** (1961), no. 1–2, 5–13; English transl. in Selected Transl. Math. Statist. and Probab., vol. 7, Amer. Math. Soc., Providence, R. I., 1968.

170. I. I. Banis [J. Banys], *Convergence for densities in the L_1 metric for a stable limit law in the two-dimensional case*, Litovsk. Mat. Sb. **17** (1977), no. 1, 13–18; English transl. in Lithuanian Math. J. **17** (1977).

171. ____, *An estimate of the rate of convergence in a local theorem in the multidimensional case*, Litovsk. Math. Sb. **19** (1979), no. 2, 13–21; English transl. in Lithuanian Math. J. **19** (1979).

172. ____, *A refinement of the rate of convergence to a stable law*, Litovsk. Mat. Sb. **16** (1976), no. 1, 5–22; English transl. in Lithuanian Math. J. **16** (1976).

173. ____, *A refinement of the rate of convergence to a stable law in a local theorem in the multidimensional case*, Litovsk. Mat. Sb. **16** (1976), no. 3, 13–20; English transl. in Lithuanian Math. J. **16** (1976).

174. ____, *A refinement of the rate of convergence of densities to a stable law with characteristic exponent $0 < \alpha < 1$ in the L_p-metric*, Litovsk. Mat. Sb. **18** (1978), no. 2, 21–27; English transl. in Lithuanian Math. Sb. **18** (1978).

175. ____, *On the rate of convergence in a multidimensional local theorem in the case of a stable limit law*, Litovsk. Mat. Sb. **13** (1973), no. 1, 17–22; English transl. in Math. Trans. Acad. Sci. Lithuanian SSR **13** (1973).

176. ____, *On a nonuniform estimate of the remainder term in the integral limit theorem*, Litovsk. Mat. Sb. **14** (1974), no. 3, 57–65; English transl. in Lituanian Math. Trans. **14** (1974).

177. ____, *On an integral limit theorem for convergence to a stable law in the multidimensional case*, Litovsk. Mat. Sb. **10** (1970), 665–672. (Russian)

178. R. Bartels, *Generating non-normal stable variates using limit theorem properties*, J. Statistical Computation and Simulation **7** (1978), 199–212.

179. Sujit K. Basu, *On a local limit theorem concerning variables in the domain of normal attraction of a stable law of index* α, $1 < \alpha < 2$, Ann. Probab. **4** (1976), 486–489.

180. Sujit K. Basu, Makoto Maejima and Nishith K. Patra, *A nonuniform rate of convergence in a local limit theorem concerning variables in the domain of normal attraction of a stable law*, Yokohama Math. J. **27** (1979), 63–72.

181. Sujit K. Basu and Makoto Maejima, *A local limit theorem for attractions under a stable law*, Math. Proc. Cambridge Philos. Soc. **87** (1980), 179–187.

182. P. L. Butzer and L. Hahn, *General theorems on rates of convergence in distribution of random variables. II: Applications to the stable limit laws and weak law of large numbers*, J. Multivariate Anal. **8** (1978), 202–221.

183. Harald Bergström, *On distribution functions with a limiting stable distribution function*, Ark. Mat. **2** (1952/53), 463–474.

184. ——, *Limit theorems for convolutions*, Almqvist & Wiksell, Stockholm, and Wiley, New York, 1963.

185. N. H. Bingham, *Maxima of sums of random variables and suprema of stable processes*, Z. Wahrsch. Verw. Gebiete **26** (1973), 273–296.

186. Virool Boonyasombut and Jesse M. Shapiro, *The accuracy of infinitely divisible approximations to sums of independent variables with application to stable laws*, Ann. Math. Statist. **41** (1970), 237–250.

187. Gerd Christoph, *Konvergenzaussagen und asymptotische Entwicklungen für die Verteilungsfunktionen einer Summe unabhängiger, identisch verteilter Zufallsgrössen im Falle einer stabilen Grenzverteilungsfunktion*, Zentralinst. Math. Mech. Akad. Wiss. DDR, Berlin, 1979.

188. ——, *Über notwendige und hinreichende Bedingungen für Konvergenzgeschwindigkeitsaussagen im Falle einer stabilen Grenzverteilung*, Z. Wahrsch. Verw. Gebiete **54** (1980), 29–40.

189. ——, *Über die Konvergenzgeschwindigkeit im Falle einer stabilen Grenzverteilung*, Math. Nachr. **90** (1979), 21–30.

190. Harold Cramér, *On asymptotic expansions for sums of independent random variables with a limiting stable distribution*, Sankhyā Ser. A **25** (1963), 13–24, 216.

191. ——, *On the approximation to a stable probability distribution*, Studies in Math. Anal. and Related Topics (Essays in Honor of George Pólya), Stanford Univ. Press, Stanford, Calif., 1962, pp. 70–76.

192. ——, *Random variables and probability distributions*, 2nd ed., Cambridge Univ. Press, 1962.

193. D. A. Darling, *The influence of the maximum term in the addition of independent random variables*, Trans. Amer. Math. Soc. **73** (1952), 95–107.

194. D. A. Darling and P. Erdös, *A limit theorem for the maximum of normalized sums of independent random variables*, Duke Math. J. **23** (1956), 143–155.

195. I. Dubinskaĭte [J. Dubinskaitė], *On the accuracy of approximation of distributions of sums of independent random variables by a stable distribution*, Litovsk. Mat. Sb. **23** (1983), no. 1, 74–91; English transl. in Lithuanian Math. J. **23** (1983).

196. V. A. Egorov, *On the rate of convergence to a stable law*, Teor. Veroyatnost. i Primenen. **25** (1980), 183–190; English transl. in Theory Probab. Appl. **25** (1980).

197. William Feller, *Fluctuation theory of recurrent events*, Trans. Amer. Math. Soc. **67** (1947), 98–119.

198. M. I. Fortus, *A uniform limit theorem for distributions which are attracted to a stable law with index less than one*, Teor. Veroyatnost. i Primenen. **2** (1957), 486–487; English transl. in Theor. Probab. Appl. **2** (1957).

199. B. V. Gnedenko, *On the theory of domains of attraction of stable laws*, Uchen. Zap. Moskov. Gos. Univ. Mat. **30** (1939), 61–81. (Russian; English summary).

200. ——, *On a local theorem for stable limit distributions*, Ukraïn. Mat. Zh. **1** (1949), no. 4, 3–15. (Russian)

201. B. V. Gnedenko and V. S. Korolyuk, *Some remarks on the theory of domains of attraction of stable distributions*, Dopovīdī Akad. Nauk Ukrain. SSR **1950**, 275–278. (Ukrainian)

202. Peter Hall, *Two-sided bounds on the rate of convergence to a stable law*, Z. Wahrsch. Verw. Gebiete **57** (1981), 349–364.

203. Yu. I. Ignat and P. V. Slyusarchuk, *On convergence to stable laws*, Dokl. Akad. Nauk Ukrain SSR Ser. A **1977**, 912–913. (Russian)

204. N. Kalinauskaitė, a) *On the upper and lower functions for stable random processes*. I, Litovsk. Mat. Sb. **5** (1965), 541–553. (Russian)

b) *Upper and lower functions for sums of independent random variables with stable limit distributions*, Litovsk. Mat. Sb. **6** (1966), 249–256; English transl. in Selected Transl. Math. Statist. and Probab., vol. 10, Amer. Math. Soc., Providence, R. I., 1972.

205. ——, *Some properties of stable random processes*, Litovsk. Mat. Sb. **4** (1964), 493–495. (Russian)

206. ——, *On attraction to stable laws of Lévy-Feldheim type*, Litovsk. Mat. Sb. **14** (1974), no. 3, 93–105; English transl. in Lithuanian Math. Trans. **14** (1974).

207. Stephen R. Kimbleton, *A simple proof of a random stable limit theorem*, J. Appl. Probab. **7** (1970), 502–504.

208. Miriam Lipschutz, *On the magnitude of the error in the approach to stable distributions.* I, II, Nederl. Akad. Wetensch. Proc. Ser. A **59** = Indag. Math. **18** (1956), 281–287, 288–294.

209. Makoto Maejima, *A nonuniform estimate in the local limit theorem for densities.* II, Yokohama Math. J. **26** (1978), 119–135.

210. Masatomo Udagawa, *On some limit theorems for sums of identically distributed independent random variables*, Kōdai Math. Sem. Rep. **8** (1956), 85–92.

211. J. David Mason, *Convolutions of stable laws as limit distributions of partial sums*, Ann. Math. Statist. **41** (1970), 101–114.

212. A. A. Mitalauskas, *On a local limit theorem in the case of a stable limit distribution*, Litovsk. Mat. Sb. **1** (1961), 131–139. (Russian)

213. ——, *On a local limit theorem for stable distributions*, Teor. Veroyatnost. i Primenen. **7** (1962), 185–190; English transl. in Theor. Probab. Appl. **8** (1962).

214. ——, *An asymptotic expansion for independent random variables in the case of a stable limit distribution*, Litovsk. Mat. Sb. **3** (1963), 189–193. (Russian)

215. ——, *On an integral limit theorem for covergence to a stable limit law*, Litovsk. Mat. Sb. **4** (1964), 235–240. (Russian)

216. ——, *On an estimate of the rate of convergence in an integral limit theorem in the case of a stable limit distribution*, Litovsk. Mat. Sb. **6** (1966), 85–90. (Russian)

217. A. A. Mitalauskas and V. A. Statulyavichus [Statulevičius], *On local limit theorems.* I, Litovsk. Mat. Sb. **14** (1974), no. 4, 129–144; English transl. in Lithuanian Math. Trans. **14** (1974).

218. ——, *An asymptotic expansion in the case of a stable approximating law*, Litovsk. Mat. Sb. **16** (1976), no. 4, 149–166; English transl. in Lithuanian Math. J. **16** (1976).

219. Thomas A. O'Connor, *Some classes of limit laws containing the stable distributions*, Z. Wahrsch. Verw. Gebiete **55** (1981), 25–33.

220. Ouyang Guang-Yirong, *On limit theorems and stable limit distributions for sums of a random number of independent random variables*, Acta Sci. Nat. Univ. Fudan **10** (1956), no. 1, 1–8. (Chinese)

221. V. I. Paulauskas, *Estimation of the remainder term in a limit theorem in the case of a stable limit law*, Litovsk. Mat. Sb. **14** (1974), no. 1, 165–187; English transl. in Lithuanian Math. Trans. **14** (1974).

222. ——, *Some nonuniform estimates in limit theorems of probability theory*, Dokl. Akad. Nauk SSSR **211** (1973), 791–792; English transl. in Soviet Math. Dokl. **14** (1973).

223. ———, *On estimates of the rate of convergence in limit theorems by means of pseudomoments*, Dokl. Akad. Nauk SSSR **199** (1971), 26–29; English transl. in Soviet Math. Dokl. **12** (1971).

224. ———, *Uniform and nonuniform estimates of the remainder term in a limit theorem with a stable limit law*, Litovsk. Mat. Sb. **14** (1974), no. 4, 171–185; English transl. in Lithuanian Math. Trans. **14** (1974).

225. ———, *On the rate of convergence in a multidimensional limit theorem in the case of a stable limit law*, Litovsk. Mat. Sb. **15** (1975), no. 1, 207–228; English transl. in Lithuanian Math. J. **15** (1975).

226. R. N. Pillai, *Semistable laws as limit distributions*, Ann. Math. Statist. **42** (1971), 780–783.

227. Yu. V. Prokhorov, *A local theorem for densities*, Dokl. Akad. Nauk SSSR **83** (1952), 797–800. (Russian)

228. G. N. Sakovich, A single form of attraction to stable laws, Teor. Veroyatnost. i Primenen. **1** (1956), 357–361; English transl. in Theor. Probab. Appl. **1** (1956).

229. K. I. Satybaldina, *On the question of estimating the rate of convergence in a limit theorem with a stable limit law*, Teor. Veroyatnost. i Primenen. **18** 211–212; English transl. in Theor. Probab. Appl. **18** (1973).

230. ———, *Absolute estimates of the rate of convergence to stable laws*, Teor. Veroyatnost. i Primenen. **17** (1972), 773–775; English transl. in Theory Probab. Appl. **17** (1972).

231. ———, *The influence of the smoothness of the distribution function of a random variable on an estimate of the rate of convergence to a stable limit law*, Theoretical and Applied Problems in Mathematics and Mechanics (T. I. Amanov, editor), "Nauka", Alma-Ata, 1979, pp. 198–202. (Russian)

232. Jesse M. Shapiro, *On the rate of convergence of distribution functions of sums of reciprocals of random variables to the Cauchy distribution*, Houston J. Math. **4** (1978), 439–445.

233. V. A. Statulyavichus [Statulevičius], *On limit theorems in the case of a stable limit law*, Litovsk. Mat. Sb. **7** (1967), 321–328; English transl. in Selected Transl. Math. Statist. and Probab., vol. 11, Amer. Math. Soc., Providence, R. I., 1973.

234. Charles Stone, *Local limit theorems for asymptotically stable distribution functions*, Notices Amer. Math. Soc. **12** (1964), 465.

235. S. G. Tkachuk, *A theorem on large deviations in R^s in the case of a stable limit law*, Random Processes and Statistical Inference, vyp. 4, "Fan", Tashkent, 1974, pp. 178–184. (Russian)

236. ——, *Local limit theorems taking into account large deviations in the case of stable limit laws*, Izv. Akad. Nauk UzSSR Ser. Fiz.-Mat. Nauk **17** (1973), no. 2, 30–33. (Russian)

237. Howard G. Tucker, *Convolutions of distributions attracted to stable laws*, Ann. Math. Statist. **39** (1968), 1381–1390.

238. P. Vaitkus, *On large deviations of sums of random variables in the case of a stable limit law*, Litovsk. Mat. Sb. **12** (1972), no. 1, 85–97. (Russian)

239. N. A. Volodin, *Estimates in a weak form of a local limit theorem for the case of a stable distribution*, Limit Theorems and Mathematical Statistics (S. A. Sirazhdinov, editor), "Fan", Tashkent, 1976, pp. 32–36. (Russian)

240. Stephen James Wolfe, *A note on the complete convergence of stable distribution functions*, Ann. Math. Statist. **43** (1972), 363–364.

241. V. M. Zolotarev, *On the choice of normalizing constants in increasing sums of independent random variables*, Trudy Moskov. Fiz.-Tekhn. Inst. **7** (1961), 158–161. (Russian)

242. ——, *On a new viewpoint of limit theorems taking in account large deviations*, Proc. Sixth All-Union Conf. Theor. Probab. and Math. Statist. (Vilnius, 1960), Gos. Izdat. Politichesk. i Nauchn. Lit. Litovsk. SSR, Vilnius, 1962, pp. 43–47; English transl. in Selected Transl. Math. Statist. and Probab., vol. 9, Amer. Math. Soc., Providence, R. I., 1971.

243. ——, *An analogue of the Edgeworth-Cramér asymptotic expansion for the case of convergence to stable distribution laws*, Proc. Sixth All-Union Conf. Theor. Probab. and Math. Statist. (Vilnius, 1960), Gos. Izdat. Politichesk. i Nauchn. Lit. Litovsk. SSR, Vilnius, 1982, pp. 49–50. (Russian)

244. ——, *An analogue of the law of the iterated logarithm for semicontinuous stable processes*, Teor. Veroyatnost. i Primenen. **9** (1964), 566–567; English transl. in Theor. Probab. Appl. **9** (1964).

V. Supplementary references

245. George W. Brown and John W. Tukey, *Some distributions of sample means*, Ann. Math. Statist. **17** (1946), 1–12.

246. J. M. Chambers, C. L. Mallows and B. W. Stuck, *A method for simulating stable random variables*, J. Amer. Statist. Assoc. **71** (1976), 340–344.

247. Wolfgang Gawronski, *On the bell-shape of stable densities*, Ann. Probab. **12** (1984), 230–242.

248. A. V. Nagaev and S. M. Shkol′nik, *An invariant estimation of the characteristic parameter of a stable law*, Manuscript No. 233 Uz-84 Dep, deposited at the UzNIINTI, Tashkent, 14 November 1984. (Russian)

249. ——, *On a family of probability distributions*, Mat. Zametki **37** (1985), 594–598; English transl. in Math. Notes **37** (1985).

250. V. I. Paulauskas, *Convergence to stable laws and their simulation*, Litovsk. Mat. Sb. **22** (1982), no. 3, 146–156; English transl. in Lithuanian Math. J. **22** (1982).

251. A. P. Prudnikov, Yu. A. Bychkov and O. I. Marichev, *Integrals and series*, "Nauka", Moscow, 1981. (Russian)

252. Hirotaka Sakasegawa, *On a generation of normal pseudo-random numbers*, Ann. Inst. Statist. Math. **30** (1978), 271–279.

253. G. Udny Yule, *A mathematical theory of evolution, based on the conclusions of Dr. J. C. Willis, F. R. S.*, Philos. Trans. Roy. Soc. London Ser. B **213** (1925), 21–87.

254. J. C. Willis, *Age and area*, Cambridge Univ. Press, 1922.

List of Notation (*)

$\mathfrak{M}$, class of M-infinitely divisible distributions

$\mathfrak{S}$, family of SD's

$\mathfrak{W}$, class of strictly SD's

$\mathfrak{W}_1$, class of distributions in $\mathfrak{W}$ with $\lambda_C = 1$

$\mathfrak{G}$, class of IDL's

$\lambda = \lambda_B, \lambda_A$, a parameter—the time in stable processes, and the scale parameter for strictly SD's (in the form (B), (A))

$\beta = \beta_B, \beta_A$, a parameter—the degree of asymmetry of an SD (in the form (B), (A))

$\gamma = \gamma_B, \gamma_A$, the bias characteristic of an SD (in the form (B), (A)) for fixed λ

$K(\alpha) = \alpha - 1 + \varepsilon(\alpha) = \operatorname{sgn}(1 - \alpha)$, functions in certain forms of expression for the CF of an SD

$\mathfrak{g}_A(t, \alpha, \beta, \gamma, \lambda) = \mathfrak{g}(t)$, the CF of an SD with parameter system in the form (A)

α, the main parameter of an SL, the same in all the forms except for the form (E)

$g(x, \alpha, \beta, \gamma, \lambda)$, the density of an SD corresponding to the form (A) or (B)

$g(x, \alpha, \beta) = g(x, \alpha, \beta, 0, 1)$, the density of a standard SD, in the form (B) beginning with §2.2

$G(x, \alpha, \beta, \gamma, \lambda)$, DF of an SD

$G(x, \alpha, \beta) = G(x, \alpha, \beta, 0, 1)$, DF of a standard SD, in the form (B) beginning with §2.2

θ, a parameter—the degree of assymetry of an SD in the forms (C) and (E)

$\rho = (1 + \theta)/2$, a parameter replacing the parameter θ in the form (C)

$\nu = 1/\alpha^2$, a parameter replacing the parameter α in the form (E)

τ, a parameter responsible for scale changes of an SD in the form (E)

$Y(\alpha, \beta, \gamma, \lambda)$, a random variable with SD in the form (B) (sometimes also in the form (A), if explicitly mentioned)

$Y(\alpha, \beta) = Y(\alpha, \beta, 0, 1)$, a random variable with a standard SD, in the form (B) beginning with §2.2

$Y(\alpha, \theta) = Y_C(\alpha, \theta)$, a random variable with strictly SD in the form (C) and with DF $G_C(x, \alpha, \theta)$

$Z(\alpha, \rho)$, a cutoff of a random variable $Y(\alpha, \theta)$

$Y(\alpha)$, abbreviated notation for a random variable $Y(\alpha, 1)$ for $0 < \alpha < 1$

$M(s, \alpha, \rho)$, the Mellin transform of the density of the SD of a variable $Z(\alpha, \rho)$

$w_k(s, \alpha, \theta)$, $k = 0, 1$, an element of the characteristic transform of the SD with density $g_C(x, \alpha, \theta, \lambda)$

E, a random variable with DF $1 - \exp(-x)$, $x \geq 0$

(*)Abbreviations: IDL = infinitely divisible law; SD (SL) = stable distribution (stable law); CF = characteristic function; DF = distribution function.

Subject Index (*)

(*)Abbreviations: IDL = infinitely divisible law; SD (SL) = stable distribution (stable law); CF = characteristic function; DF = distribution function.

ABCDEFGHIJ — 89876